T0205238

Methods in Pharmacology and Toxicology

For further volumes:
http://www.springer.com/series/7653

Apoptosis Methods in Toxicology

Edited by

Perpetua M. Muganda

Department of Biology, North Carolina A&T State University, Greensboro, NC, USA

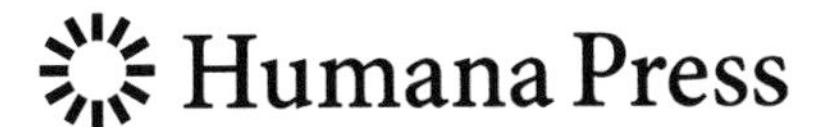

Editor
Perpetua M. Muganda
Department of Biology
North Carolina A&T State University
Greensboro, NC, USA

ISSN 1557-2153 ISSN 1940-6053 (electronic)
Methods in Pharmacology and Toxicology
ISBN 978-1-4939-8086-4 ISBN 978-1-4939-3588-8 (eBook)
DOI 10.1007/978-1-4939-3588-8

Softcover reprint of the hardcover 1st edition 2016

Cover Illustration: Taken from Chapter 5, Figure 2.

Printed on acid-free paper

This Humana Press imprint is published by Springer Nature
This registered company is Springer Science+Business Media LLC New York

Preface

Apoptosis is a highly regulated and active form of programmed cell death that is used to eliminate excess, damaged, or cancerous cells throughout life in a variety of organisms, thus maintaining normal homeostasis. Inappropriate apoptosis may occur within cells in response to various toxicological stresses, including drugs and other compounds. Since dysregulated apoptosis is likely to result in disease, it is important to quantitate the level of apoptosis and determine its mechanisms and signaling pathways through suitable apoptotic methods.

The most suitable methods for detecting and quantitating apoptosis must be able to distinguish apoptosis from other more recently discovered forms of programmed cell death, such as programmed necrosis and autophagic cell death. Apoptosis is characterized by specific morphological and biochemical changes, and is executed principally through the mitochondria and death receptor pathways. Thus, the most appropriate and useful methods for the specific analysis of and quantitation of apoptosis must measure as many apoptosis parameters as possible, in addition to distinguishing apoptosis from other forms of cell death, such as autophagy and necrosis. Mechanisms underlying apoptosis and other forms of programmed cell death have been the focus of toxicology, as well as many scientific fields, within the past decade. This has spiked recent developments as well as updates in time-tested apoptosis methods.

Apoptosis Methods in Toxicology is designed to provide a single, valuable reference source for methods that definitively identify and accurately quantify apoptosis. Experts in the field have been recruited to provide extensive reviews and references of time-tested, recently updated, and newly developed apoptosis methods in toxicology. In addition to relevant reviews, authors provide a detailed step-by-step description of their best state-of-the-art, time-tested, recently developed, and updated techniques for studying apoptosis in toxicology. Technical problems, challenges, and limitations of the methods are also discussed. This volume, which is designed for the novice as well as the expert in toxicology and other related fields, is divided into four interrelated and overlapping sections. The introduction section (Chaps. 1 and 2) contains reviews on the most common methods utilized to detect and quantitate apoptosis, as well as apoptosis signaling pathways in toxicological and other related research. This is an excellent introductory section for the novice scientist. The second section focuses on multiparametric and phased apoptosis assays for detecting early and late apoptosis (Chaps. 3 and 4), or distinguishing apoptosis from necrosis and autophagy (Chap. 5). The third section focuses on recent advances in real time and high-throughput assays to detect and quantitate apoptosis (Chaps. 6–10) and apoptosis signaling pathways (Chap. 11). This section covers apoptosis assays that utilize low-end instrumentation that can be found in most laboratories, as well as methods that rely on high-end sophisticated instrumentation. The last section of the book reviews recent developments in preclinical anticancer therapeutics targeting apoptosis. Chapter 12 focuses on the interrelationship of preclinical anticancer small molecules with apoptosis and autophagy. Chapter 13 discusses the enhancement of cancer cell death subroutine therapeutic effects of small molecules through the use of various liposome formulations. Thus, the content, as organized, can be utilized by novice scientists as well as experts, utilizing a range of instruments from common laboratory equipment to high-end expensive and automated machinery capable of performing real time apoptotic measurements.

Greensboro, NC, USA *Perpetua M. Muganda*

Contents

Contributors

WILLIAM P. BOZZA • *Office of Biotechnology Products, Center for Drug Evaluation and Research, Food and Drug Administration, Silver Spring, MD, USA*
JACK COLEMAN • *ENZO Life Sciences, Farmingdale, NY, USA*
AKAMU J. EWUNKEM • *Department of Energy and Environmental Studies, North Carolina A&T State University, Greensboro, NC, USA*
HONGBO GU • *Cell Signaling Technology, Inc., Danvers, MA, USA*
CHRISTIAN T. HELLWIG • *Centre for Systems Medicine, Department of Physiology and Medical Physics, Royal College of Surgeons in Ireland, Dublin 2, Ireland*
DEEPA INDIRA • *Cancer Research Program-1, Rajiv Gandhi Centre for Biotechnology (RGCB), Poojappura, Thiruvananthapuram, Kerala, India*
MAGISETTY JHANSILAKSHMI • *Department of Biochemistry, Central Food Technological Research Institute, Mysore, India*
JEENA JOSEPH • *Cancer Research Program-1, Rajiv Gandhi Centre for Biotechnology (RGCB), Poojappura, Thiruvananthapuram, Kerala, India*
SU-RYUN KIM • *Office of Biotechnology Products, Center for Drug Evaluation and Research, Food and Drug Administration, Silver Spring, MD, USA*
ARUN KUMAR • *ENZO Life Sciences, Farmingdale, NY, USA*
BO LIU • *State Key Laboratory of Biotherapy and Cancer Center, West China Hospital, Sichuan University, and Collaborative Innovation Center of Biotherapy, Chengdu, China*
RUI LIU • *ENZO Life Sciences, Farmingdale, NY, USA*
AGNIESZKA H. LUDWIG-GALEZOWSKA • *Centre for Systems Medicine, Department of Physiology and Medical Physics, Royal College of Surgeons in Ireland, Dublin 2, Ireland*
KRUPA ANN MATHEW • *Cancer Research Program-1, Rajiv Gandhi Centre for Biotechnology (RGCB), Poojappura, Thiruvananthapuram, Kerala, India*
PATRICIA MCGRATH • *Phylonix, Cambridge, MA, USA*
PENG MIAO • *CAS Key Lab of Bio-Medical Diagnostics, Suzhou Institute of Biomedical Engineering and Technology, Chinese Academy of Sciences, Suzhou, Jiangsu, China*
PERPETUA M. MUGANDA • *Department of Biology, North Carolina A&T State University, Greensboro, NC, USA*
ROBERT H. NEWMAN • *Department of Biology, North Carolina A&T State University, Greensboro, NC, USA*
MAGISETTY OBULESU • *Department of Materials Science, Graduate School of Pure and Applied Sciences, Tsukuba University, Tsukuba, Ibaraki, Japan*
CARL D. PARSON II • *Department of Biology, North Carolina A&T State University, Greensboro, NC, USA*
PRAKASH RAJAPPAN PILLAI • *Cancer Research Program-1, Rajiv Gandhi Centre for Biotechnology (RGCB), Poojappura, Thiruvananthapuram, Kerala, India*
INDU RAMACHANDRAN • *Cancer Research Program-1, Rajiv Gandhi Centre for Biotechnology (RGCB), Poojappura, Thiruvananthapuram, Kerala, India*
MARKUS REHM • *Centre for Systems Medicine, Department of Physiology and Medical Physics, Royal College of Surgeons in Ireland, Dublin 2, Ireland*

SANTHOSHKUMAR THANKAYYAN RETNABAI • *Cancer Research Program-1, Rajiv Gandhi Centre for Biotechnology (RGCB), Poojappura, Thiruvananthapuram, Kerala, India*
WEN LIN SENG • *Phylonix, Cambridge, MA, USA*
JEFFREY C. SILVA • *Cell Signaling Technology, Inc., Danvers, MA, USA*
MATTHEW P. STOKES • *Cell Signaling Technology, Inc., Danvers, MA, USA*
WILLIAM G. TELFORD • *Experimental Transplantation and Immunology Branch, National Cancer Institute, National Institutes of Health, Bethesda, MD, USA*
MAO TIAN • *State Key Laboratory of Biotherapy and Cancer Center, West China Hospital, Sichuan University, and Collaborative Innovation Center of Biotherapy, Chengdu, China*
JULIANNE D. TWOMEY • *Office of Biotechnology Products, Center for Drug Evaluation and Research, Food and Drug Administration, Silver Spring, MD, USA*
SHANKARA NARAYANAN VARADARAJAN • *Cancer Research Program-1, Rajiv Gandhi Centre for Biotechnology (RGCB), Poojappura, Thiruvananthapuram, Kerala, India*
JINHUI WANG • *College of Pharmacy, Xinjiang Medical University, Urumqi, China*
KATHY WANG • *ENZO Life Sciences, Farmingdale, NY, USA*
JIAN YIN • *CAS Key Lab of Bio-Medical Diagnostics, Suzhou Institute of Biomedical Engineering and Technology, Chinese Academy of Sciences, Suzhou, Jiangsu, China*
BAOLIN ZHANG • *Office of Biotechnology Products, Center for Drug Evaluation and Research, Food and Drug Administration, Silver Spring, MD, USA*
DAWEI ZHANG • *Phylonix, Cambridge, MA, USA*
LAN ZHANG • *State Key Laboratory of Biotherapy and Cancer Center, West China Hospital, Sichuan University, and Collaborative Innovation Center of Biotherapy, Chengdu, China*
SHOUYUE ZHANG • *State Key Laboratory of Biotherapy and Cancer Center, West China Hospital, Sichuan University, and Collaborative Innovation Center of Biotherapy, Chengdu, China*
YAXIN ZHENG • *State Key Laboratory of Biotherapy and Cancer Center, West China Hospital, Sichuan University, and Collaborative Innovation Center of Biotherapy, Chengdu, China*

Chapter 1

An Overview of Apoptosis Methods in Toxicological Research: Recent Updates

Perpetua M. Muganda

Abstract

Apoptosis is the most common form of programmed cell death. Apoptosis plays a critical role in many physiological functions, and its dysregulation is an underlying defect in various diseases, including cancer. In fact, many toxicants and chemotherapeutic drugs exert their mechanisms of action through modulation of the apoptosis process. Thus, interest in the apoptosis process, as well as the methods used to assess and quantify its various aspects has continued to spike. This chapter provides a brief overview of the apoptosis process, the most common apoptosis methods, and the principles upon which these methods function. Furthermore, this chapter overviews the most recent improvements and trends in apoptosis methods, and introduces *Apoptosis Methods in Toxicology* book content. The information provided is useful to novice scientists, as well as the more advanced scientist.

Key words Apoptosis methods, Toxicology, Recent improvements, Review

1 Introduction

Apoptosis is the most common form of programmed cell death (recently reviewed in [1–3]). It plays a critical role in many physiological functions; these include tissue homeostasis, embryonic development, immune system development and maintenance, as well as the disposal of damaged cells (recently reviewed in [2, 4–8]). Downregulation of apoptosis is an underlying defect in cancer, and some forms of autoimmune disease [5, 7, 9, 10]. In contrary, increased apoptosis is an underlying defect of Alzheimer's disease, Parkinson disease, and other forms of autoimmune disease [1, 5–7, 11–13]. In fact, many toxicants and chemotherapeutic drugs exert their mechanisms of action through the modulation of the apoptosis process [9, 12]. Thus, interest in the apoptosis process and methods has grown steadily among toxicologists and other scientists.

Perpetua M. Muganda (ed.), *Apoptosis Methods in Toxicology*, Methods in Pharmacology and Toxicology, DOI 10.1007/978-1-4939-3588-8_1, © Springer Science+Business Media New York 2016

1.1 Morphological and Biochemical Features of Apoptosis

The apoptosis process proceeds through a complex cascade that involves characteristic morphological, molecular, and biochemical features (recently reviewed in [2, 6]). Morphological characteristics of apoptosis frequently include chromatic condensation and margination at the nuclear membrane, nuclear condensation, and cell shrinkage. While these are early events, chromatin and nuclear fragmentation, plasma membrane blebbing, and formation of apoptotic bodies are frequently observed as late events in the apoptosis process. These morphological changes are due to characteristic molecular and biochemical events occurring in the apoptotic cell (recently reviewed in [2, 6, 12, 13]). The externalization of phosphatidylserine is one of the earliest biochemical events in the apoptotic process of most cells; this imparts "eat me" signals on the apoptotic cells, causing them to be engulfed by macrophages and nearby cells, especially in vivo [6, 13–16]. Additional biochemical events include the sequential activation of initiator and effector caspases, cleavage of poly(ADP-ribose) polymerase (PARP) and other proteins, as well as DNA fragmentation at internucleosomal sites [1, 2, 6, 17]. Some of the morphological, molecular, and biochemical events are context dependent, and sometimes transient [18–21]. Caspase activation involving initiators and effectors, however, is a universal event in the apoptosis process [13, 20]. Thus, the Nomenclature Committee on Cell Death defines apoptosis as the caspase dependent variant of programmed cell death [18, 19].

Apoptosis can be initiated through two principal apoptosis signaling pathways that converge on executioner caspases, such as caspase 3; these include the extrinsic (death receptor-mediated) and intrinsic (mitochondria) pathways [2, 6]. The extrinsic (death receptor) apoptosis pathway involves formation of the death induced signaling complex (DISC), followed by activation of initiator caspase 8 after engaging receptors, such as the FAS/CD95 death receptor [19]. Activated caspase 8 then activates caspase 3, the executioner caspase. In some cell types, amplification of the apoptotic signal is achieved by cleaving the Bcl2 family protein Bid, effectively connecting the death receptor pathway to proceed through the intrinsic mitochondria pathway. The intrinsic mitochondria pathway is initiated by various cellular stresses (such as DNA damage, oxidative stress, irradiation), and is mediated through perturbation of the balance between pro-apoptotic and anti-apoptotic Bcl2 family members (for recent review see [6, 13, 18, 22]). This leads to depolarization of the mitochondria transmembrane potential, and the release of cytochrome c and other apoptogenic factors (such as AIF, Smac/DIABLO, endoG, CAD, and HtrA2/Omi serine protease) from the mitochondria. The cytochrome c interacts with APAF1, dATP, and procaspase 9 to form the apoptosome. Procaspase 9 is activated to caspase 9 within the apoptosome, and then activates the executioner caspase 3.

Thus, both the extrinsic receptor-mediated and the intrinsic mitochondria-mediated apoptosis pathway involve the activation of the executioner caspases, such as caspase 3 [6, 17]. The late apoptotic events, such as inter-nucleosomal DNA fragmentation, PARP cleavage, and nuclear fragmentation take place after activation of the executioner caspases, such as caspase 3. The released AIF and endonuclease G can both promote caspase 3-independent apoptosis by inducing DNA fragmentation and chromatin condensation [2, 6, 20]. These features are used to determine apoptosis mechanisms and identify the apoptotic signaling pathway involved.

Apoptosis (type 1 programmed cell death) is often confused with other types of programmed cell death that include autophagy and necroptosis (programmed necrosis, type 3 programmed cell death) [13]. Necroptosis is characterized by cytoplasmic granulation, organelle and/or cellular swelling, and disruption of cell membrane [5, 6]. It is a caspase independent cell death mediated by serine/threonine Receptor Interacting Protein (RIP) and Mixed Lineage Kinase Like (MLKL) kinases [18, 19, 23]. Autophagic cell death (type II programmed cell death) involves a catabolic process that turnover cytosolic protein aggregates and damaged organelles in the lysosomes via double membrane autophagosomes (reviewed in [18, 24, 25]). Morphologically, autophagy involves increased autophagosome formation and extensive vacuolation of the cytoplasm; no chromatin condensation or other morphological features of apoptosis are involved [6]. Biochemically, autophagy is dependent on a number of proteins, including Beclin-1 and other ATG proteins, and is independent of caspases (reviewed in [18, 25–27]). Thus, apoptosis is fundamentally different from other types of cell death based on its incidence, morphology, and biochemistry [2]. Apoptosis generally requires caspases and nucleases to bring about cell death without loss of membrane integrity (reviewed in [18, 25]). It is the principal mechanism by which unwanted cells are dismantled and eliminated from organisms [24].

Although apoptosis, autophagy and necroptosis are mediated by distinct pathways, recent evidence has demonstrated that the three programmed cell death modes interconnect during cell death decisions [5, 10, 24, 25, 27, 28]. This makes it possible for multiple types of programmed cell death modes to coexist and/or overlap, depending on the cell type and nature of the stress [6, 18]. Variabilities in apoptosis biochemical properties, including caspase dependency and phosphatidylserine externalization can exist, and apoptosis morphological features are also context dependent [18–21]. It is thus recommended that a number of measurable distinct morphological, molecular, and biochemical features, including signaling pathways, be used in order to correctly identify apoptosis from other types of programmed cell death [18, 19, 21, 29].

2 Overview of Common Apoptosis Methods

Apoptosis is best assayed by utilizing methods that detect and quantitate a combination of its biochemical and molecular features, as well as its morphological features [18–21] . This is due to the fact that individual morphological, biochemical, and molecular properties of apoptosis can vary in different cellular environments [13, 19, 20]. Any apoptosis assay utilized should always be confirmed through the use of other assays measuring different features [13]. A wide variety of platforms based on immunofluorescence, immunohistochemistry, bioluminescence, flow cytometry, western blot analysis, as well as array based platforms can be utilized (for previous review see [1, 13, 30, 31]).

2.1 Morphological Methods for Determination of Apoptosis

Methods based on apoptosis morphological properties primarily seek to assess the extent of chromatic condensation and margination at the nuclear membrane, as well as nuclear condensation and fragmentation [1, 31, 32]. Although other forms of light microscopy, as well as electron microscopy can be utilized to assess these features, the use of fluorescence microscopy makes this determination simple and more definitive [31]. Cell permeable DNA binding fluorescent dyes [such as Hoechst 43332 and acridine orange] in combination with cell impermeable DNA binding fluorescent dyes [such as and propidium iodide or ethidium bromide] can be utilized for this purpose [31–33]. Normal live cell nuclei stain positive with the cell permeable dye, while necrotic cell nuclei stain positive with the cell impermeable dye. Apoptotic cells have condensed and/or fragmented nuclei, while necrotic cells have normal nuclear morphology. Thus, this method is able to distinguish normal live cells, early apoptotic cells, late apoptotic cells, as well as necrotic cells. The percentage of apoptotic cells is then determined by counting the proportion of cells with nuclear membrane marginated chromatin, as well as condensed or fragmented nuclei. These assays are inexpensive, since they can be performed on a regular fluorescent microscope accessible to most laboratories. This method is labor intensive, however, and requires training, since it is rather subjective. The presence of a specific morphological feature, however, is not sufficient to establish the occurrence of apoptosis; these assays must be combined with other biochemical and molecular assays for apoptosis [21, 31–33].

2.2 Biochemical and Molecular Methods for Determination of Apoptosis

The most common biochemical and molecular features utilized to assess the extent of apoptosis include chromosomal DNA cleavage between nucleosomes, activation of caspases, and externalization of phosphatidyl serine [13, 30].

The annexin V-FITC assay quantifies the extent of phosphatidylserine (PS) externalization [6, 13, 15, 16]. It is performed on live cells using fluorescence microscopy or flow cytometry; a number of packaged kits can be utilized to facilitate the assay. Inclusion of propidium iodide (PI) within the assay helps to distinguish and quantitate early apoptosis (Annexin-V positive, PI−), late apoptosis (Annexin-V positive, PI+), and necrotic cells (Annexin-V negative, PI+) by either fluorescence microscopy or flow cytometry. The PI within the assay can also be utilized as a second independent assay to confirm the occurrence of late apoptosis. This is based on the existence of condensed and fragmented nuclei (by fluorescence microscopy), as well as sub-G0/G1 cell population (by flow cytometry). The use of flow cytometry is preferred over the fluorescence microscopy approach, due to its ability to quantitate in a direct read out manner. The fluorescence microscopy approach is labor intensive, but is useful for laboratories without access to a flow cytometer. The annexin V-FITC assay is reviewed in Chaps. 2 and 4 of this volume, *Apoptosis Methods in Toxicology*. In recent years, improvements of this assay has resulted in the pSIVA-IANBD polarity sensitive assay based on annexin XII. This assay can be used with no wash steps, and has high signal-to-noise ratios. It is also able to detect transient PS externalization because binding to PS is polarity sensitive and reversible [13, 14, 34]. The pSIVA-IANBD assay exists in packaged kits; it is used in Chap. 3 of this *Apoptosis Methods in Toxicology* book.

The chromatin within apoptotic cells undergoes cleavage between nucleosomes; this is a hall mark of apoptosis. This can be quantified by utilizing the TUNEL assay or the DNA ladder assay. The TUNEL (**T**dt-mediated d**U**TP-biotin **N**ick-**E**nd **L**abeling) assay detects DNA fragmentation in situ after fixing cells; it can be performed by immunohistochemistry, fluorescence microscopy, flow cytometry [1, 31]. It is labor intensive, and appropriate controls should be included to avoid false positives. For the DNA ladder assay, DNA extracted from control and apoptotic cells is separated on an agarose gel to form a 200 bp DNA ladder. Care is taken to avoid damage to the DNA during the preparation step, and positive and negative controls are utilized. The DNA ladder assay is primarily a qualitative assay for apoptosis [1, 31]. This assay has recently been improved to obtain higher DNA yields by using an easy to use extraction protocol with less steps; the improved assay does not utilize commercial kits, and is cost effective and efficient [35]. Due to the fact that apoptotic cells have less DNA content, the fragmentation of DNA can also be assessed by quantitating the sub-G0/G1 cell population by flow cytometry. The extent of the types and properties of DNA binding dyes that can be used have been extensively reviewed in Chap. 4 of this volume, *Apoptosis Methods in Toxicology*. Early apoptotic cells (which stain with the cell permeable dye) are distinguished from late apoptotic cells (which stain positive with the cell impermeable dye). Apoptotic cells are distinguished from necrotic cells, since necrotic cells stain positive with the cell impermeable dye but have normal DNA content. The flow cytometer approach is quantitative, and inexpensive, since it can be performed on a low-end flow cytometer available or accessible to most laboratories. These assays must be combined with other biochemical and molecular assays for apoptosis [1, 31].

Caspase 3/7 activation is a hallmark of apoptosis in most cell systems [18, 19]. Consequently, quantification of apoptosis based on the extent of caspase 3/7 activation is preferred by most investigators [30]. Caspase 3/7 activity assays are the most frequently used apoptosis assays [30]. Caspase 3/7 activity assays can be directly quantitated utilizing fluorescence or bioluminescence substrates in microplate reader formats (Chaps. 6–8 and 11 in this volume, *Apoptosis Methods in Toxicology*). Flow cytometry readouts can also be obtained by utilizing caspase 3/7 fluorescence substrates endpoints [24, 36, 37]. Caspases 3/7 activation can also be detected and quantitated using cleaved caspase 3/7 specific antibodies; this can utilize immunofluorescence microscopy, flow cytometry, ELISA, or western blot formats

In general, apoptosis methods based on biochemical and molecular properties of apoptosis are preferred over methods based on morphological properties [18, 19, 21, 31]. This is because these methods are more quantitative, and are less prone to error dependent misinterpretations that can occur when morphological methods are used. It is still important to interpret the results with caution, and to avoid using single biochemical endpoint assays alone. This is due to the fact that some of the biochemical features may occur in non-apoptotic cells, and not every biochemical event takes place in all apoptotic cells [13, 18–21].

2.3 Apoptosis Methods for Identification of Apoptosis Signaling Pathways

The two principal pathways involved in apoptotic signaling are the death receptor and mitochondrial pathways [2, 6, 9, 12]. Apoptotic methods for detecting the activation of the mitochondria pathway quantitate levels and activation of Bcl2 family proteins, changes in mitochondrial membrane potential, the release of cytochrome c and other apoptogenic factors from the mitochondria, apoptosome formation, and caspase 9 activation [1, 19]. To ensure that these events lead to apoptosis, it is important to demonstrate that caspase 9 activation leads to caspase 3 activation and apoptosis [33]. In systems where the caspase 3 activation does not account for the entire apoptotic response, it is important to demonstrate that other apoptogenic factors, such as AIF and endoG, translocate to the nucleus to modulate the observed apoptotic response [20]. Activation of the receptor mediated pathway is identified by detecting and quantifying death induced signaling complex (DISC) formation (the rate limiting step in this pathway), as well as the activation of caspase 8, the initiator caspase in this pathway, leading to caspase 3/7 activation and apoptosis [1]. Apoptotic methods in various formats are available to detect and quantitate all these steps in both signaling pathways [1, 19, 31]. Most of these methods are already packaged into readily available commercial kits. In Chap. 9 of this book, Stokes *et al.* described a high throughput antibody-based proteomic analysis method for deducing apoptotic signaling.

2.4 Common Apoptosis Methods Described in Apoptosis Methods in Toxicology

Apoptosis Methods in Toxicology is designed to provide a single valuable reference source for methods that definitively identify and accurately quantify apoptosis. Experts in the field have been recruited to provide extensive reviews and descriptions of time-tested, recently updated, and newly developed apoptosis methods in toxicology. The volume is designed for the novice as well as the expert in toxicology and other related fields. The introductory chapters (Chaps. 1 and 2) are designed for the novice scientist. This chapter (Chap. 1) provides a brief overview of the apoptosis and its methods, as well as the book content. The content of Chap. 2 represents a review of the most common in vitro and in vivo methods utilized to detect and quantitate apoptosis; the applications of these methods in research and clinical practice are discussed. Various approaches for determining phosphatidylserine externalization, mitochondria membrane permeabilization, DNA fragmentation, and caspase activation in vitro and in vivo are discussed. Additional in vivo approaches involving magnetic resonance imaging, as well as the use of aposense molecules are also described.

3 Recent Improvements and Trends in Apoptosis Methods

Time-tested apoptosis methods have been improved over the past decade, driven by a growing interest in apoptosis and other forms of programmed cell death. The advances are based on better understanding of the apoptotic machinery, as well as parallel improvements in apoptosis detection probes and platforms [5, 13].

Apoptosis Methods in Toxicology (this volume) describes the latest developments in the use of phased and multiparametric apoptosis assays for detecting early and late apoptosis (Chaps. 3 and 4), or distinguishing apoptosis from necrosis and autophagy (Chap. 5). Chapter 3, entitled “Detecting the Extent of Toxicant-Induced Apoptosis Using Concurrent Phased Assays”, presents a protocol involving the concurrent phasing of the latest developments in apoptosis assays. The concurrent phased assays are based on phosphatidylserine externalization, caspase 3/7 activity, and nuclear morphology, and are useful for laboratories that lack access to a flow cytometer. The phosphatidylserine externalization assay protocol described in Chap. 3 utilizes a recently developed polarity-sensitive annexin-based biosensor (pSIVA-IANBD); this has advantages over the widely used annexin V sensor [13, 14, 34].

Chapter 4 reviews the latest developments in the multiparametric flow cytometric analysis of early and late apoptosis; the reviewed assays are based on various biochemical and morphological properties of apoptotic cells in a single tube. Chapter 5 discusses simple protocols for detecting and distinguishing apoptosis, necrosis, and autophagy.

The most sweeping important improvements in apoptosis methods over the last decade have resulted in high throughput, real-time readout, and/or multiplex assays. For example, Tang [38] has recently described a method that monitors cell cycle state and apoptosis in mammalian cells in a rapid one-step multiplex cell imaging assay. This method utilizes three fluorescent dyes in living cells with no wash steps, thus making it quicker, accurate, and more cost effective than conventional antibody based assays. Butterick et al. [39] has described a multiplexing method that measures apoptosis and cell number in a single tube. In Chap. 4 of this volume, Telford provides an excellent invaluable up-to-date review of a wide variety of high throughput multiplex flow cytometry apoptosis methods and detection probes.

Most of the developments in high throughput, real-time readout, and/or multiplex apoptosis assays have focused on caspase 3/7, the most common apoptosis executioner caspases [36, 37, 40]. Thus *Apoptosis Methods in Toxicology* describes protocols based on notable improvements in real time and high-throughout assays that quantitate apoptosis based on caspase 3/7 assays (Chaps. 6–8, 10 and 11). Chapter 6 is entitled "A Low-Cost Method for Tracking the Induction of Apoptosis using FRET-based Activity Sensors in Suspension Cells". It describes a relatively inexpensive protocol for monitoring the induction and progression of apoptosis; the protocol utilizes a genetically encoded fluorescence resonance energy transfer (FRET)-based biosensor of caspase-3 in live cells growing in suspension. This protocol can be done in a microplate reader, and is thus useful for laboratories that have no access to a flow cytometer or high-end imagers. Chapter 7 is entitled "FRET-Based Measurement of Apoptotic Caspase Activities by High-throughput Screening Flow Cytometry". It describes a non-invasive, highly sensitive cost efficient protocol utilizing stable adherent cell lines and minimal sample handling steps; this protocol is useful for high-throughput compound screening in a 96-well format for laboratories with access to a high end flow cytometer. Chapter 8 is entitled "Automated Ratio Imaging Using Nuclear Targeted FRET Probe Expressing Cells for Apoptosis Detection". It describes a high throughput screening (for multipoint/multidrug situations) automated microscopy based protocol for detecting caspase activation in live cells using stable cells expressing nuclear-targeted intracellular FRET probes. Chapter 11, entitled "Microplate-based Whole Zebrafish Caspase 3/7 Assay" details a whole animal microplate-based apoptosis assay based on a commercially available human specific caspase 3/7 bioluminescence; this provides a rapid, reproducible, and predictive alternative animal model for identifying potential inhibitors and activators of apoptosis. Thus, the most notable improvements have resulted in a wide range of technologies to measure caspase 3/7 activity in vitro and in vivo using colorimetric, fluorimetric, and luminescence substrates, as well as FRET based and label-free sensors (Chaps. 6–8, 10 and 11), [37].

Apoptosis has become one of the prime targets for development of new therapies and overcoming resistance in cancer treatment [9, 41, 42]. In fact, drug screening assays not based on apoptosis perform poorly [43]. Predictive tests for response to cancer chemotherapy are also being developed based on apoptosis methods; these predictive tests aim to improve the outcomes of therapy [44]. A drug induced apoptosis assay, the microculture kinetic MiCK assay, has been developed to determine which chemotherapy drugs or regimes can provide higher cell killing in vitro. The assay has undergone clinical trials, and the results of the assay have been effective in determining patient treatment plans [44]. Various developments in the design of compounds that target apoptosis has resulted in small molecules with therapeutic potential against cancer. Chapters 12 and 13 in this volume contain the latest information on preclinical small molecules and their effect on cell death through apoptosis and autophagy. Chapter 12 provides a comprehensive review of the interrelationships between the cell death subroutines (e.g., apoptosis and autophagic cell death) and relevant anticancer small-molecule compounds. This is timely information that reflects recent developments in this field. Chapter 13, entitled "Liposomes in Apoptosis Induction and Cancer Therapy" reviews enhanced cancer cell death subroutine therapeutic effects of small molecules through the use of various liposome formulations.

It is safe to speculate that additional developments in apoptosis assays will continue to take place, as judged from the progress that has been made in recent years. A number of techniques based on real time chemical sensor technology have recently been described. In Chap. 10 of this volume, Miao *et al.* has described a "Novel Electrochemical Biosensor for Apoptosis Evaluation". These investigators detail a simple, cost-effective, convenient, and sensitive real time peptide-based electrochemical biosensor method for determining the extent of apoptosis. In other studies, Chen *et al.* [45] have reported an apoptosis assay based on detection of caspase 3 using a graphene oxide-assisted electrochemical signal amplification method. Huang *et al.* [46] have also reported an apoptosis assay based on the chemiluminescent detection of caspase 3 by gold nanoparticle-based resonance energy transfer. Additional new apoptosis assays have recently been described. Pfister *et al.* [47] have described the use of a Taqman protein assay to detect and quantitate apoptosis in primary cells. Using this assay, apoptosis can be quantified in a fast and reliable manner by quantifying levels of cleaved caspase 3 at a 1000-fold higher sensitivity than western blot. In Chap. 9 of this volume Stokes *et al.* describes an "Antibody-Based Proteomic Analysis of Apoptosis Signaling" method. This novel technique utilizes an antibody-based enrichment procedure in combination with a liquid chromatography-tandem mass spectrometry analytical platform to identify apoptotic signaling pathways and networks.

Future directions in the development of apoptosis methods in toxicology will likely involve computational modeling and systems biological approaches [48, 49]. These approaches will help to determine the multifactorial determinants of cell death decisions involving apoptosis, autophagic cell death, and necroptosis [10, 24, 28]. Not all cells in a population behave the same. Thus, the study of apoptosis and other forms of cell death at the single cell level, as well as the population is necessary. This will help our understanding of the cell to cell variability in cell death [50, 51]. This will in turn help to develop combination treatment strategies for cancer, thus achieving systems based analysis of chemotherapy resistance.

References

1. Elmore S (2007) Apoptosis: a review of programmed cell death. Toxicol Pathol 35: 495–516
2. Goldar S, Khaniani MS, Derakhshan SM, Baradaran B (2015) Molecular mechanisms of apoptosis and roles in cancer development and treatment. Asian Pac J Cancer Prev 16: 2129–2144
3. Kerr JF, Wyllie AH, Currie AR (1972) Apoptosis: a basic biological phenomenon with wide-ranging implications in tissue kinetics. Br J Cancer 26:239–257
4. Monier B, Suzanne M (2015) The morphogenetic role of apoptosis. Curr Top Dev Biol 114:335–362
5. Li K, Wu D, Chen X, Zhang T, Zhang L, Yi Y, Miao Z, Jin N, Bi X, Wang H, Xu J, Wang D (2014) Current and emerging biomarkers of cell death in human disease. Biomed Res Int 2014:690103
6. Tower J (2015) Programmed cell death in aging. Ageing Res Rev 23:90–100
7. Sankari SL, Babu NA, Rajesh E, Kasthuri M (2015) Apoptosis in immune-mediated diseases. J Pharm Bioallied Scii 7:S200–S202
8. Sarvothaman S, Undi RB, Pasupuleti SR, Gutti U, Gutti RK (2015) Apoptosis: role in myeloid cell development. Blood Res 50:73–79
9. Lopez J, Tait SW (2015) Mitochondrial apoptosis: killing cancer using the enemy within. Br J Cancer 112:957–962
10. Radogna F, Dicato M, Diederich M (2015) Cancer-type-specific crosstalk between autophagy, necroptosis and apoptosis as a pharmacological target. Biochem Pharmacol 94:1–11
11. Obulesu M, Lakshmi MJ (2014) Apoptosis in Alzheimer's disease: an understanding of the physiology, pathology and therapeutic avenues. Neurochem Res 39:2301–2312
12. Koff JL, Ramachandiran S, Bernal-Mizrachi L (2015) A time to kill: targeting apoptosis in cancer. Int J Mol Sci 16:2942–2955
13. Zeng W, Wang X, Xu P, Liu G, Eden HS, Chen X (2015) Molecular imaging of apoptosis: from micro to macro. Theranostics 5: 559–582
14. Kim YE, Chen J, Chan JR, Langen R (2010) Engineering a polarity-sensitive biosensor for time-lapse imaging of apoptotic processes and degeneration. Nat Methods 7:67–73
15. Leventis PA, Grinstein S (2010) The distribution and function of phosphatidylserine in cellular membranes. Annu Rev Biophys 39: 407–427
16. Hankins HM, Baldridge RD, Xu P, Graham TR (2015) Role of flippases, scramblases and

transfer proteins in phosphatidylserine subcellular distribution. Traffic 16:35–47

17. Shalini S, Dorstyn L, Dawar S, Kumar S (2015) Old, new and emerging functions of caspases. Cell Death Differ 22:526–539
18. Galluzzi L, Bravo-San Pedro JM, Vitale I, Aaronson SA, Abrams JM, Adam D, Alnemri ES, Altucci L, Andrews D, Annicchiarico-Petruzzelli M, Baehrecke EH, Bazan NG, Bertrand MJ, Bianchi K, Blagosklonny MV, Blomgren K, Borner C, Bredesen DE, Brenner C, Campanella M, Candi E, Cecconi F, Chan FK, Chandel NS, Cheng EH, Chipuk JE, Cidlowski JA, Ciechanover A, Dawson TM, Dawson VL, De Laurenzi V, De Maria R, Debatin KM, Di Daniele N, Dixit VM, Dynlacht BD, El-Deiry WS, Fimia GM, Flavell RA, Fulda S, Garrido C, Gougeon ML, Green DR, Gronemeyer H, Hajnoczky G, Hardwick JM, Hengartner MO, Ichijo H, Joseph B, Jost PJ et al (2015) Essential versus accessory aspects of cell death: recommendations of the NCCD 2015. Cell Death Differ 22:58–73
19. Galluzzi L, Vitale I, Abrams JM, Alnemri ES, Baehrecke EH, Blagosklonny MV, Dawson TM, Dawson VL, El-Deiry WS, Fulda S, Gottlieb E, Green DR, Hengartner MO, Kepp O, Knight RA, Kumar S, Lipton SA, Lu X, Madeo F, Malorni W, Mehlen P, Nunez G, Peter ME, Piacentini M, Rubinsztein DC, Shi Y, Simon HU, Vandenabeele P, White E, Yuan J, Zhivotovsky B, Melino G, Kroemer G (2012) Molecular definitions of cell death subroutines: recommendations of the Nomenclature Committee on Cell Death 2012. Cell Death Differ 19:107–120
20. Kanemura S, Tsuchiya A, Kanno T, Nakano T, Nishizaki T (2015) Phosphatidylinositol induces caspase-independent apoptosis of malignant pleural Mesothelioma cells by accumulating AIF in the nucleus. Cell Physiol Biochem 36:1037–1048
21. Rello-Varona S, Herrero-Martin D, Lopez-Alemany R, Munoz-Pinedo C, Tirado OM (2015) "(Not) all (dead) things share the same breath": identification of cell death mechanisms in anticancer therapy. Cancer Res 75:913–917
22. Li MX, Dewson G (2015) Mitochondria and apoptosis: emerging concepts. F1000Prime Rep 7
23. Degterev A, Zhou W, Maki JL, Yuan J (2014) Assays for necroptosis and activity of RIP kinases. Methods Enzymol 545:1–33
24. Wu H, Che X, Zheng Q, Wu A, Pan K, Shao A, Wu Q, Zhang J, Hong Y (2014) Caspases: a molecular switch node in the crosstalk between autophagy and apoptosis. Int J Biol Sci 10:1072–1083
25. Nakahira K, Cloonan SM, Mizumura K, Choi AM, Ryter SW (2014) Autophagy: a crucial moderator of redox balance, inflammation, and apoptosis in lung disease. Antioxid Redox Signal 20:474–494
26. Liu B, Bao JK, Yang JM, Cheng Y (2013) Targeting autophagic pathways for cancer drug discovery. Chin J Cancer 32:113–120
27. Wu HJ, Pu JL, Krafft PR, Zhang JM, Chen S (2015) The molecular mechanisms between autophagy and apoptosis: potential role in central nervous system disorders. Cell Mol Neurobiol 35:85–99
28. Lalaoui N, Lindqvist LM, Sandow JJ, Ekert PG (2015) The molecular relationships between apoptosis, autophagy and necroptosis. Semin Cell Dev Biol 39:63–69
29. Campisi L, Cummings RJ, Blander JM (2014) Death-defining immune responses after apoptosis. Am J Transplant 14:1488–1498
30. Bucur O, Stancu AL, Khosravi-Far R, Almasan A (2012) Analysis of apoptosis methods recently used in Cancer Research and Cell Death & Disease publications. Cell Death Dis 3, e263
31. Ulukaya E, Acilan C, Ari F, Ikitimur E, Yilmaz Y (2011) A glance at the methods for detection of apoptosis qualitatively and quantitatively. Turkish J Biochem 36:261–269
32. Squier MK, Cohen JJ (2001) Standard quantitative assays for apoptosis. Mol Biotechnol 19:305–312
33. Yadavilli S, Muganda PM (2004) Diepoxybutane induces caspase and p53-mediated apoptosis in human lymphoblasts. Toxicol Appl Pharmacol 195:154–165
34. Kim YE, Chen J, Langen R, Chan JR (2010) Monitoring apoptosis and neuronal degeneration by real-time detection of phosphatidylserine externalization using a polarity-sensitive indicator of viability and apoptosis. Nat Protoc 5:1396–1405
35. Rahbar Saadat Y, Saeidi N, Zununi Vahed S, Barzegari A, Barar J (2015) An update to DNA ladder assay for apoptosis detection. Bioimpacts 5:25–28
36. McStay GP, Green DR (2014) Measuring apoptosis: caspase inhibitors and activity assays. Cold Spring Harb Protoc 2014:799–806
37. Nicholls SB, Hyman BT (2014) Measuring caspase activity in vivo. Methods Enzymol 544:251–269
38. Tang Y (2014) A one-step imaging assay to monitor cell cycle state and apoptosis in mammalian cells. Curr Protoc Chem Biol 6:1–5

39. Butterick TA, Duffy CM, Lee RE, Billington CJ, Kotz CM, Nixon JP (2014) Use of a caspase multiplexing assay to determine apoptosis in a hypothalamic cell model. J Vis Exp. doi:10.3791/51305
40. Yamaguchi Y, Kuranaga E, Nakajima Y, Koto A, Takemoto K, Miura M (2014) In vivo monitoring of caspase activation using a fluorescence resonance energy transfer-based fluorescent probe. Methods Enzymol 544:299–325
41. Bai L, Wang S (2014) Targeting apoptosis pathways for new cancer therapeutics. Annu Rev Med 65:139–155
42. Elkholi R, Renault TT, Serasinghe MN, Chipuk JE (2014) Putting the pieces together: How is the mitochondrial pathway of apoptosis regulated in cancer and chemotherapy? Cancer Metab 2:16
43. Burstein HJ, Mangu PB, Somerfield MR, Schrag D, Samson D, Holt L, Zelman D, Ajani JA, American Society of Clinical O (2011) American Society of Clinical Oncology clinical practice guideline update on the use of chemotherapy sensitivity and resistance assays. J Clin Oncol 29:3328–3330
44. Bosserman L, Rogers K, Willis C, Davidson D, Whitworth P, Karimi M, Upadhyaya G, Rutledge J, Hallquist A, Perree M, Presant CA (2015) Application of a drug-induced apoptosis assay to identify treatment strategies in recurrent or metastatic breast cancer. PLoS One 10, e0122609
45. Chen H, Zhang J, Gao Y, Liu S, Koh K, Zhu X, Yin Y (2015) Sensitive cell apoptosis assay based on caspase-3 activity detection with graphene oxide-assisted electrochemical signal amplification. Biosens Bioelectron 68:777–782
46. Huang X, Liang Y, Ruan L, Ren J (2014) Chemiluminescent detection of cell apoptosis enzyme by gold nanoparticle-based resonance energy transfer assay. Anal Bioanal Chem 406:5677–5684
47. Pfister C, Pfrommer H, Tatagiba MS, Roser F (2015) Detection and quantification of apoptosis in primary cells using Taqman(R) protein assay. Methods Mol Biol 1219:57–73
48. Wurstle ML, Zink E, Prehn JH, Rehm M (2014) From computational modelling of the intrinsic apoptosis pathway to a systems-based analysis of chemotherapy resistance: achievements, perspectives and challenges in systems medicine. Cell Death Dis 5, e1258
49. Terranova N, Rebuzzini P, Mazzini G, Borella E, Redi CA, Zuccotti M, Garagna S, Magni P (2014) Mathematical modeling of growth and death dynamics of mouse embryonic stem cells irradiated with gamma-rays. J Theor Biol 363:374–380
50. Flusberg DA, Sorger PK (2015) Surviving apoptosis: life-death signaling in single cells. Trends Cell Biol 25:446–458
51. Xia X, Owen MS, Lee RE, Gaudet S (2014) Cell-to-cell variability in cell death: can systems biology help us make sense of it all? Cell Death Dis 5, e1261

Chapter 2

Detection of Apoptosis: From Bench Side to Clinical Practice

William P. Bozza, Julianne D. Twomey, Su-Ryun Kim, and Baolin Zhang

Abstract

Apoptosis or programmed cell death is implicated in several pathological conditions, such as cancer and neurodegenerative diseases. An increasing number of therapies are developed by targeting apoptosis signaling components to either induce or inhibit apoptosis in target cells. For these reasons, it is critical to develop appropriate analytical methods for the detection of apoptotic cell death in the context of monitoring relevant disease progression and therapeutic effects of clinical treatments (e.g., chemotherapy in cancer patients). This review provides an overview of the currently used methods for detection of apoptosis and their applications in research and clinical practice.

Key words Apoptosis, Apoptosis detection, In vitro apoptosis detection, In vivo apoptosis detection, Clinical apoptosis detection, DNA fragmentation, TUNEL, Caspase activation detection, Phosphatidylserine externalization

1 Introduction

Apoptosis or programmed cell death is a highly organized cellular process for removing unwanted cells from the body during organ development, tissue remodeling, and immune responses. Apoptosis is thought to be physiologically advantageous because early apoptotic cells are cleared by phagocytosis before they lose their plasma membrane permeability barrier [1–3]. In this manner, apoptotic cells are degraded within the macrophages. Loss of control of programmed cell death (resulting in excessive apoptosis) can lead to neurodegenerative diseases, hematologic diseases, and tissue damage. For example, the progression of HIV is directly linked to excessive, unregulated apoptosis. On the other hand, insufficient or defective apoptosis is linked to the development of cancer progression and drug resistance to chemotherapy. As such, an increasing number of drugs have been approved or are under development which target specific aberrant signaling components of cell death or survival pathways [4–6], including small molecule inhibitors and

Perpetua M. Muganda (ed.), *Apoptosis Methods in Toxicology*, Methods in Pharmacology and Toxicology,
DOI 10.1007/978-1-4939-3588-8_2, © Springer Science+Business Media New York 2016

therapeutic proteins [7]. Additionally, while the primary goal of chemotherapy is to kill cancer cells by any means, a secondary goal is to have those cells die by apoptosis so that they may be cleared quickly and "quietly" by neighboring phagocytic cells. For these reasons, it is critical to develop appropriate analytical methods for the detection of apoptotic cell death for monitoring disease progression and for the effects of therapeutic intervention like chemotherapy.

Apoptosis is characterized by a discrete set of biochemical steps and morphological changes that include the activation of caspases, translocation of phosphatidylserine from the inner to the outer layer of the plasma membrane, chromatin condensation, and fragmentation of the cell into subcellular parts called apoptotic bodies [8]. There are two major apoptosis pathways that can lead to caspase activation: the mitochondria-directed intrinsic pathway and the death-receptor mediated extrinsic pathway [9–11]. The intrinsic apoptotic signaling pathway is triggered in response to various stress signals including DNA damage, γ-irradiation, hypoxia, and survival factor deprivation. Intrinsic apoptosis involves the release of mitochondrial factors (e.g., cytochrome c) that signal downstream programmed cell death events. By contrast, the extrinsic apoptosis signaling pathway is mediated through the death receptors expressed on the cell surface membrane. These receptors, including TNF receptor 1 (TNFR1), Fas, DR4, and DR5, are characterized by an intracellular death domain that can be selectively activated by their cognate ligands such TNF, Fas ligand (FasL), and TNF-related apoptosis inducing ligand (TRAIL) [12–15]. Both intrinsic and extrinsic apoptosis pathways lead to activation of a cascade of cysteine-dependent aspartyl proteases, known as caspases, which catalyze the cleavage of cellular substrates at specific amino acid sequences (e.g., DXXD for caspase 3) [16]. Another distinct feature of apoptosis is the translocation of phosphatidylserine from the inner leaflet to the outer layer of the plasma membrane [17].

2 In Vitro Apoptosis Detection Methods

Because Apoptosis entails a vast number of sequential biochemical events, thus providing many check points at which detection of apoptosis can be accomplished. This review focuses on apoptosis detection methods that are designed by exploiting detection of the flipping of phosphatidylserine to the extracellular side of the plasma membrane, mitochondrial membrane permeabilization, DNA fragmentation, and intracellular caspase activation (Fig. 1).

Fig. 1 (continued) mitochondrial intermembrane proteins (e.g., cytochrome c) through ELISA, confocal microscopy, or high performance liquid chromatography techniques. DNA fragmentation is measured either through gel electrophoresis or end point TUNEL immunohistochemical visualization.

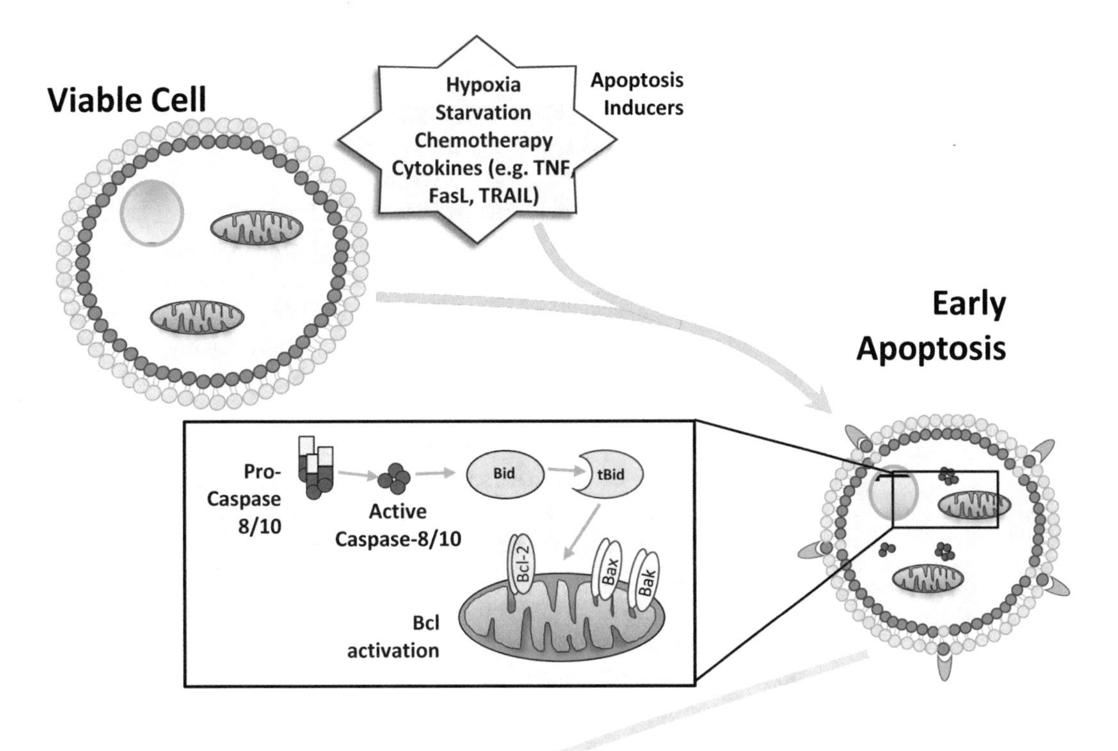

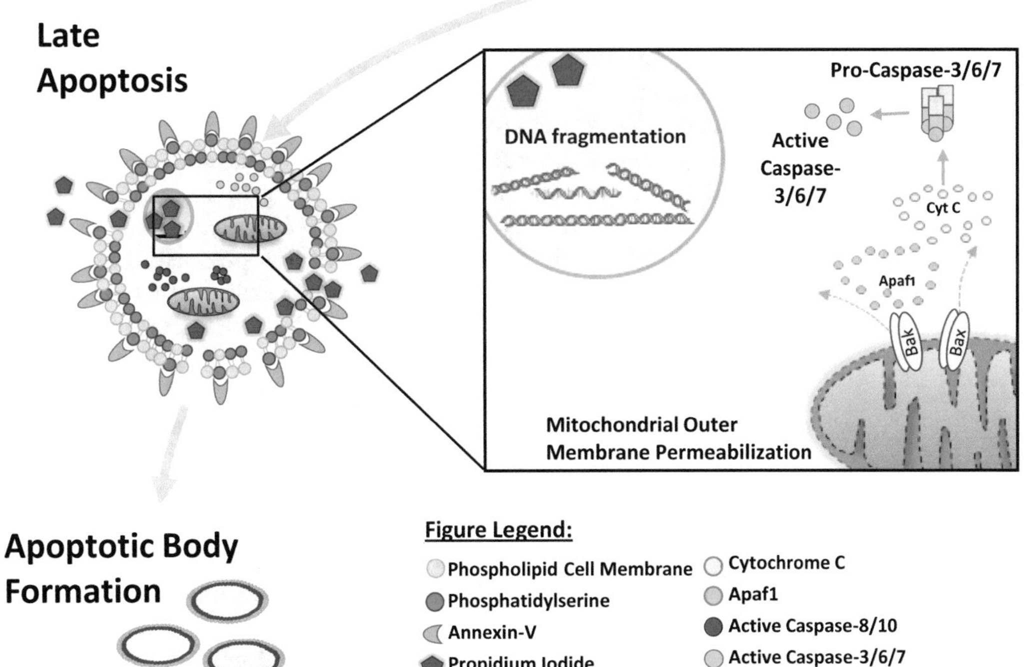

Fig. 1 An illustration of important apoptotic biochemical events that are often exploited to detect induction of apoptosis. Apoptosis is induced through varying endogenous and exogenous mechanisms such as hypoxia, starvation, chemotherapy, and targeted cytokines; and can be detected by exploiting one of the many biochemical events that occur during the programmed cell death. Loss of membrane symmetry and integrity can be visualized using labeled Annexin-V which binds to externalized phosphatidylserine on the cell membrane and propidium iodide which binds DNA after outer membrane permeabilization during late stages of apoptosis. Activated caspases can be detected and quantified using caspase labeling probes and labeled caspase substrates; while mitochondrial membrane permeabilization can be detected by identification of the release of

2.1 Phosphatidylserine Extracellular Flipping

Phosphatidylserine (PS) normally faces the inside leaflet of the cell membrane; however, during the onset of apoptosis PS flips and is exposed extracellularly. Flow cytometry has been widely used to detect cell populations with exposed PS. This detection is accomplished by exploiting the tight binding that exists between the anticoagulant protein Annexin-V and PS. In these experiments, Annexin-V is conjugated to various fluorophores [18, 19] allowing for efficient labeling of apoptotic cells. Alternatively, detection approaches utilizing Annexin-V labeled with quantum dots (QDs) have advantages in robustness and sensitivity due to their semiconductor properties [20, 21]. PS extracellular flipping is one of the earliest processes that occurs during apoptosis, and is therefore used to detect cells undergoing early stage apoptosis. PS/Annexin-V labeling is often complemented with DNA binding dyes such as 7-amino-actinomycin D (7-AAD) or propidium iodide (PI) which can only penetrate the cell membrane during late stages of apoptosis and necrosis [22]. Together this approach allows for differentiation of healthy cells from cells undergoing early or late stage apoptosis and also necrosis.

2.2 Mitochondrial Membrane Permeabilization

Mitochondrial membrane permeabilization is another cellular change often used to detect apoptosis induction. Physical alterations in mitochondrial structure during apoptosis have been visualized by electron microscopy [23, 24]. However, this approach is limited regarding automation and quantification. Alternatively, analysis of the cellular redistribution of proteins that commonly reside in the intracellular space of the mitochondria is a technique often performed to assay mitochondria permeabilization, and therefore apoptosis [25]. This has traditionally been accomplished by immunoblot and immunofluorescence detection of released mitochondrial intermembrane spaced proteins, such as apoptosis-inducing factor (AIF) and cytochrome c (cyt c) [26, 27]. To aid in assay throughput, ELISA-based immunoassays have been developed for the detection of released cytochrome c [28]. Interestingly, confocal microscopy of GFP-tagged cytochrome c revealed important kinetic information, and has indicated that cytochrome c release can precede PS extracellular flipping and loss of plasma membrane integrity [29]. Similarly, high performance liquid chromatography (HPLC) has been used to detect mitochondrial metabolites that have diffused due to membrane permeabilization [30]. Alternatively, other assays have been developed that monitor the activity of the mitochondrial respiratory chain. These assays classically involve the conversion of tetrazolium salts into colored products; only occurring in the presence functional mitochondria. This technique is routinely used as a measure of cell viability [31–33]. However, lengthy incubation times and the inability to differentiate between growth arrest and true cell death are major limitations of the mitochondrial respiratory chain assays. Lastly, many cationic fluorophores have been exploited in order to detect changes in mitochondrial transmembrane potential generated during apoptosis [34, 35].

2.3 DNA Fragmentation

DNA fragmentation is yet another key feature of apoptosis. DNA fragmentation is initiated by activated endonucleases, and sequentially yields high molecular weight DNA fragments that are further cleaved into oligonucleosomal fragments of 180–200 base pairs [36–38]. Apoptosis has historically been detected through visualization of these DNA fragmentation patterns. Specifically, conventional gel electrophoresis has been utilized in separating and visualizing low molecular weight DNA patterns, and has defined the characteristic "laddering" pattern as a hallmark of apoptosis [38, 39]. Additionally, pulse field and field-inversion gel electrophoresis have been used to resolve the larger high molecular weight DNA fragments [40, 41]. Due to its simplicity and sensitivity, single cell gel electrophoresis has been used to detect changes in DNA degradation, and therefore apoptosis at the single cell level [42]. This approach is advantageous in its ability to identify specific types of DNA damage such as single and double strand breaks. Collectively, gel electrophoresis techniques have limitations in assay time, automation, and quantification. Fragmented DNA can also commonly be detected by terminal deoxynucleotidyl transferase dUTP nick end labeling (TUNEL) [43, 44]. In TUNEL, terminal deoxynucleotidyl transferase recognizes the 3′-OH termini of DNA breaks caused by induction of apoptosis. The enzyme then catalyzes the addition of labeled dUTPs into the damaged DNA. This labeled DNA can be visualized either by fluorescence microscopy or cytometry [45]. One major advantage of this approach is when complemented with PI staining, the phase of the cell cycle where apoptosis is occurring in can be determined [46]. However, cell fixation requirements and lengthy staining protocols are experimental limitations.

2.4 Caspase Activation

Caspases play an essential role in the execution of apoptosis, and their activation is often used as a marker of apoptosis. The availability of high quality antibodies specific to pro-caspases and their cleaved active forms offers a convenient tool for monitoring apoptosis under both in vitro and in vivo settings. Although caspase activation can be easily detected by immunoblot analysis, this approach has several limitations. In efforts to aid in assay robustness, sensitivity, repeatability, and automation there has been a strong push for the development of assays that can detect (1) labeling of active caspases and (2) cleavage of caspase substrates. In regard to caspase labeling, apoptosis detection through the use of fluorochrome-labeled inhibitors of caspases (FLICA) has been rigorously demonstrated [47–49]. FLICA-based probes are cell membrane permeant fluorescent ligands that covalently bind to the caspase active site. Cells labeled with FLICAs can be detected using fluorescence microscopy, flow cytometry, or laser scanning cytometry (LSC). Recently, detection of caspase activation with FLICA has been accomplished in living lamprey brains and spinal axons, indicating the utility of this approach in an in vivo setting (vide infra) [50, 51]. However, a drawback of the FLICA approach is that in some instances these probes have actually been found to protect cells and prevent apoptosis [52, 53], complicating apoptosis detection.

Because caspases recognize and cleave specific amino acid sequences in substrate proteins, consensus peptides can serve as excellent biosensors for apoptosis detection [16]. Specifically, fluorophores such as 7-amino-4-methylcoumarin (AMC), 7-amido-4-trifluoromethylcoumarin (AFC), or rhodamine 110 have been covalently conjugated to caspase substrate peptides [54]. When covalently linked to a peptide, total fluorescence is quenched. However, in the presence of active caspase, the substrate is cleaved and the fluorophore is liberated from full length peptide yielding a pronounced fluorescence signal. In this design, high fluorescence backgrounds can often be one major obstacle. This has led to the evolution of fluorescence resonance energy transfer (FRET) as a powerful technique in detection of active caspase [55, 56]; where fluorescence donors and quenchers have been covalently attached to the termini of caspase substrate peptides. The intact substrate orients the fluorescence donor and acceptor in close proximity allowing for efficient FRET. Upon induction of apoptosis and caspase activation, the caspase substrate peptides are cleaved, allowing a distance dependent decrease in FRET. In many instances, the caspase biosensor is generated separately before addition to cells. This creates a major obstacle in cell permeability and can lead to a requirement of cell lysis, which adds undesirable complexity.

With the advancement of recombinant DNA technology, fluorescent fusion proteins have been engineered to act as excellent caspase/apoptosis FRET biosensors. In the design, one fluorescent protein acts as a donor while the other functions as an acceptor. Each protein is linked by a caspase specific peptide linker. As a cellular biosensor, this approach eliminates the need for additional substrate processing steps, such as cell penetration. Using this method, changes in FRET have been determined using many different fluorescent protein pairs [57–59]. Most commonly, FRET based detection of caspase activation relies on either microscopy or flow cytometry [58, 60–63]. While providing valuable information, this operating procedure is not ideal for high throughput formats. Notably, there are sparse examples reported which utilize a throughput plate format [64, 65].

Recently our lab has generated a novel cell-based FRET biosensor that allows for an automated detection of apoptosis induced by anticancer drugs [66]. Specifically, MDA-MB-231 breast cancer cells have been engineered to stably express a CFP-linker-YFP fusion protein, wherein CFP functions as a donor and YFP as an acceptor for fluorescence resonance energy transfer (FRET). The linker contains both caspase 3 and caspase 8 recognition sequences, DEVD and IETD, respectively, allowing for sensitivity in drugs that function via either intrinsic or extrinsic apoptosis pathways. Upon caspase activation when cells are treated with a drug, the linker connecting CFP and YFP will be cleaved and subsequently will render a loss in cellular FRET signal. Importantly, the cell-based assay has been uniquely adapted to a microplate (e.g., 96-well plate) format.

Our platform has been found successful in quantifying the bioactivity of complex therapeutic proteins, such as TNF-related apoptosis-inducing ligand (TRAIL) and death receptor agonistic antibodies that work through extrinsic apoptosis pathways. Our assay was also successful in characterization of small molecule chemotherapeutic agents (e.g., camptothecin) that are effective by inducing intrinsic mediated apoptosis. Together these results showcase how our assay allows for an unprecedented detection of both caspase 3 and caspase 8 activity in a high throughput format. From a technical aspect, our assay is superior in sensitivity, simplicity, stability, convenience, and robustness when compared to the other described apoptosis assays. Also, our methodology has been uniquely adapted to eliminate background autofluorescence associated with cell culture media; this helps to achieve an improved signal-to-noise ratio and avoids high background subtractions which can complicate FRET calculations. Additionally, our platform has been adapted to a high-throughput screening format and was engineered to be easily setup in any lab. Most importantly, this assay closely reflects the mechanism of action of cancer drugs in killing cancer cells and can distinguish between drugs that solely trigger growth inhibition.

3 In Vivo Apoptosis Detection

Development of in vivo apoptosis detection techniques can be critical in assessing progression of several diseases and also in evaluating therapeutic effects of treatments, such as chemotherapy in cancer patients. Currently many in vivo apoptosis assays utilize noninvasive imaging modalities in order to detect tissues undergoing cellular apoptosis. These assays aid in individualized treatments for clinical efficacy, and allow for earlier predictions of the patient's response to therapy.

3.1 Changes in Diffusive Properties of Tissue and Tumor

Magnetic resonance imaging (MRI) has been used to monitor fluid diffusion gradients in tissues, and can distinguish tumors from healthy tissue due to high cellularity [67].

MRI is also often performed for the evaluation of tumor response to anticancer drugs; in these cases physical changes in tumor size and diffusion properties detected by MRI have been used to identify apoptotic cell death [68]. Specifically, diffusion weighted MRI (DW-MRI) has been used to measure increases in water diffusion for tumors in response to radiation, chemotherapies, and therapeutic proteins [69]. Tissues undergoing apoptosis will demonstrate an increase in apparent diffusion coefficient (ADC), used to represent water diffusion [70–73]. Another magnetic imaging modality, ^{1}H Magnetic resonance spectroscopy (MRS), has been used to quantify lipid concentration. Lipid concentration is known to increase in apoptotic cells [74] and this change in fat-water ratio is another established biomarker of tumor cell death [75, 76]. MRS has been found successful in detecting apoptotic cell death in cervical carcinoma [77] and in triple negative breast cancer treated with anti-DR5 antibody and carboplatin combination [78].

Although MRI approaches are commonly used in clinical practice, there has been a recent push towards the development of assays that can more directly and timely confirm the induction of apoptosis. This has been accomplished by labeling of apoptotic cells by exploiting the processes of PS extracellular flipping, changes in cell membrane imprint, DNA fragmentation, and caspase activation.

3.2 Phosphatidylserine Extracellular Flipping

An early indicator of apoptosis is the loss of membrane phospholipid asymmetry, causing PS to externalize on the plasma membrane [19, 79, 80]. One of the most common ways to detect this in vivo is by imaging for labeled Annexin-V. As mentioned earlier, Annexin-V has high affinity for PS. The in vivo uptake and biodistribution of Annexin-V has been investigated. Radiolabeled Annexin-V is imaged a minimum of an hour following intravenous injection, allowing for protein biodistribution [81–86]. The radiolabeled Annexin-V is cleared from circulation as quickly as three to seven minutes post injection, allowing for a clear indication of which tissues have taken up the protein [82]. Labeled Annexin-V can be monitored in 2D using scintigraphy or in 3D by positron emission tomography (PET) and single photon-emission tomography (SPECT). Tissue reconstructions are then created from rendered projections where Annexin-V is found to localize. Various radionuclides have been conjugated to Annexin-V, such as technetium-99m (^{99m}Tc) [83, 85, 87–89], Iodine-123 (^{123}I) [90], Iodine-124 (^{124}I) [91–94], Fluorine-18 (^{18}F) [95], and Copper-64 (^{64}Cu) [96]. Each differs in biological half-life, binding affinity, and kidney, liver, and spleen uptake. Radiolabeling and imaging of exogenously administered Annexin-V has been used to detect apoptosis following an acute stroke [81], multiple sclerosis [97], allograft rejection [84, 86], arthritis [98], cardiac infarction [99, 100], cell death in patients with Alzheimer and dementia [101], and tumor response to anticancer therapies [84, 102]. Notably, in a Phase I/II clinical study, ^{99m}Tc-radiolabeled Annexin-V was successfully taken up in 7 of 15 patients that were presenting lung cancer, breast cancer, and lymphoma; and allowed for in vivo imaging of apoptosis in human cancers in response to chemotherapy treatment. Of these seven patients, four had a complete response and three had a partial response. Collectively, this work has strong implications for labeled Annexin-V functioning as a predictor of response to chemotherapy treatment.

Besides Annexin-V, synaptotagmin I [103–105] and the PS targeting monoclonal antibody fragment, PGN635 $F(ab')_2$ [94] have been used to bind PS and label apoptotic cells in vivo. Limitations of this methodology exist in pharmacokinetic problems associated with the large size of Annexin-V, nonspecific uptake, and its relatively slow clearance from the blood [85, 106, 107].

3.3 Changes in Apoptotic Membrane Imprint

Aposense molecules are low molecular weight compounds that have been developed to selectively invade, label, and image cells undergoing cell death. They target apoptotic cells by selectively binding to plasma membranes displaying changes in membrane potential, acidification of the outer leaflet and cytosol, and activation of the apoptotic scramblase system [108]. These probes then accumulate within the cytoplasm of apoptotic cells and allow for visualization due to their chemical properties. Interestingly, aposense uptake is caspase dependent [109] and has been shown to be selective for apoptotic cells, judged by TUNEL, hematoxylin and eosin (H and E) ex vivo staining, and Annexin-V staining [109, 110]. To date, several aposense molecules have been developed and reported: DDC [111], NST-732 [112], NST-729, ML-9 [113], and ML-10 [114]. Radiolabeled aposense compounds, such as ^{18}F ML-10, allow for positron emission tomography which adds tremendous value in clinical settings. Together this technology has been shown to be successful in identifying apoptosis induced by anticancer drugs, renal failure, neurodegenerative disease, and stroke [109–113]. Human studies revealed that ^{18}F ML-10 has advantages in biodistribution, stability, dosimetry, signal to noise, and safety [115]. It is currently being evaluated by clinical trials as a radiotracer for brain metastases, head and neck neoplasms, and non-small-lung carcinomas in response to radiation or chemoradiation (as provided by https://clinicaltrials.gov). Additionally, it has advantages in selectivity and can distinguish between cells dying through apoptotic and necrotic pathways [109, 114].

3.4 DNA Fragmentation

DNA fragmentation in biopsy samples is used to detect apoptosis induction through terminal deoxynucleotidyl transferase (TdT) dUTP nick end labeling (TUNEL). As previously mentioned, TUNEL staining identifies nicks within genomic DNA by detecting either fluorescence or DAB (3, 3′-diaminobenzidine) streptavidin-labeled dUTPs that are incorporated into damaged DNA by TdT. The degree of apoptosis is determined by the intensity of the TUNEL staining within apoptotic nuclei, as well as visualization of DNA fragments within the cytosol of the cell. Histology sections of biopsies are used to compare the baseline pretreatment and posttreatment samples to determine drug efficacy [116–118]. While this assay allows for direct apoptotic labeling within the tumor, it is an end-stage detection method and lacks the ability to monitor apoptosis in real-time to enhance personalized medicine. However, end point analysis of biopsy histology can be used in conjunction with imaging techniques to verify long-term responses.

3.5 Caspase Activation

Caspase activation is a well-known apoptotic biomarker that has been extensively showcased both in vitro and in vivo. Complementing detection of caspase activation with in vivo imaging provides early and selective detection capabilities for the onset of disease and therapeutic efficacy in relation to induction of apoptosis. As with traditional in vitro techniques, detection of caspase activation can be accomplished in vivo either by labeling active caspases with ligands or by detecting cleavage of caspase-based substrates.

Peptide-based caspase labeling ligands that mimic natural physiological substrates have been exploited for the in vivo detection of caspase activity, and therefore apoptosis. In one recent promising example, [^{18}F]4-fluorobenzylcarbonyl–Val–Ala–Asp(OMe)–fluoromethylketone was developed and validated in vivo as a novel PET imaging probe capable of labeling active caspases and detecting apoptotic cell death triggered by anticancer treatment in mouse models of human colorectal cancer [119]. A major drawback of peptides based ligands is their overall modest uptake. In this regard, small molecule caspase ligands offer an attractive alternative. The most prominent examples exist within the isatin sulfonamide moiety family. Radiolabeled isatin sulfonamides, such as [^{18}F]WC-II-89 [120, 121], [^{11}C]4 (WC-98) [120], [^{18}F]ICMT-11 [122], and [^{18}F]WC-4-116 [123], are competitive caspase-3 inhibitors/labelers that can be traced using PET in order to detect caspase activation and apoptosis in vivo. However, a limitation exists in selectivity where other cysteine containing proteases can be labeled by these small molecule probes.

Caspase activity can also be measured by detection of enzymatic cleavage of substrate-based probes. In one recent example, a common caspase-3 peptide-based substrate (DEVD) was conjugated to a PET imaging radionuclide (^{18}F-CP18) on one side and a cell permeating polyethylene glycol (PEG)-sequence on the other [124]. This allows the tracer to efficiently enter the cell and to be cleaved by activated caspases existing in apoptotic cells. The PEG sequence is subsequently released and causes the radionuclide labeled peptide to become trapped and accumulate within the dying cells. These tracers have advantages in selectivity and stability. Importantly, tumor bearing mouse studies revealed that in vivo PET signal images correlated well with caspase 3/7 activity and caspase-3 immunohistochemistry. This approach has significant advantages in target selectivity, signal to noise background, biodistribution, and clearance. However, chemotherapy induced apoptosis in mouse xenografts was not analyzed, limiting the knowledge of whether this probe can detect therapeutically relevant changes in apoptosis.

4 Conclusion

In summary, we outline and describe here many of the existing methods used to detect apoptosis both in vitro and in vivo. These detection methods are presented in order to show how in vitro method development has led to promising in vivo diagnostic tools. When attempting to select an appropriate apoptosis detection method, it is essential to carefully analyze the purpose and the end goal of the experiment. Our lab is primarily interested in the detection of apoptosis in the context of therapies that target apoptosis signaling components to either induce or inhibit apoptosis in target cells. In this regard, it is our opinion that selectivity and early detection is critical. Also, it is important to select a platform that can distinguish between growth arrest and true cell death. Therefore, we favor apoptosis detection methods that rely on caspase activation. For in vitro purposes, we feel that assays which are highly automated, sensitive, and quantitative stand out amongst other. Among the examples described, our cellular FRET-based detection of caspase activation in a 96-well plate format satisfies all of these criteria. For clinically relevant in vivo detection of apoptosis, it is necessary not only to correlate increases in apoptosis with the efficacy and toxicity of applied drugs, but also to track disease progression in real-time. Noninvasive imaging techniques that detect either PS binding to Annexin-V or caspase activation allow apoptosis monitoring that most closely resembles real-time. This is highly beneficial when observing patient response during drug treatment or in monitoring disease progression. Although further research and clinical studies are clearly needed, complementing PET-based imaging probes with detection of caspase activation seems highly promising in regard to selectivity, pharmacokinetic uptake, and clearance.

Disclaimer

This chapter reflects the views of the author and should not be constructed to represent FDA's views or policies.

References

1. Anderson HA, Maylock CA, Williams JA, Paweletz CP, Shu H, Shacter E (2003) Serum-derived protein S binds to phosphatidylserine and stimulates the phagocytosis of apoptotic cells. Nat Immunol 4:87–91
2. Shacter E, Williams JA, Hinson RM, Senturker S, Lee YJ (2000) Oxidative stress interferes with cancer chemotherapy: inhibition of lymphoma cell apoptosis and phagocytosis. Blood 96:307–313
3. Uehara H, Shacter E (2008) Auto-oxidation and oligomerization of protein S on the apoptotic cell surface is required for Mer tyrosine kinase-mediated phagocytosis of apoptotic cells. J Immunol 180:2522–2530
4. Fesik SW (2005) Promoting apoptosis as a strategy for cancer drug discovery. Nat Rev Cancer 5:876–885
5. Qiao L, Wong BC (2009) Targeting apoptosis as an approach for gastrointestinal cancer therapy. Drug Resist Updat 12:55–64
6. Wong KK (2009) Recent developments in anti-cancer agents targeting the Ras/Raf/MEK/ERK pathway. Recent Pat Anticancer Drug Discov 4:28–35
7. Brunelle JK, Zhang B (2010) Apoptosis assays for quantifying the bioactivity of anticancer drug products. Drug Resist Updat 13:172–179
8. Kroemer G, Galluzzi L, Vandenabeele P, Abrams J, Alnemri ES, Baehrecke EH et al (2009) Classification of cell death: recommendations of the Nomenclature Committee on Cell Death 2009. Cell Death Differ 16:3–11
9. Eum KH, Lee M (2011) Crosstalk between autophagy and apoptosis in the regulation of paclitaxel-induced cell death in v-Ha-ras-transformed fibroblasts. Mol Cell Biochem 348:61–68
10. Ouyang L, Shi Z, Zhao S, Wang FT, Zhou TT, Liu B et al (2012) Programmed cell death pathways in cancer: a review of apoptosis, autophagy and programmed necrosis. Cell Prolif 45:487–498
11. Ward TH, Cummings J, Dean E, Greystoke A, Hou JM, Backen A et al (2008) Biomarkers of apoptosis. Br J Cancer 99:841–846
12. Ashkenazi A, Dixit VM (1998) Death receptors: signaling and modulation. Science 281:1305–1308
13. Bazzoni F, Beutler B (1996) The tumor necrosis factor ligand and receptor families. N Engl J Med 334:1717–1725
14. Elmore S (2007) Apoptosis: a review of programmed cell death. Toxicol Pathol 35:495–516
15. Locksley RM, Killeen N, Lenardo MJ (2001) The TNF and TNF receptor superfamilies: integrating mammalian biology. Cell 104:487–501
16. Riedl SJ, Shi Y (2004) Molecular mechanisms of caspase regulation during apoptosis. Nat Rev Mol Cell Biol 5:897–907
17. Bratton DL, Fadok VA, Richter DA, Kailey JM, Guthrie LA, Henson PM (1997) Appearance of phosphatidylserine on apoptotic cells requires calcium-mediated nonspecific flip-flop and is enhanced by loss of the aminophospholipid translocase. J Biol Chem 272:26159–26165
18. Galluzzi L, Aaronson SA, Abrams J, Alnemri ES, Andrews DW, Baehrecke EH et al (2009) Guidelines for the use and interpretation of assays for monitoring cell death in higher eukaryotes. Cell Death Differ 16:1093–1107
19. Koopman G, Reutelingsperger CP, Kuijten GA, Keehnen RM, Pals ST, van Oers MH (1994) Annexin V for flow cytometric detection of phosphatidylserine expression on B cells undergoing apoptosis. Blood 84:1415–1420
20. Park J, Park Y, Kim S (2013) Signal amplification via biological self-assembly of surface-engineered quantum dots for multiplexed subattomolar immunoassays and apoptosis imaging. ACS Nano 7:9416–9427
21. Prinzen L, Miserus RJ, Dirksen A, Hackeng TM, Deckers N, Bitsch NJ et al (2007) Optical and magnetic resonance imaging of cell death and platelet activation using annexin a5-functionalized quantum dots. Nano Lett 7:93–100
22. Vermes I, Haanen C, Steffens-Nakken H, Reutelingsperger C (1995) A novel assay for apoptosis. Flow cytometric detection of phosphatidylserine expression on early apoptotic cells using fluorescein labelled Annexin V. J Immunol Methods 184:39–51
23. de Graaf AO, van den Heuvel LP, Dijkman HB, de Abreu RA, Birkenkamp KU, de Witte T et al (2004) Bcl-2 prevents loss of mito-

chondria in CCCP-induced apoptosis. Exp Cell Res 299:533–540
24. Terauchi S, Yamamoto T, Yamashita K, Kataoka M, Terada H, Shinohara Y (2005) Molecular basis of morphological changes in mitochondrial membrane accompanying induction of permeability transition, as revealed by immuno-electron microscopy. Mitochondrion 5:248–254
25. Galluzzi L, Zamzami N, de La Motte RT, Lemaire C, Brenner C, Kroemer G (2007) Methods for the assessment of mitochondrial membrane permeabilization in apoptosis. Apoptosis 12:803–813
26. Loeffler M, Daugas E, Susin SA, Zamzami N, Metivier D, Nieminen AL et al (2001) Dominant cell death induction by extramitochondrially targeted apoptosis-inducing factor. FASEB J 15:758–767
27. Waterhouse NJ, Trapani JA (2003) A new quantitative assay for cytochrome c release in apoptotic cells. Cell Death Differ 10:853–855
28. Anantharam V, Kitazawa M, Wagner J, Kaul S, Kanthasamy AG (2002) Caspase-3-dependent proteolytic cleavage of protein kinase Cdelta is essential for oxidative stress-mediated dopaminergic cell death after exposure to methylcyclopentadienyl manganese tricarbonyl. J Neurosci 22:1738–1751
29. Goldstein JC, Waterhouse NJ, Juin P, Evan GI, Green DR (2000) The coordinate release of cytochrome c during apoptosis is rapid, complete and kinetically invariant. Nat Cell Biol 2:156–162
30. Vander Heiden MG, Chandel NS, Li XX, Schumacker PT, Colombini M, Thompson CB (2000) Outer mitochondrial membrane permeability can regulate coupled respiration and cell survival. Proc Natl Acad Sci U S A 97:4666–4671
31. Baize S, Leroy EM, Georges-Courbot MC, Capron M, Lansoud-Soukate J, Debre P et al (1999) Defective humoral responses and extensive intravascular apoptosis are associated with fatal outcome in Ebola virus-infected patients. Nat Med 5:423–426
32. de La Motte RT, Galluzzi L, Olaussen KA, Zermati Y, Tasdemir E, Robert T et al (2007) A novel epidermal growth factor receptor inhibitor promotes apoptosis in non-small cell lung cancer cells resistant to erlotinib. Cancer Res 67:6253–6262
33. Tajeddine N, Galluzzi L, Kepp O, Hangen E, Morselli E, Senovilla L et al (2008) Hierarchical involvement of Bak, VDAC1 and Bax in cisplatin-induced cell death. Oncogene 27:4221–4232
34. Metivier D, Dallaporta B, Zamzami N, Larochette N, Susin SA, Marzo I et al (1998) Cytofluorometric detection of mitochondrial alterations in early CD95/Fas/APO-1-triggered apoptosis of Jurkat T lymphoma cells. Comparison of seven mitochondrion-specific fluorochromes. Immunol Lett 61:157–163
35. Scaduto RC Jr, Grotyohann LW (1999) Measurement of mitochondrial membrane potential using fluorescent rhodamine derivatives. Biophys J 76:469–477
36. Bicknell GR, Snowden RT, Cohen GM (1994) Formation of high molecular mass DNA fragments is a marker of apoptosis in the human leukaemic cell line, U937. J Cell Sci 107(Pt 9):2483–2489
37. Schwartzman RA, Cidlowski JA (1993) Apoptosis: the biochemistry and molecular biology of programmed cell death. Endocr Rev 14:133–151
38. Wyllie AH (1980) Glucocorticoid-induced thymocyte apoptosis is associated with endogenous endonuclease activation. Nature 284:555–556
39. Oberhammer F, Wilson JW, Dive C, Morris ID, Hickman JA, Wakeling AE et al (1993) Apoptotic death in epithelial cells: cleavage of DNA to 300 and/or 50 kb fragments prior to or in the absence of internucleosomal fragmentation. EMBO J 12:3679–3684
40. Cantor CR, Smith CL, Mathew MK (1988) Pulsed-field gel electrophoresis of very large DNA molecules. Annu Rev Biophys Biophys Chem 17:287–304
41. Carle GF, Frank M, Olson MV (1986) Electrophoretic separations of large DNA molecules by periodic inversion of the electric field. Science 232:65–68
42. Collins AR (2002) The comet assay. Principles, applications, and limitations. Methods Mol Biol 203:163–177
43. Gavrieli Y, Sherman Y, Ben-Sasson SA (1992) Identification of programmed cell death in situ via specific labeling of nuclear DNA fragmentation. J Cell Biol 119:493–501
44. Negoescu A, Lorimier P, Labat-Moleur F, Drouet C, Robert C, Guillermet C et al (1996) In situ apoptotic cell labeling by the TUNEL method: improvement and evaluation on cell preparations. J Histochem Cytochem 44:959–968
45. Darzynkiewicz Z, Galkowski D, Zhao H (2008) Analysis of apoptosis by cytometry using TUNEL assay. Methods 44:250–254
46. Gorczyca W, Gong J, Ardelt B, Traganos F, Darzynkiewicz Z (1993) The cell cycle related differences in susceptibility of HL-60 cells to apoptosis induced by various antitumor agents. Cancer Res 53:3186–3192
47. Bedner E, Smolewski P, Amstad P, Darzynkiewicz Z (2000) Activation of cas-

pases measured in situ by binding of fluorochrome-labeled inhibitors of caspases (FLICA): correlation with DNA fragmentation. Exp Cell Res 259:308–313
48. Darzynkiewicz Z, Bedner E, Smolewski P, Lee BW, Johnson GL (2002) Detection of caspases activation in situ by fluorochrome-labeled inhibitors of caspases (FLICA). Methods Mol Biol 203:289–299
49. Darzynkiewicz Z, Pozarowski P, Lee BW, Johnson GL (2011) Fluorochrome-labeled inhibitors of caspases: convenient in vitro and in vivo markers of apoptotic cells for cytometric analysis. Methods Mol Biol 682:103–114
50. Barreiro-Iglesias A, Shifman MI (2012) Use of fluorochrome-labeled inhibitors of caspases to detect neuronal apoptosis in the whole-mounted lamprey brain after spinal cord injury. Enzyme Res 2012:835731
51. Barreiro-Iglesias A, Shifman MI (2015) Detection of activated caspase-8 in injured spinal axons by using fluorochrome-labeled inhibitors of caspases (FLICA). Methods Mol Biol 1254:329–339
52. Budihardjo I, Oliver H, Lutter M, Luo X, Wang X (1999) Biochemical pathways of caspase activation during apoptosis. Annu Rev Cell Dev Biol 15:269–290
53. Earnshaw WC, Martins LM, Kaufmann SH (1999) Mammalian caspases: structure, activation, substrates, and functions during apoptosis. Annu Rev Biochem 68:383–424
54. Liu J, Bhalgat M, Zhang C, Diwu Z, Hoyland B, Klaubert DH (1999) Fluorescent molecular probes V: a sensitive caspase-3 substrate for fluorometric assays. Bioorg Med Chem Lett 9:3231–3236
55. Boeneman K, Mei BC, Dennis AM, Bao G, Deschamps JR, Mattoussi H et al (2009) Sensing caspase 3 activity with quantum dot-fluorescent protein assemblies. J Am Chem Soc 131:3828–3829
56. Lee GH, Lee EJ, Hah SS (2014) TAMRA- and Cy5-labeled probe for efficient kinetic characterization of caspase-3. Anal Biochem 446:22–24
57. Elphick LM, Meinander A, Mikhailov A, Richard M, Toms NJ, Eriksson JE et al (2006) Live cell detection of caspase-3 activation by a Discosoma-red-fluorescent-protein-based fluorescence resonance energy transfer construct. Anal Biochem 349:148–155
58. Kawai H, Suzuki T, Kobayashi T, Sakurai H, Ohata H, Honda K et al (2005) Simultaneous real-time detection of initiator- and effector-caspase activation by double fluorescence resonance energy transfer analysis. J Pharmacol Sci 97:361–368
59. Xu X, Gerard AL, Huang BC, Anderson DC, Payan DG, Luo Y (1998) Detection of programmed cell death using fluorescence energy transfer. Nucleic Acids Res 26:2034–2035
60. He L, Wu X, Meylan F, Olson DP, Simone J, Hewgill D et al (2004) Monitoring caspase activity in living cells using fluorescent proteins and flow cytometry. Am J Pathol 164:1901–1913
61. Luo KQ, Yu VC, Pu Y, Chang DC (2001) Application of the fluorescence resonance energy transfer method for studying the dynamics of caspase-3 activation during UV-induced apoptosis in living HeLa cells. Biochem Biophys Res Commun 283:1054–1060
62. Luo KQ, Yu VC, Pu Y, Chang DC (2003) Measuring dynamics of caspase-8 activation in a single living HeLa cell during TNFalpha-induced apoptosis. Biochem Biophys Res Commun 304:217–222
63. Wu X, Simone J, Hewgill D, Siegel R, Lipsky PE, He L (2006) Measurement of two caspase activities simultaneously in living cells by a novel dual FRET fluorescent indicator probe. Cytometry A 69:477–486
64. Jones J, Heim R, Hare E, Stack J, Pollok BA (2000) Development and application of a GFP-FRET intracellular caspase assay for drug screening. J Biomol Screen 5:307–318
65. Zhu X, Fu A, Luo KQ (2012) A high-throughput fluorescence resonance energy transfer (FRET)-based endothelial cell apoptosis assay and its application for screening vascular disrupting agents. Biochem Biophys Res Commun 418:641–646
66. Bozza WP, Di X, Takeda K, Rivera Rosado LA, Pariser S, Zhang B (2014) The use of a stably expressed FRET biosensor for determining the potency of cancer drugs. PLoS One 9, e107010
67. Hoff BA, Bhojani MS, Rudge J, Chenevert TL, Meyer CR, Galban S et al (2012) DCE and DW-MRI monitoring of vascular disruption following VEGF-Trap treatment of a rat glioma model. NMR Biomed 25:935–942
68. Foroutan P, Kreahling JM, Morse DL, Grove O, Lloyd MC, Reed D et al (2013) Diffusion MRI and novel texture analysis in osteosarcoma xenotransplants predicts response to anti-checkpoint therapy. PLoS One 8, e82875
69. Chinnaiyan AM, Prasad U, Shankar S, Hamstra DA, Shanaiah M, Chenevert TL et al (2000) Combined effect of tumor necrosis factor-related apoptosis-inducing ligand and ionizing radiation in breast cancer therapy. Proc Natl Acad Sci U S A 97: 1754–1759
70. Kim H, Morgan DE, Buchsbaum DJ, Zeng H, Grizzle WE, Warram JM et al (2008) Early therapy evaluation of combined anti-death receptor 5 antibody and gemcitabine in

orthotopic pancreatic tumor xenografts by diffusion-weighted magnetic resonance imaging. Cancer Res 68:8369–8376

71. Oliver PG, LoBuglio AF, Zhou T, Forero A, Kim H, Zinn KR et al (2012) Effect of anti-DR5 and chemotherapy on basal-like breast cancer. Breast Cancer Res Treat 133:417–426
72. Wang H, Galban S, Wu R, Bowman BM, Witte A, Vetter K et al (2013) Molecular imaging reveals a role for AKT in resistance to cisplatin for ovarian endometrioid adenocarcinoma. Clin Cancer Res 19:158–169
73. Zhang F, Zhu L, Liu G, Hida N, Lu G, Eden HS et al (2011) Multimodality imaging of tumor response to doxil. Theranostics 1:302–309
74. Schmitz JE, Kettunen MI, Hu DE, Brindle KM (2005) 1H MRS-visible lipids accumulate during apoptosis of lymphoma cells in vitro and in vivo. Magn Reson Med 54:43–50
75. Jagannathan NR, Singh M, Govindaraju V, Raghunathan P, Coshic O, Julka PK et al (1998) Volume localized in vivo proton MR spectroscopy of breast carcinoma: variation of water-fat ratio in patients receiving chemotherapy. NMR Biomed 11:414–422
76. Kumar M, Jagannathan NR, Seenu V, Dwivedi SN, Julka PK, Rath GK (2006) Monitoring the therapeutic response of locally advanced breast cancer patients: sequential in vivo proton MR spectroscopy study. J Magn Reson Imaging 24:325–332
77. Lyng H, Sitter B, Bathen TF, Jensen LR, Sundfor K, Kristensen GB et al (2007) Metabolic mapping by use of high-resolution magic angle spinning 1H MR spectroscopy for assessment of apoptosis in cervical carcinomas. BMC Cancer 7:11
78. Zhai G, Kim H, Sarver D, Samuel S, Whitworth L, Umphrey H et al (2014) Early therapy assessment of combined anti-DR5 antibody and carboplatin in triple-negative breast cancer xenografts in mice using diffusion-weighted imaging and (1)H MR spectroscopy. J Magn Reson Imaging 39:1588–1594
79. Martin SJ, Reutelingsperger CP, McGahon AJ, Rader JA, van Schie RC, LaFace DM et al (1995) Early redistribution of plasma membrane phosphatidylserine is a general feature of apoptosis regardless of the initiating stimulus: inhibition by overexpression of Bcl-2 and Abl. J Exp Med 182:1545–1556
80. Wood BL, Gibson DF, Tait JF (1996) Increased erythrocyte phosphatidylserine exposure in sickle cell disease: flow-cytometric measurement and clinical associations. Blood 88:1873–1880
81. Blankenberg FG, Kalinyak J, Liu L, Koike M, Cheng D, Goris ML et al (2006) 99mTc-HYNIC-annexin V SPECT imaging of acute stroke and its response to neuroprotective therapy with anti-Fas ligand antibody. Eur J Nucl Med Mol Imaging 33:566–574
82. Blankenberg FG, Vanderheyden JL, Strauss HW, Tait JF (2006) Radiolabeling of HYNIC-annexin V with technetium-99m for in vivo imaging of apoptosis. Nat Protoc 1:108–110
83. Belhocine T, Steinmetz N, Green A, Rigo P (2003) In vivo imaging of chemotherapy-induced apoptosis in human cancers. Ann N Y Acad Sci 1010:525–529
84. Blankenberg FG, Katsikis PD, Tait JF, Davis RE, Naumovski L, Ohtsuki K et al (1998) In vivo detection and imaging of phosphatidylserine expression during programmed cell death. Proc Natl Acad Sci U S A 95:6349–6354
85. Kemerink GJ, Liu X, Kieffer D, Ceyssens S, Mortelmans L, Verbruggen AM et al (2003) Safety, biodistribution, and dosimetry of 99mTc-HYNIC-annexin V, a novel human recombinant annexin V for human application. J Nucl Med 44:947–952
86. Ogura Y, Krams SM, Martinez OM, Kopiwoda S, Higgins JP, Esquivel CO et al (2000) Radiolabeled annexin V imaging: diagnosis of allograft rejection in an experimental rodent model of liver transplantation. Radiology 214:795–800
87. Belhocine T, Steinmetz N, Hustinx R, Bartsch P, Jerusalem G, Seidel L et al (2002) Increased uptake of the apoptosis-imaging agent (99m) Tc recombinant human Annexin V in human tumors after one course of chemotherapy as a predictor of tumor response and patient prognosis. Clin Cancer Res 8:2766–2774
88. Blankenberg FG, Naumovski L, Tait JF, Post AM, Strauss HW (2001) Imaging cyclophosphamide-induced intramedullary apoptosis in rats using 99mTc-radiolabeled annexin V. J Nucl Med 42:309–316
89. Luo QY, Zhang ZY, Wang F, Lu HK, Guo YZ, Zhu RS (2005) Preparation, in vitro and in vivo evaluation of (99m)Tc-Annexin B1: a novel radioligand for apoptosis imaging. Biochem Biophys Res Commun 335:1102–1106
90. Lahorte CM, Van de Wiele C, Bacher K, van den Bossche B, Thierens H, Van BS et al (2003) Biodistribution and dosimetry study of 123I-rh-annexin V in mice and humans. Nucl Med Commun 24:871–880
91. Dekker B, Keen H, Lyons S, Disley L, Hastings D, Reader A et al (2005) MBP-annexin V radiolabeled directly with iodine-124 can be used to image apoptosis

in vivo using PET. Nucl Med Biol 32: 241–252
92. Dekker B, Keen H, Shaw D, Disley L, Hastings D, Hadfield J et al (2005) Functional comparison of annexin V analogues labeled indirectly and directly with iodine-124. Nucl Med Biol 32:403–413
93. Keen HG, Dekker BA, Disley L, Hastings D, Lyons S, Reader AJ et al (2005) Imaging apoptosis in vivo using 124I-annexin V and PET. Nucl Med Biol 32:395–402
94. Stafford JH, Hao G, Best AM, Sun X, Thorpe PE (2013) Highly specific PET imaging of prostate tumors in mice with an iodine-124-labeled antibody fragment that targets phosphatidylserine. PLoS One 8, e84864
95. Murakami Y, Takamatsu H, Taki J, Tatsumi M, Noda A, Ichise R et al (2004) 18F-labelled annexin V: a PET tracer for apoptosis imaging. Eur J Nucl Med Mol Imaging 31:469–474
96. Cauchon N, Langlois R, Rousseau JA, Tessier G, Cadorette J, Lecomte R et al (2007) PET imaging of apoptosis with (64)Cu-labeled streptavidin following pretargeting of phosphatidylserine with biotinylated annexin-V. Eur J Nucl Med Mol Imaging 34:247–258
97. Assadi M, Nemati R, Nabipour I, Salimipour H, Amini A (2011) Radiolabeled annexin V imaging: a useful technique for determining apoptosis in multiple sclerosis. Med Hypotheses 77:43–46
98. Post AM, Katsikis PD, Tait JF, Geaghan SM, Strauss HW, Blankenberg FG (2002) Imaging cell death with radiolabeled annexin V in an experimental model of rheumatoid arthritis. J Nucl Med 43:1359–1365
99. Lehner S, Todica A, Brunner S, Uebleis C, Wang H, Wangler C et al (2012) Temporal changes in phosphatidylserine expression and glucose metabolism after myocardial infarction: an in vivo imaging study in mice. Mol Imaging 11:461–470
100. Lehner S, Todica A, Vanchev Y, Uebleis C, Wang H, Herrler T, et al. (2014) In vivo monitoring of parathyroid hormone treatment after myocardial infarction in mice with [68Ga]annexin A5 and [18F]fluorodeoxyglucose positron emission tomography. Mol Imaging 13
101. Lampl Y, Lorberboym M, Blankenberg FG, Sadeh M, Gilad R (2006) Annexin V SPECT imaging of phosphatidylserine expression in patients with dementia. Neurology 66:1253–1254
102. Belhocine T, Steinmetz N, Li C, Green A, Blankenberg FG (2004) The imaging of apoptosis with the radiolabeled annexin V: optimal timing for clinical feasibility. Technol Cancer Res Treat 3:23–32
103. Fang W, Wang F, Ji S, Zhu X, Meier HT, Hellman RS et al (2007) SPECT imaging of myocardial infarction using 99mTc-labeled C2A domain of synaptotagmin I in a porcine ischemia-reperfusion model. Nucl Med Biol 34:917–923
104. Wang F, Fang W, Zhang MR, Zhao M, Liu B, Wang Z et al (2011) Evaluation of chemotherapy response in VX2 rabbit lung cancer with 18F-labeled C2A domain of synaptotagmin I. J Nucl Med 52:592–599
105. Zhao M, Zhu X, Ji S, Zhou J, Ozker KS, Fang W et al (2006) 99mTc-labeled C2A domain of synaptotagmin I as a target-specific molecular probe for noninvasive imaging of acute myocardial infarction. J Nucl Med 47:1367–1374
106. Hoebers FJ, Kartachova M, de Bois J, van den Brekel MW, van Tinteren H, van Herk M et al (2008) 99mTc Hynic-rh-Annexin V scintigraphy for in vivo imaging of apoptosis in patients with head and neck cancer treated with chemoradiotherapy. Eur J Nucl Med Mol Imaging 35:509–518
107. Rottey S, van den Bossche B, Slegers G, Van BS, Van de Wiele C (2009) Influence of chemotherapy on the biodistribution of [99mTc] hydrazinonicotinamide annexin V in cancer patients. Q J Nucl Med Mol Imaging 53:127–132
108. Reshef A, Shirvan A, Akselrod-Ballin A, Wall A, Ziv I (2010) Small-molecule biomarkers for clinical PET imaging of apoptosis. J Nucl Med 51:837–840
109. Cohen A, Ziv I, Aloya T, Levin G, Kidron D, Grimberg H et al (2007) Monitoring of chemotherapy-induced cell death in melanoma tumors by N, N′-Didansyl-L-cystine. Technol Cancer Res Treat 6:221–234
110. Reshef A, Shirvan A, Grimberg H, Levin G, Cohen A, Mayk A et al (2007) Novel molecular imaging of cell death in experimental cerebral stroke. Brain Res 1144:156–164
111. Damianovich M, Ziv I, Heyman SN, Rosen S, Shina A, Kidron D et al (2006) ApoSense: a novel technology for functional molecular imaging of cell death in models of acute renal tubular necrosis. Eur J Nucl Med Mol Imaging 33:281–291
112. Aloya R, Shirvan A, Grimberg H, Reshef A, Levin G, Kidron D et al (2006) Molecular imaging of cell death in vivo by a novel small molecule probe. Apoptosis 11:2089–2101
113. Grimberg H, Levin G, Shirvan A, Cohen A, Yogev-Falach M, Reshef A et al (2009) Monitoring of tumor response to chemotherapy in vivo by a novel small-molecule detector of apoptosis. Apoptosis 14:257–267
114. Cohen A, Shirvan A, Levin G, Grimberg H, Reshef A, Ziv I (2009) From the Gla domain

to a novel small-molecule detector of apoptosis. Cell Res 19:625–637
115. Hoglund J, Shirvan A, Antoni G, Gustavsson SA, Langstrom B, Ringheim A et al (2011) 18F-ML-10, a PET tracer for apoptosis: first human study. J Nucl Med 52:720–725
116. Cazzaniga M, Decensi A, Pruneri G, Puntoni M, Bottiglieri L, Varricchio C et al (2013) The effect of metformin on apoptosis in a breast cancer presurgical trial. Br J Cancer 109:2792–2797
117. Dai G, Tong Y, Chen X, Ren Z, Ying X, Yang F et al (2015) Myricanol induces apoptotic cell death and anti-tumor activity in non-small cell lung carcinoma in vivo. Int J Mol Sci 16:2717–2731
118. Sooriakumaran P, Coley HM, Fox SB, Macanas-Pirard P, Lovell DP, Henderson A et al (2009) A randomized controlled trial investigating the effects of celecoxib in patients with localized prostate cancer. Anticancer Res 29:1483–1488
119. Hight MR, Cheung YY, Nickels ML, Dawson ES, Zhao P, Saleh S et al (2014) A peptide-based positron emission tomography probe for in vivo detection of caspase activity in apoptotic cells. Clin Cancer Res 20:2126–2135
120. Zhou D, Chu W, Chen DL, Wang Q, Reichert DE, Rothfuss J et al (2009) [18F]- and [11C]-labeled N-benzyl-isatin sulfonamide analogues as PET tracers for apoptosis: synthesis, radiolabeling mechanism, and in vivo imaging study of apoptosis in Fas-treated mice using [11C]WC-98. Org Biomol Chem 7:1337–1348
121. Zhou D, Chu W, Rothfuss J, Zeng C, Xu J, Jones L et al (2006) Synthesis, radiolabeling, and in vivo evaluation of an 18F-labeled isatin analog for imaging caspase-3 activation in apoptosis. Bioorg Med Chem Lett 16: 5041–5046
122. Nguyen QD, Smith G, Glaser M, Perumal M, Arstad E, Aboagye EO (2009) Positron emission tomography imaging of drug-induced tumor apoptosis with a caspase-3/7 specific [18F]-labeled isatin sulfonamide. Proc Natl Acad Sci U S A 106:16375–16380
123. Chen DL, Engle JT, Griffin EA, Miller JP, Chu W, Zhou D et al (2015) Imaging Caspase-3 Activation as a Marker of Apoptosis-Targeted Treatment Response in Cancer. Mol Imaging Biol 17(3):384–393
124. Xia CF, Chen G, Gangadharmath U, Gomez LF, Liang Q, Mu F et al (2013) In vitro and in vivo evaluation of the caspase-3 substrate-based radiotracer [(18)F]-CP18 for PET imaging of apoptosis in tumors. Mol Imaging Biol 15:748–757

Chapter 3

Determining the Extent of Toxicant-Induced Apoptosis Using Concurrent Phased Apoptosis Assays

Akamu J. Ewunkem and Perpetua M. Muganda

Abstract

Apoptosis is a stage-dependent process exhibiting characteristic biochemical, molecular and morphological features that vary progressively through the apoptosis process. Apoptosis induced by toxicants may activate varied features of apoptosis to different extents and kinetics. Some of the features activated may occur transiently, while others may not occur in a cell system undergoing toxicant-induced apoptosis. Thus, the best approach for quantitating the extent of toxicant-induced apoptosis involves the utilization of a combination of assays focusing on different morphological, biochemical and molecular features of apoptosis. The kinetics of the apoptosis process will also need to be studied in any cell system subjected to toxicant-induced apoptosis for the first time, especially when more than one compound (such as an apoptosis inducing agent and an inhibitor) is being utilized. The use of multiple concurrent and phased apoptosis assays will streamline the process of determining the extent of apoptosis in such systems. Thus, in this report, we describe the concurrent use of an early, intermediate, and late apoptosis assay in order to measure different biochemical and morphological properties of apoptosis in the same system. We demonstrate the usefulness of the concurrent phased apoptosis assays using human TK6 lymphoblasts undergoing diepoxbutane-induced apoptosis. Additional notes, as well as tips for modifying the protocol for adherent cells, are included.

Key words Apoptosis, Apoptosis assay, Concurrent phased apoptosis assays, Lymphoblasts, Caspase-Glo 3/7 assay, pSIVA-IANBD apoptosis assay, Nuclear morphology fluorescent dual dye staining assay

1 Introduction

Apoptosis is a process of eliminating cells that have fulfilled their biological function during development and tissue homeostasis in multicellular organisms (recently reviewed in Refs. [1–4]). Apoptosis can be initiated by a wide variety of stimuli and conditions, such as irradiation, drugs, chemicals, and hormones [3–8]. Once initiated, apoptosis functions in multiple contexts. For example, in development, apoptosis allows for the definition of shape [3, 4]. Apoptosis is also beneficial as a natural anticancer mechanism, since damaged cells undergo apoptosis, preserving the

Perpetua M. Muganda (ed.), *Apoptosis Methods in Toxicology*, Methods in Pharmacology and Toxicology,
DOI 10.1007/978-1-4939-3588-8_3, © Springer Science+Business Media New York 2016

healthy state of the organism (reviewed in Ref. [2]). Thus, apoptosis plays a critical role in the development and maintenance of organisms [1, 8, 9].

Apoptosis can be seen as a stage-dependent process that exhibits both biochemical and morphological changes (recently reviewed in Refs. [2, 4]). Early stage apoptosis morphological events include chromatin condensation and margination at the nuclear membrane, nuclear condensation, and cell shrinkage. Late stage apoptosis morphological events include nuclear fragmentation, plasma membrane blebbing, and formation of apoptotic bodies [2, 3]. The sequence of biochemical events as the apoptosis process progresses include the externalization of phosphatidylserine, changes in mitochondrial membrane potential, the sequential activation of initiator and effector caspases, DNA fragmentation at inter-nucleosomal sites, and proteolytic cleavage of various cellular substrates, thus effectively dismantling the cell (recently reviewed in Refs. [2–4, 10]). Not all cells undergo the same changes. Some of these morphological, molecular and biochemical events are context dependent, and events such as phosphatidylserine externalization can be transient [6, 7, 11–14]. Although caspase activation appears to be a universal biochemical event in the apoptosis process, caspase 3/7 independent apoptosis has also been reported [4, 11]. Since every biochemical feature of apoptosis may not occur in all toxicant–cell systems, or may be transient, the use of single biochemical apoptosis endpoint assays needs to be avoided. Thus, the best approaches for quantitating apoptosis involve the utilization of a combination of assays focusing on various different biochemical and molecular features of apoptosis, with the addition of well-known apoptosis morphological features [4, 6, 7, 11, 12]. Each assay chosen, however, should be validated for each new toxicant–cell system prior to being utilized for experiments. Some of the most common biochemical and molecular features utilized to assess the extent of apoptosis include the externalization of phosphatidylserine, activation of caspase 3/7, and the cleavage of chromosomal DNA between nucleosomes [5, 11, 15].

The most common method utilized to determine the extent of phosphatidylserine (PS) externalization in apoptotic cells is the annexin V-FITC assay [5]. This assay is performed on live cells using fluorescence microscopy or flow cytometry. Most of the annexin-V-FITC assay kits on the market include a cell impermeable DNA binding fluorescent dye, such as propidium iodide (PI). This enables the distinction of live cells (with intact membranes, and thus PI negative) and necrotic cells (with damaged membranes, and thus PI positive). The assay also enables early apoptotic cells (Annexin-V positive, PI−) to be distinguished from late apoptotic cells (Annexin-V positive, PI+), and necrotic cells (Annexin-V negative, PI+) by either fluorescence microscopy or flow cytometry. In addition to distinguishing cells with compromised membranes, the PI included in the annexin-V assay can also be utilized as a second independent assay to confirm the occurrence of apoptosis; this is detected by quantitating the percentage of condensed and fragmented nuclei (by fluorescence microscopy), or the sub-G0/G1 cell population (by flow cytometry) [5].

In recent years, the pSIVA-IANBD assay for quantitating PS externalization has been described [5, 13, 14], and packaged kits are available on the market. pSIVA (**P**olarity **S**ensitive **I**ndicator of **V**iability & **A**poptosis) is an annexin XII-based molecule that is conjugated to IANBD, a polarity sensitive probe. It can be used as a replacement for annexin-V-FITC, with several advantages. It emits fluorescence measured by using the FITC filter only when it is bound to the membrane; this is in contrast to annexin-V-FITC, which emits fluorescence even when it is not bound to the membrane. Unlike annexin-V-FITC, it can be added to cells in media with no wash steps, and is able to detect transient PS externalization, since binding to PS is polarity sensitive and reversible. This assay, thus, has high signal-to-noise ratios, and is amendable to continuous live-cell imaging techniques; this provides a means to track the progression and timing of a cell through apoptosis starting from the initial stages of the pathway [13, 14]. It is also amendable to high volume screening by using a flow cytometer, and is thus suitable for laboratories with access to flow cytometer resources. The pSIVA-IANBD assay is able to detect early apoptosis, similar to the annexin V-FITC assay [4, 5, 13, 14, 16–18].

Activation of caspase 3/7 activity takes place in the majority of toxicant–cell systems undergoing apoptosis, and is thus a hallmark of apoptosis (recently reviewed in Refs. [2, 5]). Caspase 3/7 activity assay is, thus, one of the most commonly used apoptosis assays [15]. The assay can be performed beginning with live cells, and is directly quantifiable utilizing fluorescence or bioluminescence substrates in microplate formats. Flow cytometry can also be conducted by utilizing caspase 3/7 fluorescence substrates [19–21]. Caspase 3/7 activity assays are amendable to high volume screening, and in this regard they are suitable for laboratories with access to better resources. Disadvantages of the assay include the cost of reagents, as well as the fact that the integrity of the sample is destroyed during the assay. Additional assay formats for detecting and quantitating caspases 3/7 activation include immunofluorescence microscopy, flow cytometry, ELISA, and western blot formats using cleaved caspase 3/7 specific antibodies. Assays for caspase 3/7 activation detect an early to intermediate apoptosis event.

Methods based on biochemical and molecular properties of apoptosis are preferred over methods based on morphological properties because they are more quantitative, and thus less error prone [6, 7, 12]. Some morphological apoptosis assays that assess nuclear fragmentation, a hallmark of apoptosis, quantitate the extent of late apoptosis by determining the percentage of cells with fragmented nuclei. This is achieved through fluorescence microscopy of cells stained with a combination of cell permeable and cell impermeable DNA binding fluorescent dyes. One such assay involves the use of acridine orange (AO) cell permeable and ethidium bromide (EB) cell impermeable DNA binding fluorescent dyes. In this simple and inexpensive apoptosis assay, cells are dual stained with a mixture of acridine orange and ethidium bromide DNA binding dyes. Acridine orange is taken up by all cells, including those with intact membranes, and stains the nucleus of healthy non-apoptotic cells a uniform green color. Early apoptotic cells are identified due to their small condensed green nuclei, margination of chromatin, and green fragmented nuclei. Ethidium bromide is excluded from cells with intact membranes, but is taken up by cells possessing compromised membranes. Ethidium bromide overpowers the acridine orange stain in cells with compromised membranes, and imparts an orange color to their nuclei. Thus, ethidium bromide stains and distinguishes necrotic cells (with uniform orange nuclei) from late apoptotic cells (which possess orange fragmented nuclei). This is a low throughput assay that is best suited for the determination of late stages apoptosis in laboratories with limited resources. Although nuclear fragmentation is a hallmark of apoptosis, the suitability of this assay for any system should be validated prior to its use. This can be accomplished by comparing apoptosis measurements from this assay to measurements obtained from the DNA ladder assay, the TUNNEL assay, or by quantitating the sub-G0/G1 cell fraction by flow cytometry; all these assays are able to quantitate late apoptosis [5].

Toxicant-induced apoptotic systems may activate varied biochemical and morphological features of apoptosis to different extents and kinetics; some of the biochemical features activated may be transient. In previously uncharacterized toxicant–cell systems, or in systems where the apoptosis inducing or inhibitory activity of different compounds is being compared for the first time, the use of multiple concurrent and phased apoptosis assays is needed. In this scenario, assays that detect phosphatidylserine externalization, caspase 3/7 activation, and chromatin and nuclear fragmentation work well as phased assays, and can be conducted concurrently. The externalization of phosphatidylserine occurs early, and may be transient in most systems [4, 13, 14, 16, 17], while activation of caspase 3/7 and nuclear/DNA fragmentation are intermediate and late events, respectively. The use of concurrent phased assays will ensure the detection and quantitation of all features as apoptosis progresses to different extents and kinetics in the various toxicant–cell systems being compared. The approach will in turn shed light on the apoptotic stage being affected by an inhibitor or activator of apoptosis, thus saving time, effort, and expense.

In this report, we describe the concurrent use of early, intermediate, and late apoptosis assays that measure different biochemical and morphological properties of apoptosis in the same system. The phased early, intermediate, and late assays are represented by the pSIVA-IANBD assay [5, 13, 14, 22], the Caspase 3/7 Glo® assay [23], and the nuclear morphology fluorescent dual dye staining assay [24, 25], respectively. These assays were selected because they have been extensively used in other cells lines to study apoptosis [22, 23, 25, 26], do not require sophisticated instrumentation, and are relatively less expensive when compared to other assays. We demonstrate the usefulness of the concurrent phased apoptosis assays using human TK6 lymphoblasts undergoing diepoxbutane-induced apoptosis. Diepoxybutane (DEB) is the most active metabolite of butadiene [27]. It induces apoptosis through activation of caspases mediated through mitochondrial pathway [25, 26]. The protocols provided for pSIVA-IANBD assay and the nuclear morphology fluorescent dual dye staining assay can be performed using fluorescence microscopy or modified for flow cytometry. The description here utilized fluorescence microscopy to render the protocols especially useful for laboratories without access to a flow cytometer.

2 Materials

2.1 pSIVA Apoptosis/Viability Assay

1. pSIVA™-IANBD Apoptosis/Viability Microscopy Set (Cat # IMG-6701 K, IMGENIX, now Cat # NBP2-29382 by Novus Biologicals, USA).
2. Fluorescence microscope equipped with fluorescein isothiocyanate (FITC) and tetramethylrhodamine isothiocyanate (TRITC) filters.

2.2 Caspase-Glo® 3/7 Assay

1. Caspase-Glo®3/7 assay kit (Cat# G8091 for 100 assays, Promega Corporation, USA). This is utilized to conduct the caspase 3/7 bioluminescence assay in microplate format. Store at −20 °C protected from light.
2. Micro plate reader: Any instrument capable of measuring Caspase-Glo® 3/7 assay bioluminescence can be utilized. An example of such an instrument is the SpectraMax M5 multimode microplate reader (Molecular Devices) utilized for the example experiments described in this chapter.
3. Shaker.
4. Microplate(s) suitable for cell culture and for use in luminometer.
5. Additional supplies: sterile pipet tips, sterile Eppendorf tubes, pipet aids, multichannel pipet, microscope slides, and 10 × 10 mm No.1 coverslips.

2.3 Nuclear Morphology Fluorescent Dual Dye Staining Assay

1. Acridine orange, 10 mg/ml solution (Thermo Fisher Scientific Life Technologies). This is a cell permeable nucleic acid binding dye that stains the nucleus green. It is stored in a dark container at 4 °C, since it is light sensitive. It is also a mutagen; thus every precaution should be taken to protect personnel and environment from exposure. Handle only under class 2 type A2 conditions.
2. Ethidium bromide, 10 mg/ml solution (Thermo Fisher Scientific Life Technologies). This is a cell impermeable nucleic acid binding dye that stains the nucleus orange.

   Ethidium bromide is stored in a dark container at 4 °C, since it is light sensitive. It is also a mutagen; thus every precaution should be taken to protect personnel exposure. Handle only under class 2 type A2 conditions.
3. Sterile 0.9 % saline made by adding 0.9 g of NaCl to 100 ml of distilled deionized water. This should be sterilized by autoclaving, and then kept sterile at all times.
4. Fluorescence microscope equipped with a fluorescein isothiocyanate (FITC) filter (excitation at 488 nm, emission at 520 nm).

2.4 Cells, Cell Culture Materials, and Apoptosis Inducing Agent

1. Human TK6 lymphoblasts were obtained from the American Type Culture Collection.
2. Roswell Park Memorial Institute 1640 culture media (RPMI 1640) (obtained from Thermo Fisher Scientific) supplemented with 10 % fetal bovine serum (FBS, Atlanta Biologicals, USA), and 2 mM glutamine (Sigma Chemical Company, USA).
3. Cell counting reagents and instrument: Any cell counting system capable of enumerating total cells, live cells and dead cells can be utilized. Examples include: The Tali® Image Cytometer and the Tali® Viability Kit—Dead Cell Red (Thermo Fisher Scientific Life Technologies), the Vi-CELL XR Cell Viability Analyzer and associated reagents (Beckman Coulter), as well as a manual system using a hemocytometer.
4. An apoptosis inducing agent of choice: Any suitable compound of interest to the reader can be utilized for induction of apoptosis. Since the induction of apoptosis is concentration dependent, it is best to utilize concentrations that induce substantial apoptosis without causing necrosis. In the example experiments provided in this chapter, diepoxybutane (DEB, Sigma Cat# 202533-1G) was utilized to induce apoptosis in TK6 human lymphoblasts. Diepoxybutane is purchased as an 11.27 M stock solution; this is used to make a 1000× DEB working solution of 10 mM in sterile RPMI without serum. DEB is carcinogen and mutagen. Protective clothing and gloves should thus be worn for any work involving DEB. All

work involving DEB should be carried out in a biosafety cabinet (class 2 type A2) ducted to the outside of the building.

5. Plasticware, microscope slides and coverslips, and other materials, as needed: These include Eppendorf tubes, cell culture ware, pipette tips, pipettes, and pipetting devices.
6. Data analysis software, such as GraphPad Prism 6.

3 Methods

3.1 Cell Culture

Various cell types can be used for these experiments (see Note 1), and thus growth conditions optimized for the cell type being used should be utilized. For TK6 human lymphoblasts, cells are passaged every 36 h at 2.0×10^5 cells/ml in RPMI medium supplemented with 2 mM l-glutamine and 10 % fetal bovine serum. The cells are incubated in a humidified 37 °C incubator with a 5 % CO_2 atmosphere. The cells passaged into fresh media at 12 h prior to each experiment, and can be used for an experiment only after they are certified to be healthy, with less than 1 % necrotic cells. All work should be performed under class 2 type A2 conditions to protect personnel and material.

3.2 Induction of Apoptosis

Various compounds can be utilized to induce apoptosis in different cell types. It is a good idea to include a positive control compound that is known to induce apoptosis within the cell type being used. To induce apoptosis in TK6 lymphoblasts, cells are seeded into fresh complete RPMI media at 2×10^5 cells/ml, and exposed to a final concentration of 10 μM DEB by adding 10 mM DEB working solution at 1:1000 dilution; control unexposed cells receive vehicle (RPMI media). All manipulations are performed under class 2 type A2 conditions to protect the product as well as personnel, since DEB is a toxicant, mutagen, and carcinogen.

3.3 Detection of Apoptosis

Control and toxicant exposed cells are assayed for apoptosis utilizing the three phased (early, intermediate, and late) apoptosis assays described below. The protocols outlined below utilize control and DEB-exposed cells assayed for apoptosis at various times post-DEB exposure. The Notes section (Notes 1–10) provides information for modifications needed for these assays to work in any system.

3.3.1 pSIVA-IANBD Assay, an Early Apoptosis Assay

Cells are assayed for apoptosis utilizing the pSIVA-IANBD Apoptosis/Viability Microscopy Set Kit (Cat # IMG-6701 K, IMGENIX), as well as a modification of the kit manufacturer's protocol, as described below. Steps 1–5 are carried out under class 2 type A2 conditions to protect personnel as well as the product.

1. Take approximately 2×10^5 cells each from control and toxicant (such as DEB) treated cells into two separate sterile Eppendorf tubes.

2. Centrifuge at a predetermined lowest speed and time needed to pellet the smaller apoptotic cells while avoiding cell damage (see Note 2). The conditions utilized for control and DEB-exposed TK6 lymphoblasts involve pelleting the cells at 800 g for 3 min.
3. Carefully remove the supernatant from the cell pellet, dislodge the pellet, and gently resuspend the pellet in 50 μl of complete growth media (see Note 2).
4. Add 1 μl of a 1:2 freshly diluted pSIVA-IANBD working solution in media to the resuspended cells, and mix gently (see Note 3). This translates to the 10 μl/ml concentration recommended by the manufacturer. Dual staining by the inclusion of propidium iodide (PI) found in the kit can be used to assess the extent of necrosis/viability within the control and apoptotic cell population (see Note 3).
5. Immediately load 15–20 μl of each of the control and toxicant treated samples onto two separate areas of the same microscope slide. Add coverslips on each sample.
6. Visualize each of the two samples using a fluorescence microscope and an FITC filter, and quantitate the percentage of pSIVA-IANBD positive cells (staining green) in each of four to five randomly selected fields per sample.
7. Additional paired control and test samples are assayed by repeating steps 1–6. Fresh 1:2 diluted pSIVA-IANBD working solution in media is made for each time point in a time course experiment. In the example given in this chapter, the time course experiment was performed at 0 h, 3 h, 8 h, 14 h, 24 h, and 36 h post DEB-exposure (see Note 4).

Data analysis for the pSIVA-IANBD assay: Apoptotic cells are quantified by dividing the number of pSIVA-IANBD positive cells by the total number of cells in each field, and multiplying by 100 to get the percentage of apoptotic cells in each field. The mean percentage of apoptotic cells in each sample is then obtained by averaging the percentage of apoptotic cells in all four to five fields per sample. The standard deviation of the percentage of apoptotic cells in each sample is also calculated. Graphs can then be plotted from this information (for example see Fig. 1). The extent of necrotic/dead cells can also be assessed within the samples by quantifying the percentage of PI positive cells in a similar manner.

3.3.2 Caspase-Glo®3/7 Assay, an Early to Intermediate Apoptosis Assay

Caspase 3/7 is measured by utilizing the Caspase-Glo®3/7 assay kit (Promega, Madison USA), and a modified kit manufacturer's protocol. The protocol described below is for control and DEB exposed lymphoblasts that grow as suspension cells. The notes [5–8] provide explanations, tips, and modifications for other cell

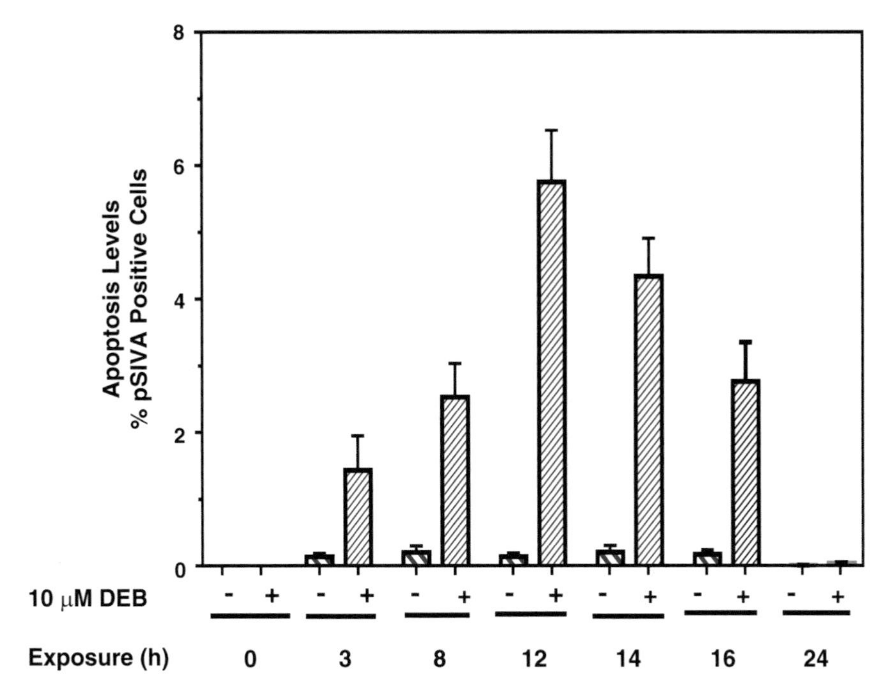

Fig. 1 Detection of apoptosis in DEB-exposed TK6 cells using the pSIVA-IANBD assay: Control and 10 μM DEB-exposed TK6 cells were stained with pSIVA at 3 h, 8 h, 12 h, 14 h, 24 h, and 36 h post DEB-exposure, as described in the protocol. The percentage of pSIVA-IANBD positive cells is plotted against the DEB exposure time (h)

types and scenarios. Steps 1–5 are carried out under class 2 type A2 conditions to protect personnel as well as the product.

1. Determine the cell count for all control and toxicant exposed samples that constitute the experiment. Perform calculations needed to deliver 1×10^4 cells in 100 μl of media for each sample within the experiment in designated wells within a 96-well plate (see Note 5).
2. Remove from the freezer and equilibrate the Caspase-3/7-Glo® substrate and buffer to room temperature; place in a room temperature shallow water bath for approximately 30 min. Be sure to wipe the container with laboratory wipes dampened with 70 % ethanol before opening the container.
3. Meanwhile (during the ~30 min substrate is equilibrating to room temperature), deliver 1×10^4 cells in 100 μl media to designated wells within the 96-well plate for each control and toxicant-exposed sample. This should be done as efficiently as possible in order to minimize the amount of time cells are at room temperature, thus avoiding interfering with the ongoing apoptosis process. Blank wells should receive 100 μl of media. The loaded plate should be returned to the CO_2 incubator at 37 °C till shortly before the assay is conducted (see Note 6).

4. Preparation of the Caspase Glo® 3/7 working reagent: This is performed by following the protocol in the kit. Briefly, the room temperature equilibrated Caspase Glo® 3/7 buffer (10 ml) is added to the room temperature equilibrated bottle of Caspase Glo® 3/7 lyophilized substrate. The substrate is dissolved by inverting or swirling, and making sure that the substrate on the bottle walls is dissolved as well. Retain enough working reagent at room temperature for the experiment. The rest of the Caspase Glo® 3/7 working reagent can be kept at 4 °C; it is best if it is used within a week (See Note 7).
5. Remove the loaded 96-well plate from the incubator, and equilibrate it at room temperature for approximately 5 min.
6. Add 100 μl of Caspase Glo® 3/7 working reagent to the wells of the 96-well plate using a multichannel pipette (see Note 8), and immediately shake at room temperature for 30 s at 300–500 rpm using the microplate reader or on a separate shaker. Incubate at 22 °C a temperature controlled microplate reader for 30 min to 2 h.
7. Measure Caspase Glo® 3/7 bioluminescence in a microplate reader beginning at 30 min incubation, and take readings every 30 min up to 2 h.
8. If performing a time course experiment, repeat steps 1–7 for each time point. In this case, steps 2 and 4 are replaced by the warming up of the Caspase Glo® 3/7 working reagent to room temperature before use.

Data Analysis for Caspase Glo® 3/7 assay:

The data are analyzed by plotting the obtained caspase 3/7 relative luminescence units on the *y*-axis, against the various samples assayed on the *x*-axis (see example in Fig. 2). By plotting the readings obtained after incubating the assay at different times from 30 min to 2 h, an optimal incubation time for the assay can be identified.

3.3.3 The Nuclear Morphology Fluorescent Dual Dye Staining Assay, a Late Apoptosis Assay

Steps 1–5 are carried out under class 2 type A2 conditions to protect personnel as well as the product. The assay is carried out as follows (see Notes 2, 9 and 10 for modifications, tips and explanations):

1. Preparation of the acridine orange–ethidium bromide (AO-EB) stain working solution: At approximately 15 min prior to conducting the apoptosis assay, make the acridine orange–ethidium bromide (AO-EB) stain working solution by adding 1 μl each of the 10 mg/ml dyes to 100 μl of sterile 0.9 % saline solution in a sterile Eppendorf tube; mix well, but gently. This prepares an AO-EB solution that contains 100 μg/ml each of acridine orange and ethidium bromide; this is enough for 24 samples. All solutions containing acridine orange and ethidium

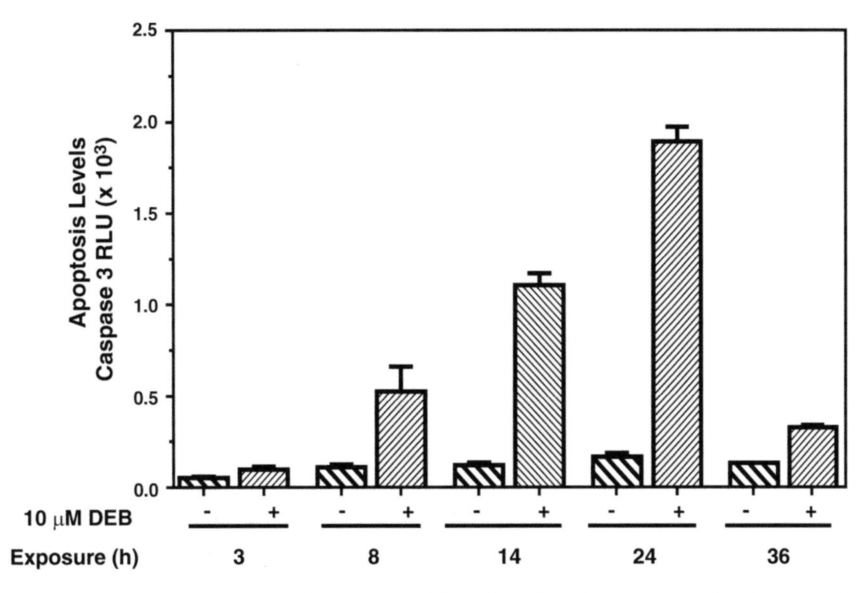

Fig. 2 Monitoring caspase 3 activity in DEB-exposed TK6 cells using the Caspase-Glo 3/7 assay: The control and 10 μM DEB-exposed TK6 cells (same batch of cells assayed in Fig. 1) were subjected to Caspase-Glo 3/7® assay at 3 h, 8 h, 12 h, 14 h, 24 h, and 36 h post DEB-exposure, as described in the protocol. Caspase 3/7 activity expressed as relative bioluminescence units is plotted against the DEB exposure time (h)

bromide should be protected from light; this can be achieved by wrapping a piece of foil around any tubes.

2. Add approximately 2×10^5 cells from control and toxicant (such as DEB) treated cells into two separate sterile Eppendorf tubes, and spin them at a predetermined lowest speed and time needed to pellet the smaller apoptotic cells (see Note 2).
3. Carefully and immediately remove the supernatant from the cell pellets, dislodge the pellets by flickering the tubes, and gently resuspend the pellet in 50 μl of complete growth media (such as the RPMI medium used in these experiments) (see Note 2).
4. Add 4 μl of AO-EB working solution (100 μg/ml of AO and 100 μg/ml of EB in 0.9 % saline) made in step 1, and mix gently.
5. Immediately load 15–20 μl of each of the control and toxicant treated stained samples onto two separate areas of the same microscope slide. Add coverslips on each sample.
6. Visualize each of the two samples using a fluorescence microscope and an FITC filter. Identify and count early apoptotic, late apoptotic, necrotic cells, and total cells; record this information for each of four to five fields for each sample (see Fig. 3a as a visual aid, and Note 9 for description).

7. Additional paired control and test samples are assayed by repeating using steps 2–6 (See Note 10). Fresh AO-EB working solution is made (step 1) for each time point in a time course experiment. In the example given in this chapter, the time course experiment was performed at 0 h, 3 h, 8 h, 14 h, 24 h, and 36 h post DEB-exposure (see Fig. 3b), and fresh AO-EB working solution was made for each time point.

Data analysis for the nuclear morphology fluorescent dual dye staining assay: Apoptotic cells are quantified by dividing the combined number of early and late apoptotic cells by the total number of cells in each field, and multiplying by 100 to get the percentage of apoptotic cells in each field. The percentage of apoptotic cells in each sample is then obtained by averaging the percentage of apoptotic cells in all four to five fields per sample. The standard deviation of the percentage of apoptotic cells in each sample is also calculated. Graphs can then be plotted from this information (for example see Fig. 3b).

Overall Integration of Data Analysis for the Early, intermediate, and Late Apoptosis Assays: Overall, all three assays can be compared to determine the optimal time point (i.e., early, intermediate or late phase) within the apoptosis time course each assay performs. The optimal assay time point is the time at which the assay displays peak parameter values during a time-course experiment. The three assays can also be compared to determine the identity of the most sensitive assay (one which demonstrates maximum fold change in experimental samples with toxicant compared to control unexposed cells) at each time point tested. In our example experiment, the pSIVA-IANBD assay peaks at 12 h post DEB exposure, and it is the most sensitive assay at this point within the apoptosis time course. This is a point prior to changes in mitochondria transmembrane potential in this experimental system [28]. Likewise, the caspase 3/7 assay peaks at 24 h post DEB exposure, and its functionality decreases considerably by 36 h post DEB exposure. The nuclear morphology fluorescent dual dye staining assay is the most sensitive apoptosis assay at 36 h post-DEB Exposure (Fig. 4). Thus collectively, the Caspase-Glo 3/7 assay is more reliable at 14 h and 24 h post-DEB exposure, but should not be used at 36 h post-DEB exposure. The nuclear morphology fluorescent dual dye staining assay can be used at 14, 24 and 36 h, and it is most useful at 36 h, a time point when the other two assays are unreliable. The pSIVA-IANBD assay is useful at 8 h and 12 h post-DEB exposure (see Figs. 1, 2, 3, and 4). Thus, the three assays can be utilized concurrently in phased experiments designed to investigate the effect on inhibitors of toxicant induced apoptosis, especially in cases where the apoptosis phase to be inhibited is unknown.

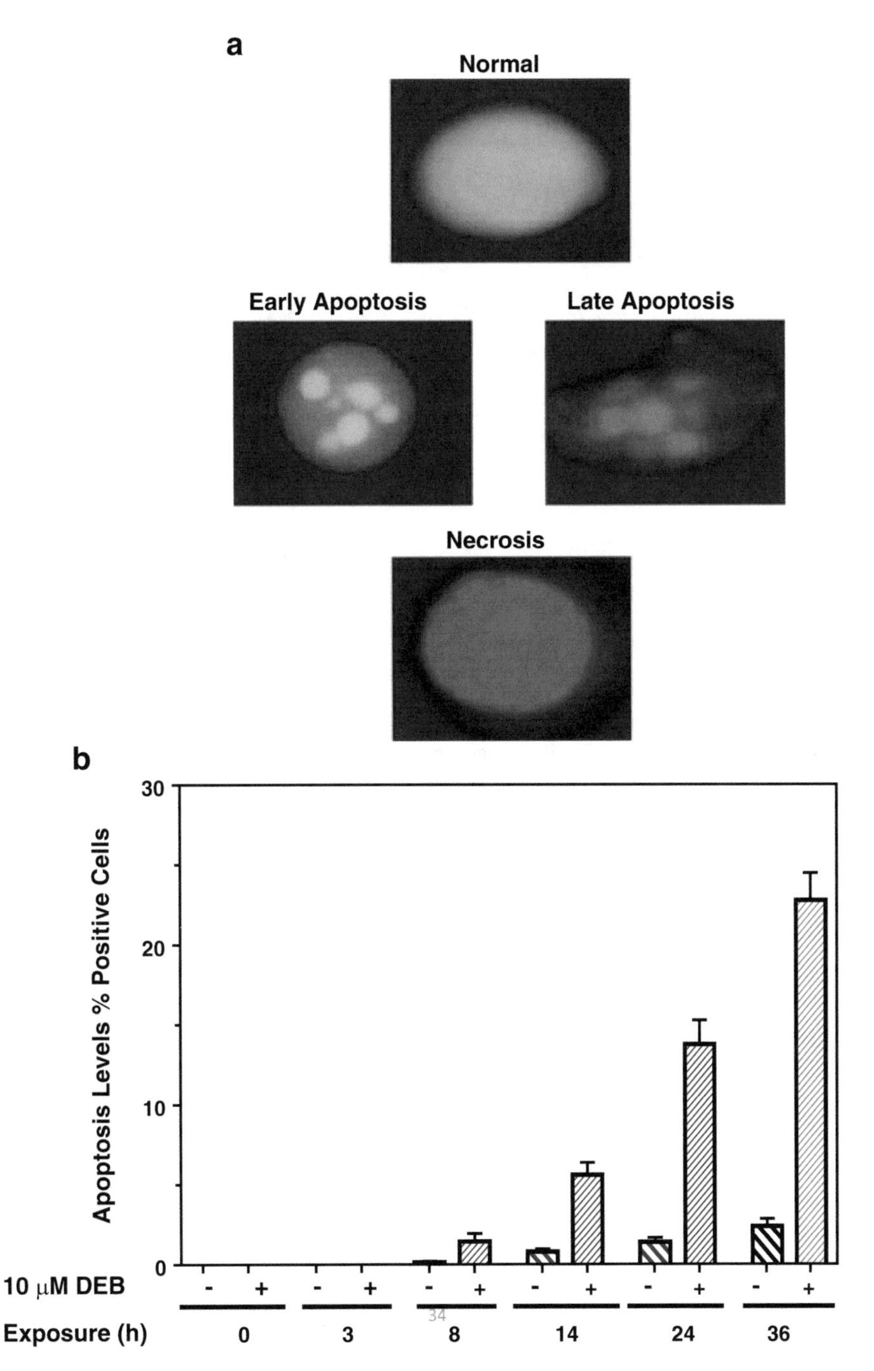

Fig. 3 Detection of apoptosis in DEB-exposed TK6 cells using the nuclear morphology fluorescent dual dye staining assay: Control and 10 μM DEB-exposed TK6 cells (same batch of cells assayed in Figs. 1 and 2) were stained with acridine orange and ethidium bromide dyes at 3 h, 8 h, 12 h, 14 h, 24 h, and 36 h post DEB-exposure, as described in the nuclear morphology fluorescent dual dye staining assay protocol. (**a**) Representative microscopy pictures for normal, early apoptotic, late apoptotic, and necrotic cells (figure taken from Yadavilli et al. [25], with permission. (**b**) The percentage of apoptotic cells obtained from the assay is plotted against the DEB exposure time (h)

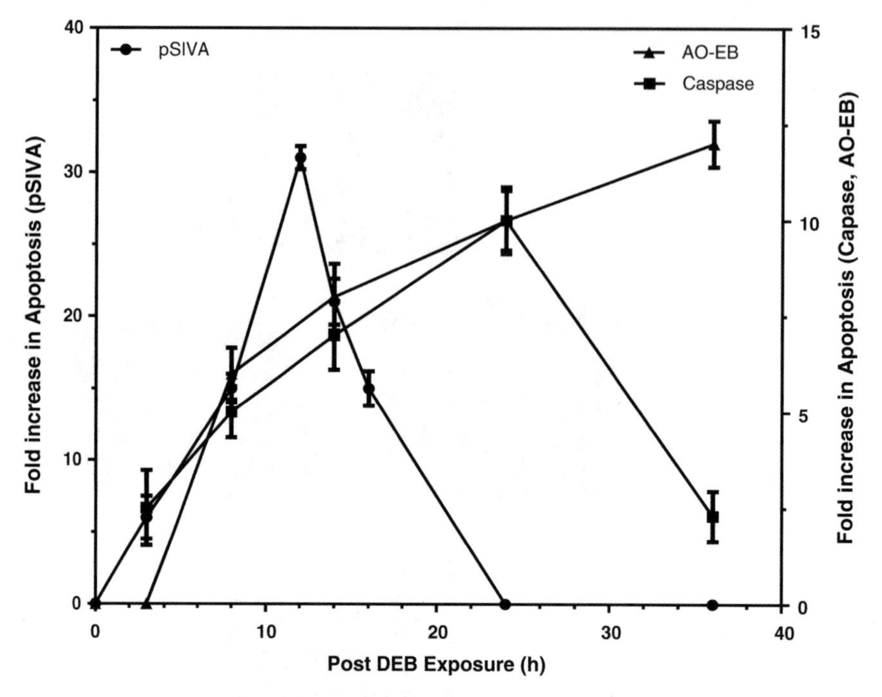

Fig. 4 The fold-increase in the extent of apoptosis in DEB-exposed cells as compared to control unexposed cells at each time point for each of the three assays. Data shown in Figs. 1, 2, and 3 were repackaged and plotted as shown

4 Notes

1. Although suspension cells were utilized for the example experiments shown here, the protocol can be adapted for any cell type, including adherent cells. The notes below will provide modifications for adherent cells at appropriate steps within the protocol.
2. It is important to predetermine the lowest speed and time needed to pellet the both the regular cells as well as the smaller apoptotic cells while avoiding or minimizing damage to all (both non-apoptotic and apoptotic) cells. This can be performed by determining this property for non-apoptotic cells first, followed by the fine tuning of the parameters by including cells undergoing apoptosis. The correct speed and centrifugation time will recover the highest quantity of apoptotic cells, with the minimum amount of cell damage (assayed by a live and dead cell assay) to all cells. The results should be compared to control cells that were not centrifuged. It is also important to avoid disturbing the cell pellet while removing media from the centrifuged cells. This is accomplished by marking the position of the pellet to avoid disturbing it. Furthermore, cell suspension to single cells is more effective if

the pellet is dislodged by gently flicking the tube with the index finger to dislodge the pellet prior to resuspending it in the 50 μl of media.

Adherent Cells: Steps 1–3 can be replaced by the following modification. Cells can be grown and exposed to a toxicant on coverslips. After removal of media and toxicant, the cells can be rinsed once with media, and then the minimum volume of pSIVA-IANBD working solution needed to cover the cells can be added. The cells can then be visualized immediately. For a time course experiment, this step can be repeated for each time point. Care must be taken to make sure that the cells are not too confluent for effective fluorescent microscopy.

3. In step 4 of the pSIVA-IANBD protocol, the reagent was diluted to avoid pipetting errors associated with pipetting less than 1 μl reagent volume. The 50 μl total cell resuspension volume was chosen in order to minimize reagent use; this volume works very well for the amount of cells used per sample. The actual final concentration of pSIVA-IANBD and propidium iodide (PI) working solutions utilized should be predetermined for the cells and system in question. This is accomplished by using negative and positive controls that represent cells not undergoing apoptosis, and cells whose percentage of apoptosis is already known. A good starting concentration is 10 μl/ml for pSIVA-IANBD and 5 μl/ml for propidium iodide (PI).

   Adherent Cells: The description provided under the modification for the adherent cells in Note 2 can be followed.

4. It is important to work quickly during the acquisition of the data by fluorescence microscopy to keep the mounted cells alive. Each specimen and its control can be processed through the microscopy component of the experiment within 5 min.

5. The calculations needed to deliver 1×10^4 cells in 100 μl of media for each sample are best performed as follows: For three replicates for each sample, where a total of 300 μl is needed, it is best to make 375–400 μl cell suspension of each sample at a concentration of 1×10^5 cells/ml; this saves as stock cell suspension for delivery of 1×10^4 cells/100 μl of media.

6. Normalization of cells numbers per well prior to the caspase assay is important because most suspension cells exposed to a toxicant or different experimental conditions can grow at different rates compared to unexposed cells. Loading the 96-well plate with cells is conducted as fast as possible to avoid equilibrating the cells at room temperature; this minimizes any interference with the ongoing apoptosis process.

7. The amount of working caspase 3/7 substrate solution retained at room temperature for use within the planned experiment is calculated as the total number of control and experimental

samples with replicates + total number of blanks+ extra 2 samples times 100 μl/sample. For example, for 11 control and experimental samples + 1 blank= 12 total samples, including the blank. 12 samples × 3 replicates = 36 + 2 extra samples (to avoid any shortages due to handling) = 38 samples × 100 μl/sample = 3.8 ml of working substrate retained at room temperature for the planned experiment. The unused stored substrate can be aliquoted and frozen at −20 °C, if it cannot be used within a week.

8. If a suitable small multichannel pipetting reservoir is not available, sterile PCR strips can be adapted for use as a pipetting reservoir for a multichannel pipette.
9. Early apoptotic, late apoptotic, necrotic cells, and total cells can be identified as follows ([25]; see Fig. 3a): Early apoptotic cells have condensed small nuclei, marginated chromatin, as well as green fragmented nuclei. Late apoptotic cells have orange fragmented nuclei. Non-apoptotic live cells have bright uniform green nuclei, and necrotic cells have orange uniform nuclei. Total cells stain with green or orange colors. See Fig. 3a for example of the expected nuclear morphology.
10. The contents of Note 4 apply to this assay as well. More than investigator may want to score the extent of apoptosis within the experimental system, especially during the training phase of the assay; different investigator should get the same results on the same samples while scoring the samples independently.

References

1. Li K, Wu D, Chen X, Zhang T, Zhang L, Yi Y, Miao Z, Jin N, Bi X, Wang H, Xu J, Wang D (2014) Current and emerging biomarkers of cell death in human disease. Biomed Res Int 2014:690103
2. Goldar S, Khaniani MS, Derakhshan SM, Baradaran B (2015) Molecular mechanisms of apoptosis and roles in cancer development and treatment. Asian Pac J Cancer Prev 16:2129–2144
3. Elmore S (2007) Apoptosis: a review of programmed cell death. Toxicol Pathol 35: 495–516
4. Tower J (2015) Programmed cell death in aging. Ageing Res Rev 23:90–100
5. Zeng W, Wang X, Xu P, Liu G, Eden HS, Chen X (2015) Molecular imaging of apoptosis: from micro to macro. Theranostics 5: 559–582
6. Galluzzi L, Vitale I, Abrams JM, Alnemri ES, Baehrecke EH, Blagosklonny MV, Dawson TM, Dawson VL, El-Deiry WS, Fulda S, Gottlieb E, Green DR, Hengartner MO, Kepp O, Knight RA, Kumar S, Lipton SA, Lu X, Madeo F, Malorni W, Mehlen P, Nunez G, Peter ME, Piacentini M, Rubinsztein DC, Shi Y, Simon HU, Vandenabeele P, White E, Yuan J, Zhivotovsky B, Melino G, Kroemer G (2012) Molecular definitions of cell death subroutines: recommendations of the Nomenclature Committee on Cell Death 2012. Cell Death Differ 19:107–120
7. Galluzzi L, Bravo-San Pedro JM, Vitale I, Aaronson SA, Abrams JM, Adam D, Alnemri ES, Altucci L, Andrews D, Annicchiarico-Petruzzelli M, Baehrecke EH, Bazan NG, Bertrand MJ, Bianchi K, Blagosklonny MV, Blomgren K, Borner C, Bredesen DE, Brenner C, Campanella M, Candi E, Cecconi F, Chan FK, Chandel NS, Cheng EH, Chipuk JE, Cidlowski JA, Ciechanover A, Dawson TM, Dawson VL, De Laurenzi V, De Maria R, Debatin KM, Di Daniele N, Dixit VM, Dynlacht BD, El-Deiry WS, Fimia GM, Flavell RA, Fulda S, Garrido C, Gougeon ML, Green DR, Gronemeyer H, Hajnoczky G, Hardwick JM, Hengartner MO, Ichijo H, Joseph B, Jost

PJ et al (2015) Essential versus accessory aspects of cell death: recommendations of the NCCD 2015. Cell Death Differ 22:58–73
8. Ulukaya E, Acilan C, Yilmaz Y (2011) Apoptosis: why and how does it occur in biology? Cell Biochem Funct 29:468–480
9. Fuchs Y, Steller H (2011) Programmed cell death in animal development and disease. Cell 147:742–758
10. Shalini S, Dorstyn L, Dawar S, Kumar S (2015) Old, new and emerging functions of caspases. Cell Death Differ 22:526–539
11. Kanemura S, Tsuchiya A, Kanno T, Nakano T, Nishizaki T (2015) Phosphatidylinositol induces caspase-independent apoptosis of malignant pleural mesothelioma cells by accumulating AIF in the nucleus. Cell Physiol Biochem 36:1037–1048
12. Rello-Varona S, Herrero-Martin D, Lopez-Alemany R, Munoz-Pinedo C, Tirado OM (2015) "(Not) all (dead) things share the same breath": identification of cell death mechanisms in anticancer therapy. Cancer Res 75:913–917
13. Kim YE, Chen J, Chan JR, Langen R (2010) Engineering a polarity-sensitive biosensor for time-lapse imaging of apoptotic processes and degeneration. Nat Methods 7:67–73
14. Kim YE, Chen J, Langen R, Chan JR (2010) Monitoring apoptosis and neuronal degeneration by real-time detection of phosphatidylserine externalization using a polarity-sensitive indicator of viability and apoptosis. Nat Protoc 5:1396–1405
15. Bucur O, Stancu AL, Khosravi-Far R, Almasan A (2012) Analysis of apoptosis methods recently used in cancer research and cell death & disease publications. Cell Death Dis 3, e263
16. Hankins HM, Baldridge RD, Xu P, Graham TR (2015) Role of flippases, scramblases and transfer proteins in phosphatidylserine subcellular distribution. Traffic 16:35–47
17. Leventis PA, Grinstein S (2010) The distribution and function of phosphatidylserine in cellular membranes. Annu Rev Biophys 39:407–427
18. Schlegel RA, Williamson P (2001) Phosphatidylserine, a death knell. Cell Death Differ 8:551–563
19. Wu H, Che X, Zheng Q, Wu A, Pan K, Shao A, Wu Q, Zhang J, Hong Y (2014) Caspases: a molecular switch node in the crosstalk between autophagy and apoptosis. Int J Biol Sci 10:1072–1083
20. Nicholls SB, Hyman BT (2014) Measuring caspase activity in vivo. Methods Enzymol 544:251–269
21. McStay GP, Green DR (2014) Measuring apoptosis: caspase inhibitors and activity assays. Cold Spring Harb Protoc 2014:799–806
22. Novus-Biologicals (2014) pSIVA-IANBD Apoptosis/Viability Microscopy Set Product Information and Manual
23. Promega (2015) Caspase-Glo 3/7 Technical Bulletin.
24. Squier MK, Cohen JJ (2001) Standard quantitative assays for apoptosis. Mol Biotechnol 19:305–312
25. Yadavilli S, Muganda PM (2004) Diepoxybutane induces caspase and p53-mediated apoptosis in human lymphoblasts. Toxicol Appl Pharmacol 195:154–165
26. Yadavilli S, Chen Z, Albrecht T, Muganda PM (2009) Mechanism of diepoxybutane-induced p53 regulation in human cells. J Biochem Mol Toxicol 23:373–386
27. Jackson MA, Stack HF, Rice JM, Waters MD (2000) A review of the genetic and related effects of 1,3-butadiene in rodents and humans. Mutat Res 463:181–213
28. Yadavilli S, Martinez-Ceballos E, Snowden-Aikens J, Hurst A, Joseph T, Albrecht T, Muganda PM (2007) Diepoxybutane activates the mitochondrial apoptotic pathway and mediates apoptosis in human lymphoblasts through oxidative stress. Toxicol In Vitro 21:1429–1441

Chapter 4

Measurement of Apoptosis by Multiparametric Flow Cytometry

William G. Telford

Abstract

Apoptosis remains a critical phenomenon in cell biology, playing a regulatory role in virtually every tissue system. It is particular crucial in the immune system, ranging from immature immune cell development and selection to downregulation of the mature immune response. Apoptosis is a primary mechanism in the action of antitumor drugs, and is thus an important phenomenon in pharmacology, drug discovery, and toxicology. Flow cytometry is the primary technique for measuring apoptosis in suspension cells; many flow cytometry assays have been developed to measure the entire apoptotic process, from the earliest signal transduction events to the late morphological changes in cell size, proteolysis, and DNA degradation. These assays become even more powerful when they can be combined into single multiparametric assays that can document the process of apoptosis in a single tube. The ability of flow cytometry to measure multiple structural and fluorescent characteristics in single cells is uniquely suited to this task. In this methods review, we show how multiple individual assays can be combined in this fashion. Combining early biochemical and late morphological assays together gives a comprehensive and detailed picture of the apoptotic process.

Key words Apoptosis measurement, Flow cytometry, Apoptosis assays, Multiparametric, Caspase 3/7 substrate, Caspase 3 immunolabeling, Annexin V, FLICA (fluorescence linked inhibitor of caspase activity), Covalent binding viability probes, DNA binding dyes

Abbreviations

7-AAD	7-aminoactinomycin D
APC	Allophycocyanin
FBS	Fetal bovine serum
FLICA	Fluorescence linked inhibitor of caspase activity
HBSS	Hanks balanced salt solution
PBS	Phosphate buffered saline
PE	Phycoerythrin
TUNEL	Terminal deoxynucleotidyl transferase dUTP nick end labeling

Perpetua M. Muganda (ed.), *Apoptosis Methods in Toxicology*, Methods in Pharmacology and Toxicology,
DOI 10.1007/978-1-4939-3588-8_4, © Springer Science+Business Media New York 2016

1 Introduction

Apoptosis or signal-directed cell death is a critical process in the regulation of cell and tissue homeostasis. This is particularly true in the immune system, where all but a small fraction of immune cells are ultimately destined to be cleared by apoptosis, both to prevent subsequent autoimmunity during immature immune cell development, and to downregulate immune responses following response to an foreign antigen or pathogen. The action of chemotherapeutic drugs also largely centers around the ability to induce apoptosis in target cell types. It is therefore necessary to have techniques that can measure this important cellular process both in vivo and in vitro.

Apoptosis is a morphological phenomenon, and the earliest assays to detect it involved recognition of apoptotic morphology by microscopy. While the process of apoptosis is highly variable depending on the cell type and the inducer, several relatively common morphological changes are associated with most forms of cell death. Most cells undergo dramatic degradation of their intracellular components during apoptosis, including perturbation of the cell membrane, collapse of the intracellular support matrix, degradation of intracellular organelles, protein degradation, and chromatin condensation and degradation. These features could often be identified morphologically. Identification of apoptosis as a central regulatory mechanism in the immune system in the 1980s made it necessary to develop more quantitative assays with higher throughput. Like morphological analysis, these assays also relied on the physical changes cells underwent during apoptosis. Internucleosomal DNA degradation, a common (although by no means universal) characteristic of cell death, could be measured by purifying DNA from cell suspensions and running it on an agarose gel. By mid-1980s, morphology and DNA degradation were the primary means by which apoptosis was measured in the immune cells [1].

Fortunately by this time flow cytometry technology was starting to achieve maturity and was starting to see widespread usage in the study of immune cells. A flow cytometer can detect individual cells in a single cell suspension, and can use a laser light sources to measure both cell light scattering properties (termed forward and side scatter). It can also measure the presence and amount of fluorescent probes incorporated into cells. Measurement of DNA content in permeabilized individual immune and tumor cells using DNA binding dyes like ethidium bromide and propidium iodide was being widely applied to immune and tumor cells by the mid 1980s. Several laboratories observed that apoptotic cells analyzed by flow cytometry displayed distinct changes in laser light forward scatter and 90° laser light side scatter, rough measures of cell size and cell granularity, respectively. Apoptotic cells were shrinking in size and becoming more light dense, a phenomenon that could be readily measured by cytometry. In addition, degradation and loss of chromatin could be measured by labeling with a DNA binding dye, and noting the decrease in the stoichiometric DNA dye binding associated with viable cells [2–4]. By the late 1980s and early 1990s, flow cytometry had largely replaced earlier methods for detecting cell death.

Flow cytometry had (and still has) several crucial advantages for measuring apoptosis over earlier techniques. Unlike cell lysate assays for DNA damage, flow cytometry analyzed cells on a per cell basis. This provided a much more accurate measure of cell death, since actual percentages of apoptotic activity could be measured. Very small numbers of cells could also be analyzed. In addition, flow cytometry is a multiparametric technique; multiple cell characteristics can be measured simultaneously. Even in the early days, two changes in cell morphology (forward and side scatter) and DNA content perturbations could be analyzed at the same time, and correlated on a per cell basis [5].

Since that time, the variety of techniques for measuring apoptosis by flow cytometry has exploded. Many flow cytometry assays now exist for detecting the later morphological manifestations of apoptosis. The TUNEL assay labeled 3′-OH terminated strand breaks that occur during chromatin degradation with a fluorochrome conjugated deoxynucleotide for detection by cytometry [6, 7]. While DNA binding dyes can still be used to measure DNA loss in apoptotic cells, they can also be used at lower concentrations to detect cells with cell membrane perturbation. DNA dyes like propidium iodide, 7-actinomycin D and DAPI will cross the membrane in advanced apoptotic cells and can be easily distinguished from viable cells by flow cytometry [8]. Covalent protein binding dyes (including the Live/Dead dye series) similarly cross the cell membrane in apoptotic cells and label these cells at higher levels than viable. Changes in mitochondrial function associated with apoptosis can be observed using mitochondrial transmembrane potential probes [9]. Fluorochrome conjugated lipids that intercalate into the plasma membrane at varying levels depending on its lipid disposition and charge are also used to measure apoptosis associated plasma membrane disruption [10]. Annexin V, which binds to phosphatidylserine (PS) residues, is probably the most common apoptosis assay in use today. PS residues normally segregate to the inner plasma membrane leaf in many cells; during apoptosis, they "flip" to the outer leaflet, where fluorochrome conjugated annexin V can identify their presence. A series of DNA and protein binding dyes (the SYTO) dyes are now being used as viable cell markers for continuous measurement of apoptosis over extended periods of time [11–13]. Most of these assays all measure the physical manifestations of apoptosis, which occur late in the process.

A more thorough understanding of the signal transduction pathways has led to assays that can identify the biochemical aspects of apoptosis, many of which occur well before the morphological manifestations. Most important of these are caspases, enzymes that play both a signaling and effecter role in many forms of apoptosis [14–16]. The early or proximal caspases (including but not limited to caspase 8, 9 and 10) are activated early in the apoptotic process; they act as signal transduction intermediates, as well as activators for the pro- form of enzymes important for cell death, including other caspases. The later caspases (including 1, 3, 6 and 7) mediate many of the morphological changes in apoptosis, including cytoskeletal collapse and chromatin degradation [14]. Like all enzymes, caspases preferentially cleave proteins at specific target sites. These target peptides have been integrated into a variety of cell-permeant fluorogenic caspase substrates that can be used to detect caspase activity in unfixed cells. The best known of these are the FLICA substrates (*F*luorescence *L*inked *I*nhibitors of *C*aspase *A*ctivation) available from multiple sources [17–20]. These probes are composed of a caspase consensus cleave domain, a fluorochrome molecule and a fluoromethylketone (FMK) reactive group. They are cell permeable, and bind to caspase activation sites in apoptotic cells [21]. The PhiPhiLux series of substrates (Oncoimmunin, Inc., Gaithersburg, MD USA) are also cell permeable, and use a pair of fluorochromes held in a quenched state by a caspase cleavage peptide. Upon cleavage, the fluorochromes are unquenched and fluoresce [22–24]. The CellEvent Green (Thermo Fisher Life Technologies, Carlsbad, CA, USA) and NucView 488 (Biotium) substrates, also cell permeable, are DNA binding dyes inhibited by a cleavage peptide. Upon cleavage, the dye migrates to the nucleus and binds to DNA [25]. All of these reagents have been used successfully to detect apoptosis. It is also possible to immunolabel active caspases in fixed cells using monoclonal antibodies. Other assays have also been developed to detect Bax translocation, cytochrome C release, histone associated proteins associated with early DNA strand breaks, and other early biochemical indicators of apoptosis.

One of the outstanding features of flow cytometry is its ability to measure many cell parameters simultaneously. This is particularly true of modern instruments, which can readily measure 12 or more scatter and fluorescent parameters. This makes flow cytometry an especially powerful tool for analyzing apoptosis; rather than relying on a single assay (which many investigators still do), we can combine multiple apoptosis assays in the same tube, giving a powerful multiparametric picture of the apoptotic process. In addition to providing multiple verifications that cell death is truly occurring, it also becomes possible to track the apoptotic process from early to late steps. This is particularly true if we combine both biochemical (early) and morphological (late) apoptosis measurements into a single assay. Fortunately, many apoptosis assays are compatible with one another, and the judicious selection of spectrally compatible fluorescent probes allows their simultaneous analysis [26]. In this guide, we describe the combination of two different fluorogenic caspase assays (a biochemical early step) with annexin V binding and DNA dye permeability (two morphological later steps) into a single tube assay. Since some experimental conditions also require cell fixation, we also describe the combination of caspase immunolabeling (early) with a covalent binding viability marker (late), an assay that does not require "viable" cells. While a more modern cytometry will give greater flexibility with regard to fluorochrome selection, even a simple cytometer can be used for these assays.

2 Materials

2.1 FLICA Fluorogenic Caspase 3/7 Substrate

This reagent is available from several sources, including Immunochemistry Technologies, LLC (trade name FLICA) and Thermo Fisher Life Technologies (trade name Vybrant FAM caspase 3/7). This probe employs fluorescein (FITC) or a fluorescein-like fluorochrome, and can be analyzed on any flow cytometer that can detect fluorescein. The FLICA caspase 3/7 substrate uses the DEVD recognition and cleavage. The reagent is diluted according to the manufacturer's instructions, and is added to the cell suspension. The substrate enters a cell by passive diffusion. When it encounters a caspase molecule that binds to the consensus sequence, the FMK portion of the molecule covalently cross-links to the caspase protein. The caspase becomes inactivated, and the probe is immobilized in the cell. The cell suspension is then washed extensively to remove unbound substrate.

1. The conventional form of FLICA can be excited with a standard blue-green 488 nm laser found on most flow cytometers, with detection in the detector reserved for fluorescein. It is spectrally compatible with propidium iodide or 7-aminoactinomycin D (which will be subsequently used for measuring apoptotic cell permeability) and a variety of fluorochrome conjugated annexin V reagents (for detection of phosphatidylserine "flipping" during apoptotic death).
2. The FLICA reagents are also available in fluorochromes other than fluorescein. Sulforhodamine 101 (SR101) labeled FLICA reagents are available from Immunochemistry Technologies (Magic Red substrates). SR101 is not well-excited at 488 nm, and requires a green (532 nm) or yellow (561 nm) laser source for best results. This fluorochrome is primarily used for imag-

ing, since these wavelengths are available from mercury arc lamp sources. However, modern cytometers are often equipped with green and yellow laser sources, so this fluorochrome is an option. FLICA reagents are also available in a red fluorochrome conjugated form from Immunochemistry Technologies (FLICA 660). This conjugate requires a red laser source, either a HeNe 633 nm or a red laser diode ~640 nm. Red lasers are common fixtures on cytometers. The detector usually reserved for allophycocyanin (APC) or Alexa Fluor 647 is used to detect this conjugate.

3. The assay described below uses the FLICA reagent specific for caspase 3 and 7, with the DEVD target peptide sequence. Both caspase 3 and 7 recognize this sequence, so they cannot be distinguished using this reagent. FLICA substrates are also available for other caspases, including 1, 6, 8, 9, and 10. A poly caspase reagent using a pan-specific caspase substrate (ZVAD) is also available from both Thermo Fisher Life Technologies and Immunochemistry Technologies.
4. FLICA reagents are usually supplied lyophilized. They are initially diluted in DMSO, with a subsequent dilution in aqueous buffer immediately prior to use. They are very hygroscopic and will inactivate quickly upon hydrolysis. The dry form should be stored over desiccant, and the initial dilution should be done with dry DMSO. The DMSO diluted form can be stored at −20 °C. The subsequent dilution in aqueous buffer should be used within 4 h of preparation, and should not be stored. Make only enough of the second aqueous dilution required for the assay at hand.
5. Unlike the PhiPhiLux and CellEvent/NucView 488 nm reagents, the FLICA reagents are fluorescent in the unreacted state and are very bright. It is therefore critical to thoroughly wash the cells after labeling with a significant volume of labeling buffer to remove any unreacted substrate. High levels of background in the flow cytometric analysis may be due to insufficient removal of unreacted substrate.
6. Some primary cell cultures may show somewhat lower levels of caspase activation than cell lines, with subsequent lower levels of substrate fluorescence (see Note 1). Cell lines may also show higher fluorescence background levels.
7. FLICA labeled cells can be fixed AFTER labeling. If fixed, they are no longer compatible with annexin V and DNA binding dye labeling as described below (Sects. 2.4 and 2.5). However, fixation FLICA is compatible with covalent binding viability probes (Sect. 2.6) if labeling is done prior to fixation.
8. While FLICA has some demonstrated specificity for caspase activity, no caspase substrate is completely specific for its target

enzyme. While specific for apoptotic cells, the FLICA reagents can bind intracellularly in cells with no caspase 3 activity [21]. See Note 3 for more information.

9. An in vivo form of FLICA, termed FLIVO, is also commercially available.

2.2 CellEvent Green or NucView 488 Caspase Substrate

This reagent is an alternative fluorogenic substrate to FLICA. This reagent is available from two sources: Thermo Fisher Life Technologies (trade name CellEvent Green caspase 3/7), and Biotium, Inc. (trade name NucView 488) This probe also employs fluorescein (FITC) or a fluorescein-like fluorochrome, and can also be analyzed on any flow cytometer than can detect fluorescein. These substrates are also specific for caspases 3 and 7 substrate, and use the same DEVD recognition and cleavage sequence as FLICA [27].

1. CellEvent Green and NucView 488 are also excited by the blue-green 488 nm laser found on most flow cytometers, and are detected as with fluorescein. They are also spectrally compatible with propidium iodide or 7-aminoactinomycin D (which will be subsequently used for measuring apoptotic cell permeability) and the same fluorochrome conjugated annexin V reagents.
2. At this time, both CellEvent Green and NucView 488 are only available in forms specific to caspase 3 and 7, and only conjugated to fluorescein (unlike FLICA, which is available in several specificities and fluorochromes).
3. Both CellEvent Green and NucView 488 are supplied already suspended in DMSO, and should be stored at −20 °C over desiccant. The stock is usually good for about one year. Initial thawing and preparation of small aliquots is recommended to prevent excessive freeze-thawing.
4. CellEvent Green and NucView 488 labeling are NOT compatible with fixation, and should be analyzed promptly following the labeling procedure.
5. While CellEvent Green and NucView 488 have some demonstrated specificity for caspase activity, no caspase substrate is completely specific for its target enzyme. See Note 3 for more information.
6. Biotium, Inc. has recently released a violet laser excited caspase substrate, termed NucView 405.

2.3 PhiPhiLux Fluorogenic Caspase Substrates

These caspase substrates from Oncoimmunin, Inc. can also be substituted for FLICA or CellEvent Green/NucView 488. Their use has been extensively described in previous publications [22–24, 26].

2.4 Fluorochrome Conjugated Annexin V

Annexin V is available conjugated to a variety of fluorochromes from many manufacturers, both in kit form and as a single reagent. For this assay, a kit is not required, and the less expensive individual form is recommended. Annexin V binds tightly to apoptotic cells with “flipped” phosphatidylserine residues on their extracellular membrane leaflet. Damaged or necrotic cells with a high degree of membrane permeability will also bind annexin V to their intracellular membrane leaflet, in spite their questionable apoptotic status; therefore, a DNA binding dye (Sect. 2.5) as a cell permeability indicator should always be included with annexin V binding assays.

1. Annexin V is available as many fluorochrome conjugates, including fluorescein (FITC) and Alexa Fluor 488, phycoerythrin (PE), PE-Cy5, allophycocyanin (APC), Alexa Fluor 647, and Pacific Blue. The selection of reagent will depend of what other apoptosis assays and fluorochromes it will be combined with. If annexin V is being combined with the fluorogenic caspase substrates above, the fluorescein or Alexa Fluor 488 nm annexin V conjugates will be spectrally incompatible and should not be combined. A PE conjugate would be compatible, but will require color compensation to separate the overlapping fluorescein and PE signals. To minimize the compensation requirement, use an annexin V conjugate with minimal spectral overlap with both the caspase substrate and DNA binding dye in your multiparametric assay.
2. Annexin V binding is absolutely dependent on the presence of calcium and magnesium ions in the labeling buffer. Your incubation buffer must therefore contain these divalent cations. If using a complete medium such as RPMI-1640 or D-MEM as an incubation buffer as recommended in Sects. 2.4 and 2.7, these cations will already be present. If using a simple buffer like Hanks Balanced Salt Solution (HBSS), make sure it contains calcium and magnesium. Many simple buffers are formulated as calcium and magnesium free.
3. Annexin V binding infrequently occurs in viable cells with significant PS levels on the extracellular membrane leaflet. See Note 5 for more information.

2.5 DNA Binding Dyes

This is the “final” step in this multicolor labeling, and is used as an indicator of cell membrane permeability. While annexin V binding is an indication of plasma membrane perturbation, it can often occur in advance of total membrane integrity loss. A DNA binding dye therefore detects one of the latest stages of apoptosis. Like annexin V, DNA binding dyes are available in a wide variety of colors and permeability properties. Some options are listed below, with selection guidelines. Specific combinations will be listed in Sect. 3.

1. *Propidium iodide* (PI) is the most common DNA binding dye viability probe. It is an intercalating DNA binding dye, and is available from a wide variety of sources. It is highly impermeant to viable cells, and will be excluded from all but the most advanced apoptotic cells. Propidium iodide is excited at 488 nm and emits in the 570–630 nm range, and is detected in the PE or PE-Cy5 detector on most cytometers. It can be used with the fluorescein based caspase substrates, but it possesses a wide emission profile and will require compensation. As a result, it is not the best dye to combine with caspase detection from a spectral standpoint. It should be dissolved in deionized water at 1 mg/ml and stored in the dark at 4 °C for up to 3 months.
2. *7-Aminoactinomycin D* is a common alternative to propidium iodide. Available from Sigma Chemical Co., St. Louis, MO USA) and Molecular Probes, Eugene, OR). It is slightly more cell permeant than PI, but is still excluded by all live cells. 7-aminoactinomycin D (7-AAD) is a DNA binding dye that excites at 488 nm and emits in the far red, with an emission peak at approximately 670 nm. 7-AAD can be used in place of propidium iodide where a longer wavelength cell permeability probe is desired. It is typically detected in the PE-Cy5 or PerCP detector on commercial cytometers. It shows much less spectral overlap with fluorescein, and is therefore more spectrally compatible with the above caspase substrates. 7-AAD should be dissolved in EtOH at 1 mg/ml and stored at –20 °C. Solubilized stocks in EtOH are good for 6 months. Diluted stocks should be used within 24 h. Thermo Fisher Life Technologies also provides a variant of 7-AAD under the name SYTOX AADvanced; it has a lower molecular weight and may be more soluble. It has spectral properties similar to 7-AAD, is somewhat more soluble than 7-AAD, and is prepared and stored similarly.
3. *DAPI, Hoechst 33258 and SYTOX Blue*. DAPI and Hoechst 33258 are available from several manufacturers. SYTOX Blue is available from Thermo Fisher Life Technologies. Although structurally dissimilar, these dyes have similar spectral properties. All are excited by ultraviolet or violet lasers, and emit in the blue 440–480 nm range. DAPI and Hoechst 33258 have similar cell permeability characteristics to PI, while SYTOX Blue is somewhat more permeable. Since all of these dyes show virtually no excitation at 488 nm, they have almost no spectral overlap into either fluorescein caspase substrates or almost any annexin V substrate (except Pacific Blue, which cannot be used with any of them). They therefore make excellent viability dyes. They do require a flow cytometer equipped with either an ultraviolet laser or a violet laser diode. Ultraviolet lasers are

not common on flow cytometers, but violet laser diodes are frequent additions to multi-laser instruments. Hoechst 33258 should not be confused with the structurally similar dye Hoechst 33342, which is highly cell permeant and will label both viable and apoptotic cells.

4. *TO-PRO-3, SYTOX Red and DRAQ7.* TO-PRO-3 and SYTOX Red are available from Thermo Fisher Life Technologies, DRAQ7 from Biostatus, Ltd., Leicestershire, UK. TO-PRO-3 and DRAQ7 show cell permeability similar to PI, while SYTOX Red shows somewhat greater cell permeability. These dyes excite using a red laser source, including HeNe 633 nm and red laser diode 640 nm units. These lasers are common on multi-laser flow cytometers. As with the violet-excited dyes, these probes show little excitation at 488 nm and are spectrally compatible with fluorescein caspase substrates. They can also be used with Pacific Blue annexin V on a three laser cytometer. They are analyzed using the detector normally reserved for APC or Alexa Fluor 647. The APC and Alexa Fluor 647 annexin V conjugates should therefore not be used with these DNA dyes.
5. *Other probes.* There are literally hundreds of DNA binding dyes available, with different spectral properties and varying degrees of cell permeability. An experienced flow cytometrist should be able to use some of these for multiparametric apoptosis detection once they are familiar with their properties. While possibly useful for imaging applications, DNA dyes with strong cell permeability, including Hoechst 33342, DRAQ5 (Biostatus), and DyeCycle Violet, Green, Orange, and Deep Red (Thermo Fisher Life Technologies) are not recommended for apoptosis detection by flow cytometry. Their permeability to viable cells, while useful for cell cycle analysis, makes them less useful for apoptotic cell discrimination. However, labeling the nuclei of apoptotic cells may be informative in imaging applications.
6. *Washing and fixation.* Cells labeled with DNA binding dyes should be analyzed by flow cytometry promptly. Since these molecules do not covalently bind to DNA, the dye needs to be present in solution to maintain labeling. Samples labeled with DNA dyes should not be fixed with paraformaldehyde or any other fixative. This will be covered in detail in Sect. 3.

2.6 Covalent Binding Viability Probes

In place of the DNA binding dyes described above, a covalent binding viability probe may be substituted. These probes are composed of a fluorochrome conjugated to an amine-reactive succinimidyl ester moiety. When incubated with viable cells in the absence of soluble protein, these probes bind to any available protein on the surface of the cell. If apoptotic cells are present, the probe will also cross the plasma membrane and bind to intracellular proteins as well. Apoptotic cells will therefore bind far more probe than viable cells, allowing this reagent to distinguish between the two.

1. There are two main suppliers for these reagents. The Live/Dead probes from Thermo Fisher Life Technologies are available in a wide variety of fluorescent probes, enabling them to be integrated into multiparametric apoptosis assays. Particularly useful variants include the Near IR conjugate, which is red laser excited and emits in the far red range (~800 nm), giving excellent spectral separation from most of the probes described above. The Aqua and Yellow variants are excited with a violet laser, and emit in the green (~510–520 nm) and yellow (550–570 nm) range, also easy to integrate with other apoptosis reagents.
2. The Zombie dyes (BioLegend, San Diego, CA) are similar to the Live Dead Reagents. Zombie Near IR, Violet and Yellow have the same spectral properties as the Near IR, Aqua and Yellow variants above. Other fluorochromes are also available.
3. Like the DNA binding dyes, these reagents are added at the end of the assay. However, they have special labeling requirements. Since they bind to any amine residue containing protein, the labeling must be carried out in protein-free buffer. A final wash step with protein-free buffer is included prior to probe labeling in Sect. 3. If cells have been labeled with annexin V, however, this buffer will still need to contain calcium and magnesium.
4. Unlike DNA binding dyes, covalent binding viability probes can be fixed following labeling, if (and only if) other labels are similarly compatible with fixation.

2.7 Fluorochrome Conjugated Antibody Against Active Caspases

If cells can (or must be) fixed and permeabilized prior to labeling, the substrates above can be replaced with antibody labeling directed against the caspase of interest. Several antibodies, either unconjugated or conjugated to fluorochrome, are available for this purpose. Since this is an intracellular labeling procedure, the cells must be both fixed and permeabilized prior to labeling, using fixatives and detergents similar to those used for intracellular cytokine labeling. Since the samples are fixed, annexin V and DNA dye binding reagents are NOT compatible with this labeling method. However, covalent binding viability probe labeling is compatible, if done prior to fixation.

1. *Antibodies.* Several commercially available antibodies against active caspases are available. BD Biosciences Pharmingen (San Diego, CA) provides a rabbit monoclonal antibody against the cleaved active form of caspase 3 (product number 561011) that works very well for intracellular labeling. It is available conjugated to fluorescein, PE and biotin. It should not be confused with the rabbit polyclonal antibody against caspase 3

from the same manufacturer; this reagent works, but shows less binding affinity. Cell Signaling Technologies (Danvers, MA, USA) also manufacturers several antibodies against caspase 3 and 7 that have been validated for flow cytometry.

2. When looking for antibodies against caspase molecules, it is critical that the reagent be certified for flow cytometry. Many antibodies against caspases are available for Western blotting; unfortunately, most of these are epitope-specific polyclonal antibodies, and are usually not suitable for flow cytometry.
3. The intracellular labeling method will be detailed in Sect. 3, using both buffers from specific manufacturers and generic reagents. Any fixation/permeabilization buffer system used for cytokine and other intracellular immunolabeling applications will usually work well. Examples include the Fix/Perm and Perm/Wash system from BD Biosciences Pharmingen.

2.8 Incubation Buffer

The entire labeling procedure should be carried out in buffers that are compatible with all aspects of the combined assay, including loading with caspase substrate, annexin V binding and DNA dye permeability. Using annexin V will require the buffer to have calcium and magnesium present throughout the entire assay. If using covalent binding viability probes, a final wash in a protein-free buffer will be required prior to and during labeling. The following buffers are recommended and will be referenced in the Methods in Sect. 3.

1. *Complete medium.* For FLICA, CellEvent Green and NucView 488 caspase substrates, as well as annexin V and DNA binding dyes, using the media the cells are normally grown in as a wash buffer can be a good strategy. Apoptosis is an ongoing process, and will continue to occur throughout the assay duration, particularly if the cells are nutritionally stressed. Using a complete medium, including RPMI-1640, D-MEM, etc. with serum) will minimize stress to the cells during the labeling process. This media will contain calcium and magnesium, ensuring compatibility with annexin V, and it will not interfere with caspase substrate loading or DNA dye activity. If using a covalent binding viability probe, a final wash in protein-free medium prior to labeling should be done; however, all prior steps can be done in complete medium.
2. *Incubation buffer.* If the cell growth media cannot be used, colorless RPMI-164 supplemented with 2 % fetal bovine serum (FBS) makes a good substitute. It contains calcium and magnesium. In the Sect. 3 methods, this medium is referred to as incubation buffer.
3. *Protein-free wash buffer.* For covalent binding viability probe, colorless RPMI-1640 or Hanks Balanced Salt Solution (HBSS) with calcium and magnesium can be used.

2.9 Control Cells

As with any assay, a positive control cell type or cell line is very useful for verifying if the assay is working. Since it needs to be available continuously, this is most likely a cell line that can be maintained in culture and grown continuously. While it may not be necessary to include a control cell line in every experiment, it should be done in the beginning of assay development, and at any point when the assay's functionality is in doubt.

1. *Cells. Many suspension cell lines are useful.* It is advantageous if the control cell line resembles the target cell type in some way. For example, T leukemia and lymphoma cell lines are relevant controls for T cell immunology.

2.10 Induction of Apoptosis

There is considerable variability in the sensitivity of cell lines to apoptotic stimuli. Some cell lines are high resistant to drugs and receptor-mediated induction of apoptosis, while others are very sensitive. However, several drug agents have been found to be useful inducers in many cell lines. Transcriptional and translational inhibitors like cycloheximide and actinomycin arrest cell cycle, and often induce apoptosis. Topoisomerase II inhibitors like camptothecin, topotecan and etoposide induce apoptosis in many cell types. The calcium mobilization inhibitor staurosporine is a frequent inducer. Receptor mediation of apoptosis is also possible; Fas expressing cells and cell lines can often be induced using an anti-Fas antibody. For drugs, low microgram to high nanogram per milliliter concentrations are usually sufficient for induction; higher concentrations may induce necrotic death. Both time and concentration titrations are critical. Specific examples of cell lines and inducers are listed in the **Note 1**.

3 Methods

3.1 Cell Presentation

Prior to the experiment, the individual indicators (caspase substrate, annexin V, DNA binding dye or covalent binding viability probe, etc.) must be chosen. The choices will be based on the required state of the cells (fixed or unfixed), what the investigator specifically wants to know, and on the flow cytometry instrumentation available. The possible multiparametric assays available are listed below.

Unfixed cells. When the investigator requires or is able to analyze the cells promptly after labeling, reagent combinations using unfixed cells can be used. The cells remain "viable" for the entire assay, and are analyzed this way. There are several options in this format.

1. *Either caspase substrate, annexin V and a DNA binding dye.* In this option, either caspase substrate can be used, followed by annexin V and a DNA binding dye. This combination will not tolerate fixation.

2. *Either caspase substrate, annexin V and a covalent binding viability probe.* In this option, either caspase substrate can be used, followed by annexin V and a covalent binding viability probe. This combination will also not tolerate fixation.

   Fixed cells. If cell fixation is desirable or required (due to longer delays following preparation, or for biohazard control purposes), the following combinations can be used

3. *FLICA only and a covalent binding viability probe.* In this option, the FLICA caspase substrate can be used, followed by a covalent binding viability probe. No annexin V or DNA binding dyes are used here.
4. *A covalent binding viability probe, followed by anti-caspase 3 immunolabeling.* The cells are initially labeled with a covalent binding viability probe, then fixed, permeabilized and labeled with a fluorochrome conjugated anti-caspase 3 antibody. Again, no annexin V or DNA binding dyes are used here.

3.2 Cytometry Instrumentation

The cytometer available to the investigator will dictate what reagents can be used. However, even simple single laser cytometers can usually manage at least two and usually three fluorescent parameters as well as scatter.

1. *Single laser instruments.* Very basic cytometers are usually equipped with a blue-green 488 nm laser, and two to four fluorescent detectors. Older instruments of this type include the BD Biosciences FACScan and Beckman Coulter XL. Newer instruments include the basic models of the BD Biosciences FACSVerse and single laser Accuri and the EMD Millipore Guava easyCyte 5 and 5HT. For any of these instruments, all probes must be excited at 488 nm. For unfixed cell assays, any of the fluorescein based caspase reagents can be combined with PE conjugated annexin V and the DNA binding dye 7-AAD. Unfortunately, most covalent binding viability probes are not well-excited at 488 nm, except Live Dead and Zombie Green, which will overlap with the caspase assays.
2. *Dual blue-green and red laser systems.* Cytometers with both a blue-green and red laser source are now very common. The BD Biosciences FACSort and FACSCalibur, the BD Biosciences Accuri and FACSVerse systems, basic BD LSR II, and Beckman-Coulter FC500 systems are usually equipped with both lasers. The EMD Millipore Guava EasyCyte 6 and 8 series cytometers are also equipped with two lasers. A second additional laser greatly increases the reagent options. Examples include:

   FLICA, CellEvent Green, or NucView 488

   APC or Alexa Fluor 647 annexin V

   7-AAD

FLICA, CellEvent Green, or NucView 488
PE-Cy5 annexin V
TO-PRO-3, SYTOX Red, or DRAQ7

FLICA, CellEvent Green, or NucView 488
APC or Alexa Fluor 647 annexin V
Live/Dead Near IR or Zombie Near IR

FLICA
Live/Dead Near IR or Zombie Near IR

Live/Dead Near IR or Zombie Near IR
PE anti-caspase 3 antibody

The first three examples are not fixable; the fourth one can be fixed. The last one requires fixation/permeabilization.

3. *Three laser blue-green, red, and violet systems.* Cytometers with three lasers are now very common. Equipped with blue-green, red, and violet laser diodes (~405 nm), virtually any of the probes listed above can be excited. Some possible combinations include:

FLICA, CellEvent Green, or NucView 488
APC or Alexa Fluor 647 annexin V
DAPI, Hoechst 33258, or SYTOX Blue

FLICA, CellEvent Green, or NucView 488
Pacific Blue annexin V
TO-PRO-3, SYTOX Red, or DRAQ7

FLICA, CellEvent Green, or NucView 488
Pacific Blue annexin V
Live/Dead or Zombie Near IR

FLICA, CellEvent Green, or NucView 488
APC or Alexa Fluor 647 annexin V
Live/Dead Aqua or Yellow, or Zombie Violet or Yellow

Live/Dead Aqua or Yellow, or Zombie Violet or Yellow
PE anti-caspase 3 antibody

The first two examples are not fixable; the third and fourth can be fixed. The last one requires fixation/permeabilization.

Of course, the combinations described in Sect. 3.2, step 1 can also be run on a three laser system.

4. Dye combinations should be planned carefully before proceeding, and checked for spectral compatibility. Using a multi-laser system and distributing the fluorescent probes among the individual lasers is a good way to minimize the requirement for

fluorescence compensation, as there will be minimal spectral overlap between fluorochromes.

3.3 Preparation of Cells Prior to Labeling

Cells can be removed from culture after apoptosis induction, centrifuged at 400 × g to pellet, and resuspended either in complete medium or incubation buffer (Sect. 2.8). Alternately, cells can be labeled directly in tissue culture medium, thus eliminating a centrifugation step. This alternative will minimize damage to cells by reducing the need for centrifugation

3.4 Preparation of Caspase Substrates

1. The FLICA reagent from Thermo Fisher Life Technologies and Immunochemistry Technologies, LLC is supplied as a lyophilized product in a glass vial, stored at −20 °C with desiccant. Prior to use, warm the vial to room temperature, and add 50 μl of dry DMSO to the vial. Gently mix, and allow time to reconstitute; the FLICA reagent will enter solution slowly. This stock (termed the 150× stock) can be stored at −20 °C, and should be stable for at least 3 months. To prevent freeze–thaw cycles, aliquot this stock into small quantities and store at −20 °C over desiccant. Immediately prior to use, dilute the above 150× stock at 1 volume stock to 4 volumes PBS, for a final concentration of 30×. This stock should be kept at 4 °C, and used within 4 h. Prepare only enough 30× stock for use in one day. Discard any unused material at the end of the experiment.
2. The CellEvent Green and NucView reagents are already reconstituted in DMSO at a stock concentration of 500 μM (100 μl total volume). This stock should be stored at −20 °C and is stable for a least 1 year. Separation into small aliquots to prevent repeated freeze/thawing is also advisable.

3.5 Preparation of Annexin V

1. Annexin V is usually provided as an aqueous solution in concentrations ranging from 100 to 500 μg/ml. Annexin V solutions should not be frozen but should be stored at 4 °C. This is especially true for PE and APC conjugates. Freeze/thaw cycles should be avoided. As a general rule, 10 μl of stock per 300 μl and 20 μl per 1 ml medium or incubation buffer should suffice for labeling.

3.6 Preparation of DNA Binding Dyes

1. Propidium iodide is supplied as a dry powder. Solutions of propidium iodide are also available, but these can have uncertain concentrations, as well as the inclusion of RNase for DNA cell cycle applications. These solutions should not be used for viability measurement. PI can be prepared as a stock solution of 1 mg/ml in PBS, which can be stored for 6 months at 4 °C. Add 2 μl PI stock per 1 ml medium for a final concentration of 2 μg/ml.

2. 7-Actinomycin D is usually supplied as a dry reagent in 1 mg amounts. Prepare a 1 mg/ml stock with 95 % EtOH or DMSO, and store at −20 °C. Add 2 μl 7-AAD stock per 1 ml medium for a final concentration of 2 μg/ml. The Thermo Fisher Life Technologies variant SYTOX AADvanced is also supplied as a dry reagent. Add 100 μl of DMSO, and add 1 μl of dye per 1 ml medium.
3. DAPI and Hoechst 33258 are both supplied as dry powders. Prepare as 1 mg/ml stock solutions in distilled water, and store for up to 3 months at 4 °C. Add 2 μl DAPI or Hoechst 33258 stock per 1 ml medium for a final concentration of 2 μg/ml.
4. TO-PRO-3, SYTOX Red and SYTOX Blue are supplied as prepared stocks in DMSO. Add 1 μl stock per 1 ml medium.
5. DRAQ7 should be used according to the manufacturer's instructions (Biostatus).

3.7 Preparation of Covalent Binding Viability Dyes

The Thermo Fisher Life Technologies Live/Dead reagents and the BioLegend Zombie dyes are both supplied as lyophilized stocks in single use vials. Prior to use, reconstitute the contents of each vial with 100 μl DMSO and vortex. Add 1 μl reconstituted stock per 1 ml sample. Unused vial contents can be restored at −20 °C, but should be stored over desiccant.

3.8 Fixation for FLICA Labeling

Following FLICA and covalent binding viability probe labeling, cells can be fixed with a final concentration of 2 % paraformaldehyde. 10 % aqueous stocks of formaldehyde (Thermo Fisher) can be diluted for this purpose. Alternately, powdered paraformaldehyde can be dissolved in distilled water at 10 % w/v, with gentle (non-boiling) heating and continuous stirring in a fume cabinet. WARNING! Paraformaldehyde and its degradation products are carcinogens. Exposure and inhalation should be avoided. The 10 % solution can be stored at 4 °C for up to 3 months, and can be diluted with additional distilled water and 10× PBS to a final 2 % concentration.

3.9 Fixation and Permeabilization Buffers for Caspase Immunolabeling

Immunolabeling with anti-caspase antibodies requires initial fixation and permeabilization of the cell sample. The BD Pharmingen Fix/Perm and Perm/Wash reagents work well for this application. However, any commercial buffer preparation for intracellular cytoplasmic labeling should work equally well. The fixation/permeabilization process consists of initial resuspension in the Fix/Perm buffer; this is supplied as a 1× solution and can be used directly from the container. After a brief incubation, the cells are washed with the Perm/Wash buffer, which can be used for both antibody labeling and final washing. This buffer is supplied as a 10× stock; a 1× stock using distilled water will need to be prepared prior to use.

3.10 Single Color Controls for Fluorescence Compensation

As will all multicolor labeling methods, some fluorescence compensation will be required between the fluorescent probes used in this method. For some labeling combinations (i.e., FITC and PE, or FITC and PI), this compensation may be considerable. Single color controls for each probe are therefore recommended to allow the user to calculate this compensation. While they might not need to be used routinely, they should be used in the beginning of assay development, until the compensation requirements are understood.

If using a multi-laser instrument, it is advisable to separate each fluorescent probe to an individual laser. For example, FLICA can be excited at 488 nm, Pacific Blue annexin V would be excited with a violet laser source, and TO-PRO-3 would be excited with a red laser source. By using independent excitation sources, fluorescence compensation should be minimal.

3.11 Labeling Method for Unfixed Cells

Below are listed four labeling procedures using either the FLICA or CellEvent Green/NucView 488 reagent, and either DNA binding dyes or covalent binding viability dyes.

3.11.1 FLICA, Annexin V and DNA Binding Dye

1. Resuspend the cells to be analyzed at 0.5–1 million per ml in either complete media or incubation buffer (see Sect. 2.8).
2. Transfer 300 μl of cell suspension to assay tubes.
3. Add 10 μl of the final FLICA solution to the cells. See Sect. 3.4 for FLICA stock solution preparation.
4. Incubate at 37 °C for 30 min.
5. Remove the tubes from the incubator, and add the desired annexin V conjugate at 10 μl per tube. Return the tubes to the incubator and continue incubation at 37 °C for an additional 15 min.
6. Remove tubes from incubator, and add 3 ml complete medium or incubation buffer per tube. Centrifuge at 400 × g for 5 min.
7. Decant, resuspend in an additional 3 ml complete medium or incubation buffer per tube, and centrifuge again at 400 × g for 5 min.
8. Decant and resuspend in 1 ml complete medium or incubation buffer.
9. Add the DNA binding dye at the desired concentration. See Sect. 3.6 for guidelines.
10. Analyze cells within 1 h. If analysis is to be delayed, do not add the DNA dye until ~15 min prior to analysis.

3.11.2 CellEvent Green or NucView 488, Annexin V and DNA Binding Dye

1. Resuspend the cells to be analyzed at 0.5 to 1 million per ml in either complete media or incubation buffer (see Sect. 2.8).
2. Transfer 1 ml of cell suspension to assay tubes.
3. Add 1 μl of the thawed CellEvent Green or NucView 488 solution to the cells. See Sect. 3.4 for CellEvent Green and NucView 488 stock solution preparation.

4. Incubate at 37 °C for 15 min.
5. Remove the tubes from the incubator, and add the desired annexin V conjugate at 20 μl per tube. Return the tubes to the incubator and continue incubation at 37 °C for an additional 15 min.
6. Remove tubes from incubator, and add 3 ml complete medium or incubation buffer per tube. Centrifuge at 400 × g for 5 min.
7. Decant and resuspend in 1 ml complete medium or incubation buffer.
8. Add the DNA binding dye at the desired concentration. See Sect. 3.6 for guidelines.
9. Analyze cells within 1 h. If analysis is to be delayed, do not add the DNA dye until ~15 min prior to analysis.

3.11.3 FLICA, Annexin V and Covalent Binding Viability Probe

1. Resuspend the cells to be analyzed at 0.5 to 1 million per ml in either complete media or incubation buffer (see Sect. 2.8).
2. Transfer 300 μl of cell suspension to assay tubes.
3. Add 10 μl of the final FLICA solution to the cells. See Sect. 3.4 for FLICA stock solution preparation.
4. Incubate at 37 °C for 30 min.
5. Remove the tubes from the incubator, and add the desired annexin V conjugate at 10 μl per tube. Return the tubes to the incubator and continue incubation at 37 °C for an additional 15 min.
6. Remove tubes from incubator, and add 3 ml complete medium or incubation buffer per tube. Centrifuge at 400 × g for 5 min.
7. Decant, resuspend in an additional 3 ml complete medium or incubation buffer per tube, and centrifuge again at 400 × g for 5 min.
8. Decant and resuspend in 3 ml RPMI or HBSS with calcium and magnesium, no protein. Centrifuge at 400 × g for 5 min.
9. Add the reconstituted desired Live/Dead or Zombie dye at 1 μl per 1 ml buffer. See Sect. 3.7 for guidelines. Incubate at 37 °C for 30 min.
10. Analyze cells within 1 h.

3.11.4 CellEvent Green or NucView 488, Annexin V and DNA Binding Dye

1. Resuspend the cells to be analyzed at 0.5 to 1 million per ml in either complete media or incubation buffer (see Sect. 2.8).
2. Transfer 1 ml of cell suspension to assay tubes.
3. Add 1 μl of the thawed CellEvent Green or NucView 488 solution to the cells. See Sect. 3.4 for CellEvent Green and NucView 488 stock solution preparation.
4. Incubate at 37 °C for 15 min.

5. Remove the tubes from the incubator, and add the desired annexin V conjugate at 20 µl per tube. Return the tubes to the incubator and continue incubation at 37 °C for an additional 15 min.
6. Remove tubes from incubator, and add 3 ml complete medium or incubation buffer per tube. Centrifuge at 400 × g for 5 min.
7. Decant and resuspend in 3 ml RPMI or HBSS with calcium and magnesium, no protein. Centrifuge at 400 × g for 5 min.
8. Add the reconstituted desired Live/Dead or Zombie dye at 1 µl per 1 ml buffer. See Sect. 3.7 for guidelines. Incubate at 37 °C for 30 min.
9. Analyze cells within 1 h.

3.12 Labeling Method for Fixed Cells

Below is listed the labeling procedure using FLICA and covalent binding viability dyes, followed by fixation at the end.

3.12.1 FLICA and Covalent Binding Viability Probe

1. Resuspend the cells to be analyzed at 0.5 to 1 million per ml in either complete media or incubation buffer (see Sect. 2.8).
2. Transfer 300 µl of cell suspension to assay tubes.
3. Add 10 µl of the final FLICA solution to the cells. See Sect. 3.4 for FLICA stock solution preparation.
4. Incubate at 37 °C for 45 min.
5. Remove tubes from incubator, and add 3 ml complete medium or incubation buffer per tube. Centrifuge at 400 × g for 5 min.
6. Decant, resuspend in an additional 3 ml complete medium or incubation buffer per tube, and centrifuge again at 400 × g for 5 min.
7. Decant and resuspend in 3 ml RPMI or HBSS with calcium and magnesium, no protein. Centrifuge at 400 × g for 5 min.
8. Add the reconstituted desired Live/Dead or Zombie dye at 1 µl per 1 ml buffer. See Sect. 3.7 for guidelines. Incubate at 37 °C for 30 min.
9. Add sufficient paraformaldehyde solution stock solution for a final concentration of 2 % (i.e., 250 µl of a 10 % aqueous solution).
10. Analyze.

3.13 Immunolabeling for Caspase 3 Using BD Biosciences Pharmingen Fix/Perm and Perm/Wash Reagents

In this method, cells are initially labeled with a covalent binding viability probe, followed by fixation, permeabilization and labeling with anti-caspase 3 monoclonal antibody.

1. Resuspend the cells to be analyzed at 0.5 to 1 million per ml in either complete media or incubation buffer (see Sect. 2.8).
2. Centrifuge at 400 × 8 for 5 min, and decant.

3. Decant and resuspend in 3 ml RPMI or HBSS with calcium and magnesium, no protein. Centrifuge at 400×g for 5 min.
4. Add the reconstituted desired Live/Dead or Zombie dye at 1 μl per 1 ml buffer. See Sect. 3.7 for guidelines. Incubate at 37 °C for 30 min.
5. Gently shake tube to disperse pellet, and add 1 ml of BD Pharmingen Fix/Perm buffer from 1× stock. Incubate at 4 °C for 30 min.
6. Centrifuge at 400×8 for 5 min, and decant. Shake tube to disperse pellet, and add 3 ml BD Pharmingen Perm/Wash buffer from a pre-prepared 1× stock (buffer is supplied as a 10× stock solution; see Sect. 3.7). Buffer and samples should be kept at 4 °C.
7. Centrifuge at 400×8 for 5 min, and decant. Add 200 μl of BD Pharmingen Perm/Wash buffer and gently shake to disperse pellet.
8. Add 10 μl of anti-caspase 3 antibody from stock (either FITC, PE or biotin conjugate as desired). See Sect. 2.7 for details.
9. Incubate for a minimum of 2 h at room temperature. For better results, incubate overnight at 4 °C.
10. Add 3 ml BD Pharmingen Perm/Wash buffer to each tube. Centrifuge at 400×8 for 5 min, and decant.
11. If a direct antibody conjugate was used, analyze.
12. If the biotin conjugated antibody was used, add a fluorochrome conjugated streptavidin secondary reagent. Incubate for 2 h.
13. Add 3 ml BD Pharmingen Perm/Wash buffer to each tube. Centrifuge at 400×8 for 5 min, and decant.
14. Resuspend in 1 ml PBS and analyze.

3.14 Flow Cytometric Data Acquisition

Unfixed samples should be analyzed promptly (within 1–2 h). If there will be a brief delay, store samples at 4 °C until analysis. Fixed samples should be analyzed within 1–2 days.

1. *Forward and side scatter.* Forward scatter should be analyzed with linear scaling. Side scatter can be analyzed in either linear o log scaling, although linear generally gives better cell distribution. For many cell types, "viable" and apoptotic cells will show distinct scatter profiles in forward and side scatter, which forward scatter decreasing and side scatter increasing (Fig. 1). However, this is not always the case. In particular, fixed cells often show less distinction between viable and apoptotic by scatter.
2. *Gating on forward and side scatter.* For initial gating, drawing a gate around (1) all the intact cells, both "viable" and

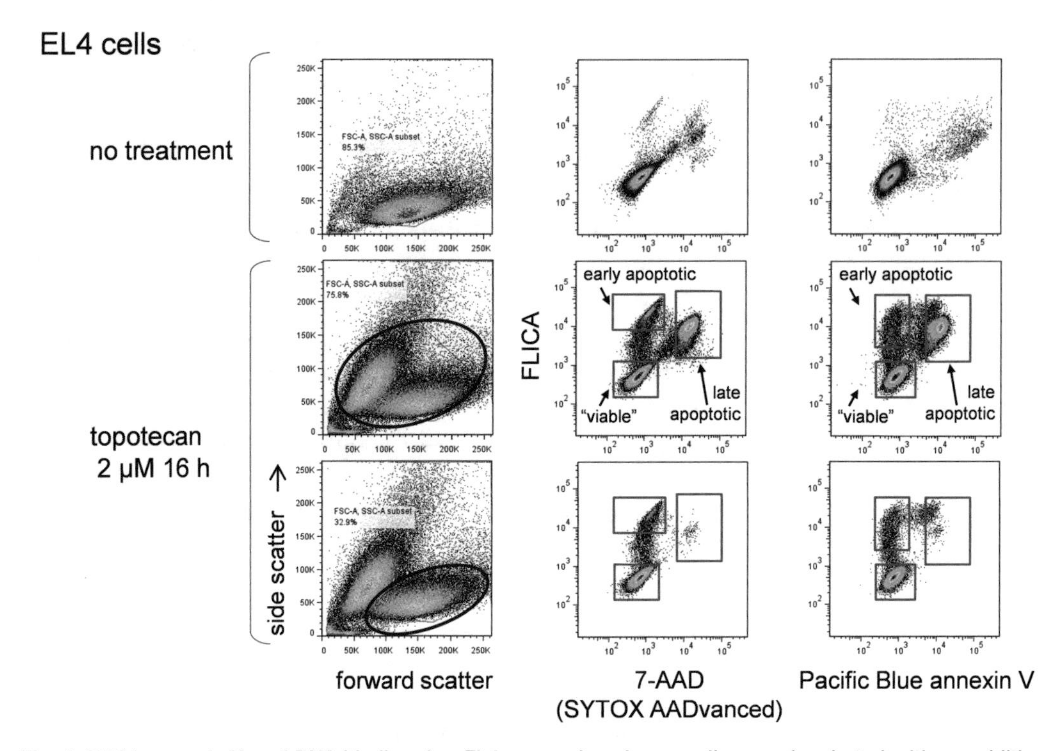

Fig. 1 *FLICA, annexin V, and DNA binding dye.* EL4 mouse lymphoma cells were incubated with no additions (*top row*) or with topotecan at 2 μM for 16 h (*bottom two rows*). Cells were analyzed on a BD Biosciences LSR II flow cytometer equipped with blue-green, red, and violet lasers. Forward versus side scatter plots (*left column*) show the distinctive shift in scatter during apoptosis. *Middle column* shows SYTOX AADvanced versus FLICA fluorescence, and the *right column* shows Pacific Blue annexin V versus FLICA fluorescence. The *middle row* shows cells gated for all single cell events; the *bottom row* shows cells gated for the scatter "viable" population only. Regions show viable, early and late apoptotic subpopulations

apoptotic, and (2) just the "viable" cells, is recommended. Gating on all intact cells (excluding debris) will provide an overall picture of the entire apoptotic process, including the very advanced or late apoptotic cells. Gating on only the scatter "viable" cells will highlight the earlier stages; as noted in Figs. 1, 2, 3, and 4, even cells that appear "viable" are in fact starting to demonstrate early stage apoptotic phenotype, including caspase activity and some early annexin V binding. By excluding the late apoptotic cells, scatter "viable" gating pinpoints the initial entry into cell death.

3. *Fluorescence detector settings.* Fluorescence detectors should be set so both "viable" and apoptotic events are on scale. Keep in mind that the fluorogenic caspase substrates can demonstrate high backgrounds in the "viable" population. A detector setting for unlabeled cells may therefore be too high to accommodate both the background and specific fluorescence signal. Run an apoptotic sample initially to make certain no signal is off scale.

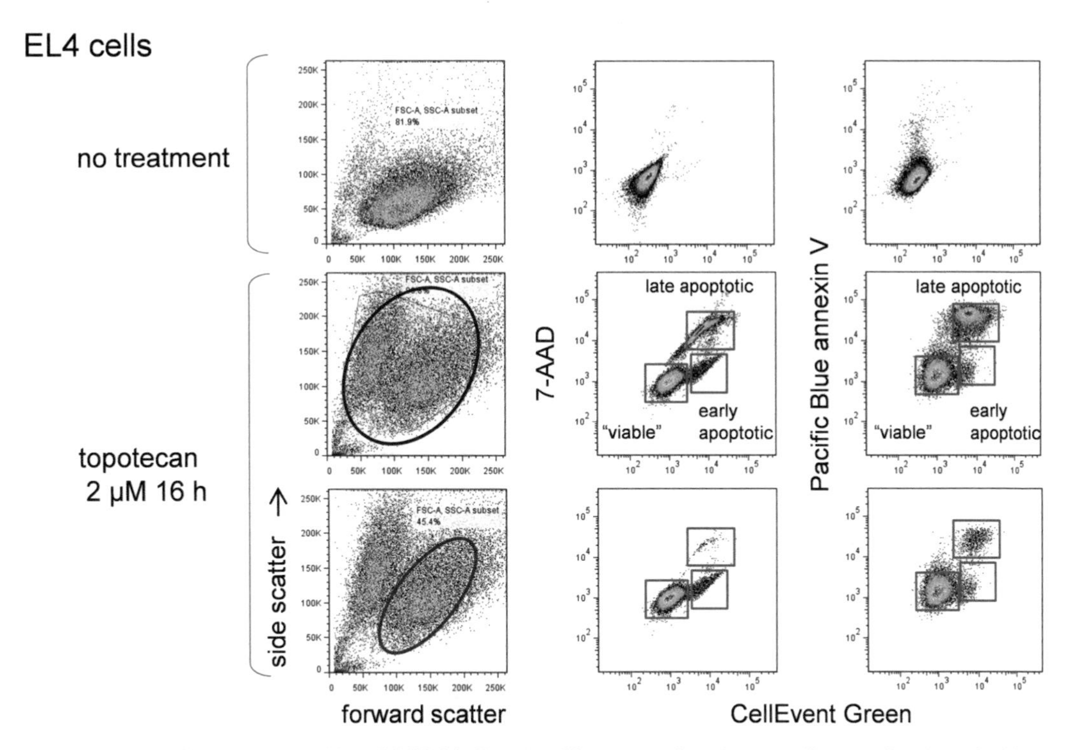

Fig. 2 *CellEvent Green, annexin V, and DNA binding dye.* EL4 mouse lymphoma cells were incubated with no additions (*top row*) or with topotecan at 2 μM for 16 h (*bottom two rows*). Cells were analyzed on a BD Biosciences LSR II flow cytometer equipped with blue-green, red, and violet lasers. Forward versus side scatter plots (*left column*) show the distinctive shift in scatter during apoptosis. *Middle column* shows CellEvent Green versus SYTOX AADvanced fluorescence, and the *right column* shows CellEvent Green versus Pacific Blue annexin V fluorescence. The *middle row* shows cells gated for all single cell events; the *bottom row* shows cells gated for the scatter "viable" population only. Regions show viable, early and late apoptotic subpopulations

4. *Compensation.* Preparing and analyzing single color compensation controls is important for identifying and minimizing fluorescence overlap between fluorochromes. This is particularly important when probes with close emission spectra are used (i.e., FITC and PE, or FITC and PI). Again, using a multi-laser system and distributing the fluorescent probes among the individual lasers is a good way to minimize the requirement for fluorescence compensation, as there will be minimal spectral overlap between fluorochromes (Sect. 3.2, step 4).

3.15 Data Analysis Examples

1. *FLICA, annexin V and DNA binding dye.* Figure 1 shows EL4 cells that have been treated with the topoisomerase II inhibitor topotecan and labeled with FLICA, Pacific Blue conjugated annexin V and the 7-AAD variant SYTOX AADvanced. The shift of the apoptotic subpopulation in forward and side scatter is very pronounced. When both "viable" and apoptotic cells are gated, the labeling with all three probes is clearly defined. "Viable" cells (all negative), an early apoptotic stage (FLICA

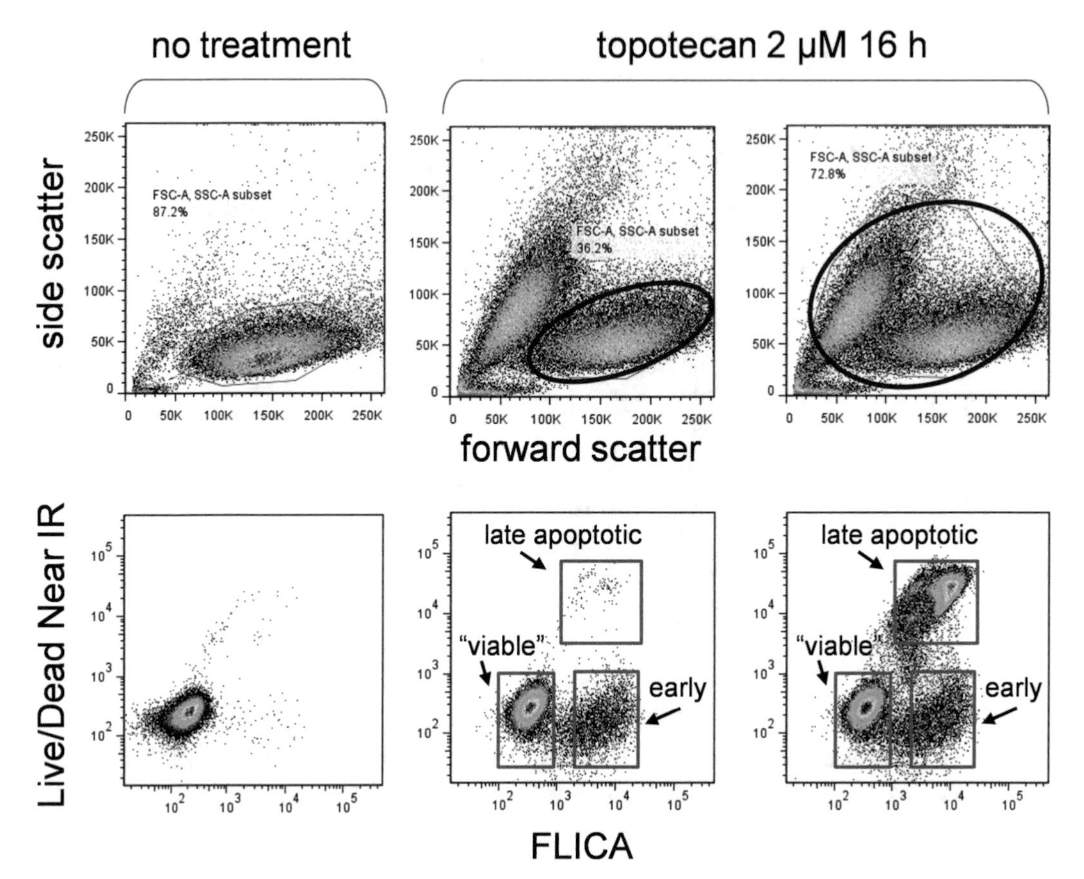

Fig. 3 *FLICA, annexin V, and DNA binding dye.* EL4 mouse lymphoma cells were incubated with no additions (*left column*) or with topotecan at 2 μM for 16 h (*middle* and *right columns*). Cells were analyzed on a BD Biosciences LSR II flow cytometer equipped with blue-green, red, and violet lasers. Forward versus side scatter plots (*top row*) show the distinctive shift in scatter during apoptosis. *Middle column* shows SYTOX AADvanced versus FLICA fluorescence, and the *bottom row* shows FLICA versus Live/Dead Near IR covalent binding viability probe. The *middle column* shows cells gated for all single cell events; the *right column* shows cells gated for the scatter "viable" population only. Regions show viable, early and late apoptotic subpopulations

positive, annexin V and DNA negative) is observed, as is an intermediate stage (FLICA positive, annexin V positive, DNA negative), and a late apoptotic stage (all positive). If only scatter "viable" cells are gated, the earliest apoptotic cells predominate. A multidimensional picture of apoptosis results, with the appearance of each probe occurring sequentially.

2. *CellEvent Green, annexin V and DNA binding dye.* Figure 2 also shows EL4 cells that have been treated with topotecan, but with CellEvent Green used in place of FLICA. Pacific Blue annexin V and SYTOX AADvanced were also used. Again, the scatter shift is very clear, and viable (all negative), early (CellEvent Green positive, annexin V and DNA negative) intermediate (CellEvent green positive, annexin V positive, DNA negative), and late (all positive) stages are all present.

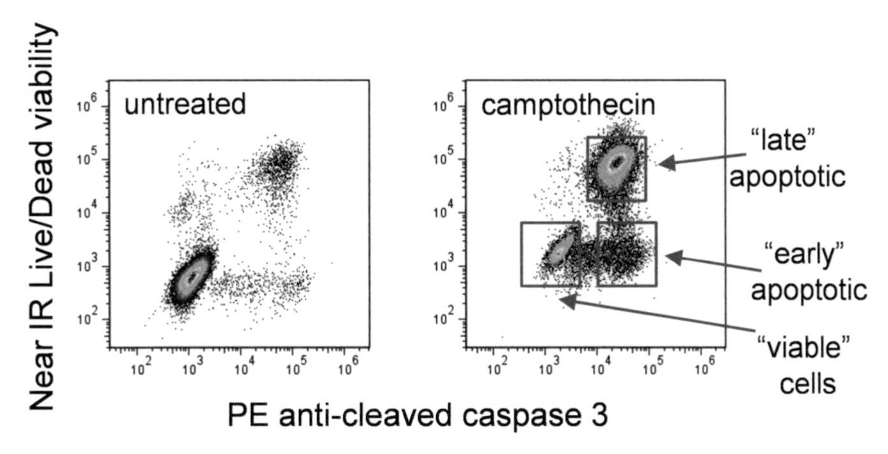

Fig. 4 EL4 mouse lymphoma cells were incubated with no additions (*left* scatter plot) or with camptothecin at 2 μM for 16 h (*right* scatter plot). Cells were analyzed on a BD Biosciences LSR II flow cytometer equipped with blue-green, red, and violet lasers. Immunolabeling for active caspase 3 using a PE conjugated antibody is shown versus Live/Dead Near IR covalent binding viability probe labeling. Regions show viable, early and late apoptotic subpopulations

3. *FLICA and covalent binding viability probe (fixed)*. Figure 3 also shows EL4 cells that have been treated with topotecan, followed by labeling with the covalent binding viability probe Live/Dead Near IR and FLICA. The viable (both negative, early (FLICA positive, Near IR negative) and late (all positive) stages are all present.
4. *Covalent binding viability probe and PE anti-caspase 3 immunolabeling (fixed)*. Figure 4 shows EL4 cells that have been treated with the topoisomerase II inhibitor camptothecin, followed by labeling with the covalent binding viability probe Live/Dead Near IR, fixation, permeabilization and immunolabeling with PE conjugated anti-active caspase 3 antibody. The viable (both negative, early (caspase positive, Near IR negative) and late (all positive) stages are all present.

4 Notes

1. *Controls.* Internal "viable" and apoptotic controls are both critical to ensure a working assay, especially in early assay development. If possible, an untreated negative control and a positive control should be included, preferably induced by an agent other than that under study (such as a cytotoxic drug). Some actual examples include: EL4 mouse lymphoma cells (ATCC clone TIB-39) induced with actinomycin D at 5 μg/ml or cycloheximide at 10 μg/ml (rapid induction at 6 h) or camptothecin or topotecan at 5 μM (slower induction at 16 h); Jurkat T cells (ATCC clone TIB-152) induced with staurosporine at

5 μM at 8 h; L1210 cells (ATCC clone CCL-219) expressing Fas antigen (CD95), induced with anti-Fas antibody (clone Jo-2, BD Pharmingen) at 1:1000 dilution, 12 h; L929 fibroblast cell line (ATCC clone CCL-1), induced with TNFα at 1000 U/ml and cycloheximide at 10 μg/ml for 12 h. These are only a few examples.

2. *Variability in apoptosis.* Apoptosis is a highly variable process involving a variety of biochemical pathways; therefore, there are no universal morphological or biochemical characteristics that are common to apoptosis in all cells. Apoptosis in different cell types (even in physiologically or morphologically similar ones) may present very different phenotypes, and may not necessarily be detectable by the same assays. In addition, new cell death processes (termed necroptosis, necrobiosis, entosis, etc.) that result in cell death by entirely different pathways are constantly being defined. Some of these novel pathways seem to show independence from caspase activity, for example. Multiparametric assays for apoptosis are very useful for measuring apoptosis for this precise reason, since the investigator is not limiting their assay to one characteristic of cell death. However, the picture of cell death illustrated here may differ significantly in other tissues.
3. *Caspase substrate specificity.* While the FLICA and CellEvent Green/NucView 488 substrates demonstrate reasonable specificity for their target caspases, no synthetic substrate is exclusively specific for any particular enzyme. This should be kept in mind for any assay involving specific proteolytic activity. The FLICA reagents in particular have been shown to bind intracellular proteins in the absence of caspase activity, although their specificity for apoptotic cells appears to be good [26]. An excess of substrate will encourage low levels of nonspecific cleavage, increasing the non-caspase background of the assay. Titration of the substrate to the lowest concentration able to distinguish activity may be necessary when the specificity of the assay is in doubt.
4. *Caspase substrate background.* "Viable" cells with no caspase activity will nonetheless show some fluorescence background when labeled with caspase substrates. Ensure that you do not interpret this background as positive signal. True caspase activation should show at least one log decade increase in comparison to background. Negative and positive controls are critical in this situation. For FLICA, additional washes may be necessary to reduce background. Increased incubation times may also cause higher backgrounds.
5. *Annexin V.* Some cell types normally segregate significant amounts of phosphatidylserine to the outer leaflet of the plasma

membrane, giving a false positive signal for annexin V binding. Megakaryocytes, platelets, and some myeloid lineage cells show this phenomenon. Phosphatidylserine can also temporarily flip in adherent cells removed from their growth matrix with scraping or trypsin treatment. Annexin V binding should be interpreted with caution in these situations. For these cell types and removed adherent cells, replacement of annexin V with a membrane intercalating probe, such as the F12N2S violet ratiometric membrane asymmetry probe from Thermo Fisher Life Technologies (product number A-35137) is recommended [10].

6. *Simultaneous immunophenotyping of "viable" and early apoptotic cells.* This technique is theoretically compatible with the incorporation of antibody immunophenotyping along with the cell death markers, resulting in a very sophisticated "screening out" of dead cells for measurement of receptor expression in "viable cells". A potentially exciting extension of this method would appear to be the phenotyping of early apoptotic cells, positive for caspase expression but negative for later markers. This method may be possible but should be approached with caution; from a standpoint of the apoptotic cell, caspase activation is probably not an "early" event in cell death, and many alterations in the plasma membrane may have occurred by this timepoint, resulting in aberrant antibody binding. Any cell surface marker expression results obtained by such methodology should be interpreted with great caution.

References

1. Telford WG, King LE, Fraker PJ (1994) Rapid quantitation of apoptosis in pure and heterogeneous cell populations using flow cytometry. J Immunol 172:1–16
2. Afanasev VN et al (1986) Flow cytometry and biochemical analysis of DNA degradation characteristic of two types of cell death. FEBS Lett 194:347
3. Darzynkiewicz Z, Juan G, Li X, Gorczyca W, Murakami T, Traganos F (1997) Cytometry in cell necrobiology: analysis of apoptosis and accidental cell death (necrosis). Cytometry 27:1–20
4. Telford WG, King LE, Fraker PJ (1991) Evaluation of glucocorticoid-induced DNA fragmentation in mouse thymocytes by flow cytometry. Cell Prolif 24:447–459
5. Vermes I, Haanen C, Reutelingsperger C (2000) Flow cytometry of apoptotic cell death. J Immunol Methods 243:167–190
6. Darzynkiewicz Z, Galkowski D, Zhao H (2008) Analysis of apoptosis by cytometry using TUNEL assay. Methods 44:250–254
7. Pozarowski P, Grabarek J, Darzynkiewicz Z (2003) Flow cytometry of apoptosis. In: Robinson JP et al (eds) Current protocols in cytometry. John Wiley and Sons, New York, NY, pp 18.8.1–18.8.34
8. Ormerod MG, Sun X-M, Snowden RT, Davies R, Fearhead H, Cohen GM (1993) Increased membrane permeability in apoptotic thymocytes: a flow cytometric study. Cytometry 14:595–602
9. Castedo M, Hirsch T, Susin SA, Zamzami N, Marchetti P, Macho A, Kroemer G (1996) Sequential acquisition of mitochondrial and plasma membrane alterations during early lymphocyte apoptosis. J Immunol 157:512–521
10. Shynkar VV, Klymchenko AS, Kunzelmann C, Duportail G, Muller CD, Demchenko AP, Freyssinet JM, Mely Y (2007) Fluorescent biomembrane probe for ratiometric detection of apoptosis. J Am Chem Soc 129:2187–2193
11. Wlodkowic D, Skommer J, Darzynkiewicz Z (2009) Flow cytometry-based apoptosis detection. Methods Mol Biol 559:19–32

12. Wlodkowic D, Skommer J, Darzynkiewicz Z (2010) Cytometry in cell necrobiology revisited. Recent advances and new vistas. Cytometry A 77:591–606
13. Wlodkowic D, Telford WG, Skommer J, Darzynkiewicz Z (2011) Apoptosis and beyond: cytometry in studies of programmed cell death (invited book chapter). In: Darzykiewicz Z et al (eds) Recent advances in cytometry, methods in cell biology volume 103. Academic, New York, NY, pp 55–99
14. Earnshaw WC, Martins LM, Kaufmann SH (1999) Mammalian caspases: structure, activation, substrates and functions during apoptosis. Ann Rev Biochem 68:383–424
15. Henkart PA (1996) ICE family proteases: mediators of all cell death? Immunity 14:195–201
16. Lazebnik Y, Kaufmann SH, Desnoyers S, Poirier GG, Earnshaw WC (1994) Cleavage of poly(ADP-ribose) polymerase by proteinase with properties like ICE. Nature 371:346–347
17. Bedner E, Smolewski P, Amstad P, Darzynkiewicz Z (2000) Activation of caspases measured in situ by binding of fluorochrome-labeled inhibitors of caspases (FLICA); correlation with DNA fragmentation. Exp Cell Res 260:308–313
18. Belloc F, Belaund-Rotureau MA, Lavignolle V, Bascans E, Braz-Pereira E, Durrieu F, Lacombe F (2000) Flow cytometry of caspase-3 activation in preapoptotic leukemic cells. Cytometry 40: 151–160
19. Koester SK, Bolton WE (2001) Cytometry of caspases. Methods Cell Biol 63:487–504
20. Overbeek R, Yildirim M, Reutelingsperger C, Haane C (1998) Early features of apoptosis detected by four different flow cytometry assays. Apoptosis 3:115–120
21. Darzynkiewicz Z, Pozarowski P, Lee BW, Johnson GL (2010) Fluorochrome-labeled inhibitors of caspases: convenient in vitro and in vivo markers of apoptotic cells for cytometric analysis. Methods Mol Biol 682:103–114
22. Komoriya A, Packard BZ, Brown MJ, Wu ML, Henkart PA (2000) Assessment of caspase activities in intact apoptotic thymocytes using cell-permeable fluorogenic caspase substrates. J Exp Med 191:1819–1828
23. Packard BZ, Topygin DD, Komoriya A, Brand L (1996) Profluorescent protease substrates: intramolecular dimers described by the exciton model. Proc Natl Acad Sci U S A 93: 11640–11645
24. Packard BZ, Komoriya A, Brotz TM, Henkart PA (2001) Caspase 8 activity in membrane blebs after anti-Fas ligation. J Immunol 167:5061–5066
25. Huang TC, Chen JY (2013) Proteomic analysis reveals that pardaxin triggers apoptotic signaling pathways in human cervical carcinoma HeLa cells: cross talk among the UPR, c-Jun and ROS. Carcinogenesis 34:1833–1842
26. Telford WG, Komoriya A, Packard BZ, Bagwell CB (2011) Multiparametric analysis of apoptosis by flow cytometry. In: Hawley TS, Hawley RG (eds) Methods in molecular biology volume 699, flow cytometry protocols, 4th edn. Humana Press, London, pp 203–228
27. Cen H, Mao F, Aronchik I, Fuentes RJ, Firestone GL (2008) FASEB J online article fj.07-099234

Chapter 5

Detecting Apoptosis, Autophagy, and Necrosis

Jack Coleman, Rui Liu, Kathy Wang, and Arun Kumar

Abstract

There are many commercially available kits to identify specific types of cell death, but at the present time, there is no simple assay that can distinguish apoptosis, necrosis, and autophagy. Autophagy and apoptosis are highly conserved processes that maintain organism and cellular homeostasis. They are also prime targets for the design of tumor therapeutics. Apoptosis is a highly regulated process involved in removing unwanted or unhealthy cells. Autophagy is a metabolic process, in which proteins and organelles are targeted for degradation in the lysosome. Necrosis is initiated by external factors, such as toxins, infection, or trauma, and results in the unregulated digestion of cell components. We discuss the tools we have developed for a simple protocol for detecting apoptosis or necrosis, as well as a simple technique for detecting autophagy. We discuss the potential pitfalls of the methods, suggest guidelines for designing experiments, and describe step by step protocols to identify apoptotic, necrotic and autophagic cell death of any cell line in response to effector.

Key words Autophagy, Apoptosis, Necrosis, Cell death, Annexin V, Phospholipidosis

1 Introduction: Types of Cell Death

Cell death is associated with at least three different morphologically different mechanisms: apoptosis, necrosis, and autophagic cell death. There are some questions in the literature whether autophagy is a true mechanism of cell death, or cells that are dying by another pathway have also induced autophagy [1]. Apoptosis is programmed cell death, activated by either external or internal stimuli. Necrosis is cell death due to injury or disease. Autophagy is the catabolic mechanism used by the cell to degrade dysfunctional or extraneous cellular components. Apoptosis and autophagy are highly conserved in evolution, and there is cross-talk between the two processes [2]. There have been other less common types of cell death [3] that will not be discussed further in this article. There are many dye based fluorescent assays for apoptosis and necrosis. Most current assays for autophagy, however, require assays for specific proteins in cell extracts, or transfection of the cells to express fluorescent protein fusions with LC3; LC3 is a

Perpetua M. Muganda (ed.), *Apoptosis Methods in Toxicology*, Methods in Pharmacology and Toxicology, DOI 10.1007/978-1-4939-3588-8_5, © Springer Science+Business Media New York 2016 protein that is modified and localized to autophagic vesicles during induction of autophagy. We describe a novel fluorescent dye that targets to the autophagic vesicle.

1.1 Apoptosis Stages

Apoptosis has several morphological characteristics. As cells first become apoptotic, they change their refractive index [4], followed by shrinkage of the cell and chromosomal condensation. The cell membrane starts to show blebs or protrusions from the membrane that can separate from the dying cell. During apoptosis, cells lose the asymmetry of phospholipids, exposing phosphatidylserine to the outer surface of the cell. The mitochondria lose membrane potential, and leak cytochrome C into the cytoplasm.

Apoptosis is induced by either the extrinsic pathway or the intrinsic pathway. The extrinsic pathway is usually activated through cell surface receptors, such as FAS, TNF-RI (tumor necrosis factor receptor I) or TRAIL-RI (TNF related apoptosis inducing ligand receptor I). The extrinsic pathway typically involves activation of caspase 8, which in turn activates other proteins. The intrinsic pathway is induced by methods not involving cell surface receptors, such as DNA damage, topoisomerase inhibition [5]. The intrinsic pathway involves the mitochondria, and is often referred to as the mitochondrial pathway. This pathway involves the activation of caspase 9, followed by caspase 3 in a tightly regulated mechanism [6].

In this chapter we describe materials and methods used to study apoptosis, necrosis, and autophagy, followed by detailed methods for the more relevant assays needed.

2 Materials

2.1 Apoptosis Assays

2.1.1 TUNEL

Many apoptosis assays exist. One of the more popular assays is the TUNEL (Terminal deoxynucleotidyl transferase dUTP nick end labeling) assay developed by Gavrieli and coworkers in 1992 [7]. The reagents for the TUNEL assay are available from several sources, including ENZO Life Sciences (APO BRDU™ kit), Life Technologies (ApoBrdU kit), Abcam (in situ Direct DNA Fragmentation (TUNEL) Assay Kit), EMD-Millipore (TUNEL Apoptosis Detection kit), Roche (in situ Cell Death Detection Kit), Promega (DeadEnd™ Colorimetric TUNEL Sytem), and R&D Systems (TUNEL Labeling Kits, TdT in situ DAB). This assay detects DNA breaks associated with apoptosis. Terminal deoxynucleotidyl transferase is added to the cells in the presence of a labeled nucleotide, such as biotin-dUTP. The biotin (or other label) is only added if the DNA has been nicked, healthy cells remain unlabeled. This method of detection detects the late stages of apoptosis, but in some instances also detects necrosis.

2.1.2 Comet Assay

The Comet assay is another method for the detection of apoptosis [8]. The Comet assay also looks at fragmented DNA, but in this case apoptotic cells are identified by the tail of DNA that is electrophoresed out of cells embedded in agarose. The reagents for this assay are available from many sources, including ENZO Life Sciences (Comet SCGE assay kit), Trevigen (CometAssay® kit), Cell Biolabs (OxiSelect™ Comet Assay Kit), R&D Systems (CometAssay Single Cell Gel Electrophoresis Assay), and Amsbio (CometAssay® Kit 96 wells). The DNA is stained with a fluorescent dye, such as ethidium bromide, and the length of the tail of DNA from the cell corresponds to the degree of apoptosis.

2.1.3 Caspase Assays

During the process of apoptosis, caspases 2, 8, 9, and 10 are activated first, and these initiators activate the executioner caspases 3, 6, and 7. A third method used for detection of apoptosis thus relies on the detection of caspases. During most types of apoptosis, various caspases are activated, which in turn cleave other proteins within the cell to activate apoptosis. The reagents for Caspase assays are available from many companies, including ENZO Life Sciences (Carboxyfluorescein multi-caspase activity kit), Life Technologies (CellEvent™ Caspase-3/7 Green ReadyProbes® Reagent), Abcam (Caspase Family Colorimetric Substrate Kit Plus), Promega (Caspase-Glo® 3/7 Assay Systems), and Cell Signaling Technology (Caspase-3 Activity Assay Kit). These companies also offer kits for different caspases and specific caspase enzyme substrates that are inactive until they are cleaved by the caspase enzyme. Upon cleavage, the resulting product either develops a strong color or a bright fluorescence or luminescence. Caspase assays are used to monitor stages of apoptosis (early or late stages).

2.1.4 Chromosome Condensation

A fourth method for the detection of apoptosis assesses the condensed chromosomes that occur during apoptosis. During apoptosis, the chromosome goes through distanced stages of chromosome condensation [9]. The reagents to detect chromosome condensation are available from many companies, including ENZO Life Sciences (NUCLEAR-ID® Green chromatin condensation detection kit for fluorescence microscopy, flow cytometry, and microplate assays), Life Technologies (Chromatin Condensation & Membrane Permeability Dead Cell Apoptosis Kit with Hoechst 33342, YO-PRO®-1, and PI dyes, for flow cytometry), and Abcam (Nuclear Condensation Assay Kit—Green Fluorescence). Some cell-permeable DNA staining dyes, such as those mentioned above, fluoresce brighter as the chromosome condenses, differentiating apoptotic cells with condensed nuclei from normal cells.

2.1.5 Annexin V Assays

A fifth method commonly used to detect apoptosis relies on the fact that healthy cells maintain an asymmetrical distribution of phospholipids in their membranes. Reagents for this assay can be purchased from a number of companies, including ENZO Life Sciences (Annexin V-FITC apoptosis detection kit), Clontech (Annexin V Apoptosis Assay), Life Technologies (Annexin V, Alexa Fluor® 488 conjugate), Abcam (Annexin V-FITC Apoptosis Detection Kit), and Phoenix Flow Systems (V Annexin V Assay). Normally phosphatidylserine is not exposed on the cell surface. During apoptosis, the membrane loses the asymmetry, and phosphatidylserine is exposed to the cell surface. Annexin V is a protein that specifically binds to phosphatidylserine, but it is not cell-permeable. Using fluorescently tagged Annexin V, apoptosis can be visualized in cells in culture by microscopy or flow cytometry. Annexin V is often used in combination with a cell-impermeable DNA staining dye, such as propidium iodide to differentiate necrotic cells from apoptotic cells.

2.1.6 PARP Cleavage Assays

There are two well-described pathways of apoptotic cell death; the caspase pathway, and the caspase-independent pathway triggered by Poly-ADP Ribose Polymerase (PARP) activation [10]. PARP triggered apoptosis is well known in ischemic tissues [10]. PARP-1 binds to damaged DNA and cleaves NAD^+ into nicotinamide and ADP-ribose, which forms ADP-Ribose polymers (PAR) on histones and other proteins. The excessive polymer formation is toxic to cells. During apoptosis, the levels of PARP-1 activity initially increases, but in the later stages the activity drops due to automodification and caspase cleavage [11].

There are a few kits on the market now, including Trevigen (HT Chemiluminescent PARP/Apoptosis Assay Kit) and R&D Systems (HT PARP In Vivo Pharmacodynamic Assay II). These kits quantify the amount of PAR in the cell using an ELISA format. First PAR or Histones are captured on a surface, then, using labeled antibodies to PAR, the level of PAR in the cell extract can be determined.

2.2 Necrosis Assays

The hallmark of necrosis is the permeabilization of the cell membrane. The two most common assays for necrosis rely on this permeability.

2.2.1 Lactate Dehydrogenase Activity

One assay measures the amount of lactate dehydrogenase (LDH, or other enzymes) that leaks out of the cell. LDH catalyzes the interconversion of pyruvate and lactate with concomitant interconversion of NADH and NAD+. There are several kits in the market for this assay, including Cayman Chemical (LDH Cytotoxicity Assay Kit), Life Technologies (Pierce® LDH Cytotoxicity Assay Kit), and Abcam (LDH-Cytotoxicity Assay Kit II). The LDH activity can be measured by the quantification of NADH, which catalyzes the reduction of various substrates and can be specifically detected by colorimetric assays.

2.2.2 Cell-Impermeable DNA Binding Dye

The other more common assay uses a cell-impermeable DNA binding dye, such as propidium iodide, that fluoresces only if it enters the necrotic cell. This assay is often used in multiplex with apoptosis assays. There are many companies that sell the reagents for this type of test, among them are: ENZO Life Sciences (GFP CERTIFIED® Apoptosis/Necrosis detection kit for microscopy and flow cytometry), AAT Bioquest (Cell Meter™ Apoptotic and Necrotic Detection Kit), Biotium (Apoptosis and Necrosis Quantification kit), Promokine (PromoKine Apoptotic/Necrotic Cells Detection Kit), and Life Technologies (Alexa Fluor® 488 Annexin V/Dead Cell Apoptosis Kit). These kits all depend on a dye that cannot enter intact cells, but as the membrane ruptures during apoptosis, the dye enters and binds DNA causing fluorescence.

2.3 Autophagy

Autophagy is a catabolic process by which intracellular components are delivered to lysosomes for degradation by their resident hydrolases. Three different types of autophagy have been described in mammalian cells depending on the mechanisms used for the delivery of cargo to lysosomes: macroautophagy, microautophagy, and chaperone-mediated autophagy (CMA) [12–15]. In microautophagy, the lysosomal membrane invaginates or tubulates to engulf whole regions of the cytosol, but its components and regulation are poorly understood in mammals [16].

Macroautophagy, hereafter referred to as autophagy, is a pathway that delivers a portion of a cell's cytoplasm to the lysosome for degradation. Macroautophagy and CMA activity are maximally activated under stress conditions such as starvation, oxidative stress, or conditions leading to enhanced protein misfolding [17–22]. In CMA, selective cytosolic proteins are recognized by a chaperone/co-chaperone complex that delivers them to the lysosomal surface [13, 23].

Turnover of proteins and long-lived organelles by autophagy are a normal part of the homeostatic activities of a cell. During periods of metabolic stress, such as nutrient limitation, autophagy is upregulated to generate the necessary macromolecules for new synthesis. In addition to its importance in the stress response, autophagy is known to play a role in neurodegenerative, liver, muscle, and cardiac diseases, as well as cancer, aging, and clearance of infections. The involvement of autophagy in a wide variety of disorders has made it an attractive target in drug discovery.

Autophagy initiates with the formation of a phagophore, which is a membrane that begins to surround a portion of the cytoplasm (Fig. 1). The phagophore develops into a fully enclosed vesicle called an autophagosome that completely sequesters a portion of the cytoplasm. The autophagosome then docks and fuses with a lysosome, creating an autolysosome, and the contents of the vesicle and the vesicle itself are degraded.

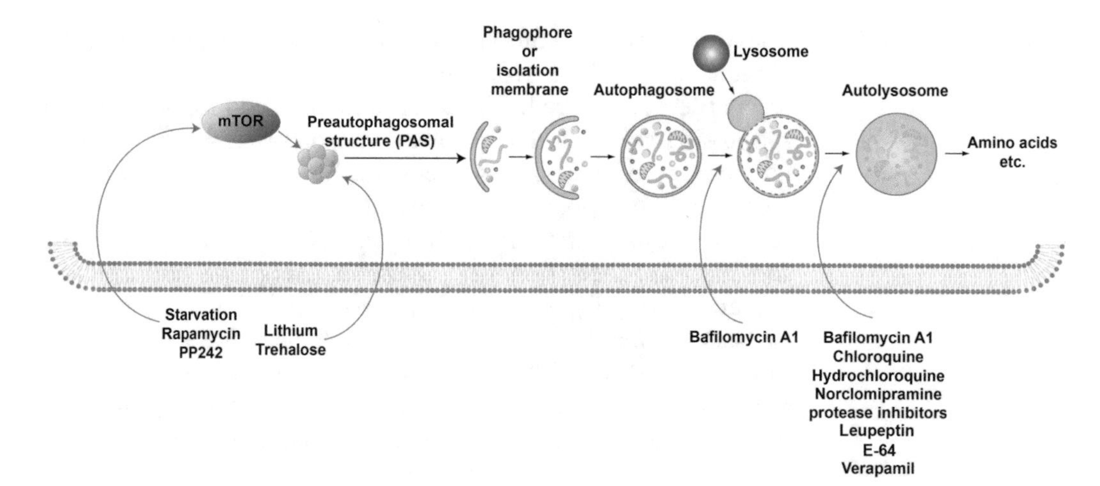

Fig. 1 Pathway of autophagy. When autophagy is induced, either through mammalian Target of Rapamycin (mTOR) or other means, a pre-autophagosomal structure is formed, and then expands to form the phagophore that surrounds the material to be degraded. The engulfed material surrounded by the phagophore is the autophagosome. The autophagosome then fuses with the lysosome to form the autolysosome, where the contents of the autophagosome are degraded

In a general sense, autophagy, or "self-eating" refers to a variety of complex and tightly regulated mechanisms by which cells sense nutrient levels and degrade cellular components to obtain molecules essential for cell homeostasis [24].

2.3.1 LC3 Assays

Most assays for autophagy rely on the microtubule associated protein LC3. When autophagy is induced, cytosolic LC3-I is modified with phosphatidylethanolamine to make LC3-II, which is localized to the autophagosomal membrane. LC3 fusions to fluorescent proteins such as Green Fluorescent Protein (GFP) or Red Fluorescent Protein (RFP) can be detected using fluorescent microscopy. When autophagy is induced, the fluorescent signal goes from a general cytoplasmic fluorescent to punctate signals.

LC3 Gene Fusions

Several companies make fusion proteins (genes) of LC3 and a fluorescent protein, including EMD-Millipore (FlowCellect™ GFP-LC3 Reporter Autophagy Assay Kit), which sells cell lines with the fusion, and Gen Target Inc (RFP-LC3 fusion lentiviral particles) and Life Technologies (Premo™ Autophagy Tandem Sensor RFP-GFP-LC3B Kit) which sell viruses with the fusion, that can be used to transfect most any cell line of interest. The GFP-RFP loses the green fluorescence after the autophagosome fuses with the acidic lysosome. Because these assays require transfection, they are susceptible to artifacts induced by transfection, and not all cells are transfected [25].

LC3-GFP fusion proteins can also be used in flow cytometry to detect autophagy. If the cells are permeabilized, cytosolic LC3-GFP leaks out, but LC3-GFP attached to the autophagosome remains in the cell. A second method using LC3-GFP fusions to detect autophagy using flow cytometry relies on the fact that the LC3 gets degraded in the lysosome during autophagy. In this case, the signal from the LC3-GFP fusion would decrease if autophagy has been induced.

Autophagy Specific Antibodies for Western or IHC

Western blot analysis for LC3-II, or other autophagy specific proteins, such Beclin-1 or Atg 5 is also used to measure the level of induction of autophagy in cell lysates. LC3 and other autophagy specific antibodies are available from many companies, including; ENZO Life Sciences, Life Technologies, Qiagen, Zyagen, Cell Signaling Technologies, AbD Serotec, Epitomics, Abgent, Novus, and Abcam. Because autophagy is a continuous pathway, autophagosomes being formed, then fusing and being degraded in the lysosome (see Fig. 1), most researchers add an inhibitor of lysosomal degradation, such as bafilomycin A1, chloroquine or pepstatin A/E64d [26].

Autophagy PCR

PCR assays for autophagy related genes have also been used to detect induction of autophagy. Qiagen makes the "RT^2 Profiler™ PCR Array Human Autophagy" and other organism targeted autophagy PCR arrays.

Autophagy Specific Dyes

Another method to detect the induction of autophagy is to use a dye that targets the acidic autophagosomal compartment, such as monodansylcadaverine (MDC [27]) and acridine orange, sold by many companies, including: Sigma-Aldrich, Cayman Chemical, and Abcam. In some works, MDC staining is indistinguishable from Lysosomal staining [25]. The autophagy specific dyes are thought to accumulate in acidic compartments due to charge groups on the dye. They are thought to fluoresce brighter in the hydrophobic environment of the autophagosome [28].

A brighter dye, Cyto-ID® Autophagy Green Detection Reagent [29] (ENZO Life Sciences) has been developed that is more specific to the autophagosome [25]. Upon induction of autophagy by amino acid starvation, the Cyto-ID Autophagy dye accumulates in compartments that are not stained by lysosomal probes, but it has a high correlation with compartments containing LC3-RFP fusions [25]. These dyes detect the autophagosome, but also suffer from the rapid degradation of the autophagosome. Inclusion of an inhibitor of lysosomal degradation helps this assay [30].

It should be noted that if autophagy is severely inhibited genetically or with bafilomycin A, monensin, or hydroxychloroquine, apoptosis is induced [31]. Cationic amphiphilic drugs that inhibit lysosomal function induce phospholipidosis, which is accompanied by cellular responses such as growth arrest and autophagy [32] under mild conditions. Severe phospholipidosis leads to autophagic cell death [33].

3 Methods

In this section we describe the necessary steps for carrying out a combined apoptosis/necrosis assay, followed by steps for carrying out autophagy assays. For both assays, we describe methods for detection using flow-cytometry, followed by methods for detection by microscopy. Finally, we describe a method for analyzing apoptosis/necrosis and autophagy. The choice of an assay for cell death studies depends on the types of samples being tested. Individual assays can use any colored dye or fluorescence available. If multiple assays, such as apoptosis, necrosis, and autophagy, are performed at the same time, several dyes that do not interfere with each other must be chosen.

3.1 Detection of Apoptosis/Necrosis by Flow Cytometry

The transition from apoptosis to necrosis is a loosely defined continuum that necessitates recognition of the various stages of the process. The display of phosphatidylserine (PS) on the extracellular face of the plasma membrane remains a unifying hallmark of early apoptosis. Phospholipid-binding proteins such as Annexin V bind with a high affinity to PS, in the presence of Ca^{2+}. Given that Annexin V is not cell-permeable, the binding to externalized PS is selective for early apoptotic cells. Similarly, the loss of plasma membrane integrity, as demonstrated by the ability of a membrane-impermeable DNA intercalating dye to label the nucleus, represents a straightforward approach to demonstrate late stage apoptosis and necrosis.

Our description is for the Apoptosis/Necrosis kit from ENZO Life Sciences (ENZ-51002), but is applicable to kits from other vendors mentioned in the materials section, as well as “home-brewed” kits.

3.1.1 Treatment and Growth of Cells

Treatment of cells is performed prior to staining. We use Staurosporine as a model system for inducing apoptosis. Cells should be maintained using standard tissue culture practices. For each sample, there should be between 2×10^5 and 1×10^6 cells. Control untreated cells should be grown standard medium with vehicle for the test compound (in our example of Staurosporine, this is DMSO at 0.2 %). For treated cells, add the compound of interest to the medium (in our example this is 0.2 % of 1 mM Staurosporine to make a final solution of 2 μM). The cells should be grown under normal conditions (37°, 5 % CO_2) for 1–18 h.

3.1.2 Staining and Analyzing Cells

After treatment, attached cells are detached as usual using trypsin, minimizing cellular damage that may be caused by over trypsinization. Detached cells or non-adherent cells are collected by centrifugation at 400 × g for 5 min at room temperature. The medium is carefully removed by aspiration, and the cells are resuspended in 5 ml cold PBS [34]. We centrifuge this at 400 × g for 5 min at room temperature, and carefully aspirate the supernatant.

The detection reagent (labeled Annexin V and cell-impermeable DNA staining dye) is prepared shortly before staining. The detection reagent must be prepared in binding buffer. Commercial kits provide this binding buffer typically as a 10× concentration. This buffer must contain calcium for Annexin V to bind, typically at 2.5 mM (final concentration). The binding buffer must be of neutral pH, using a non-phosphate buffer and it must be isosmotic to keep the cells healthy. Fluorescently labeled Annexin V is added to the binding buffer as recommended by the manufacturer, or you may titrate your own Annexin V, starting around 1 μg/ml. A cell-impermeable DNA staining dye, such as propidium iodide or 7-aminoactinomycin D (7-AAD) is added to the Annexin V solution as recommended by the manufacturer. Note that this staining solution is stable for only about 1 h at 4 °C.

The cells are stained by resuspending cells in 500 μl of the staining solution described above. The cells are stained for 15 min at room temperature in the dark. After staining, the cells are analyzed immediately using flow cytometry. It is important to remember that cells must remain in binding buffer for the Annexin V to remain bound. Samples and controls should be kept on ice before the assay is run and analyzed via flow cytometry within 1 h of staining. For the ENZO ENZ-51002, it is suggested to run the following controls to determine how much compensation is required for each channel.

(a) Unstained cells.

(b) Cells stained with Apoptosis Detection Reagent (without Necrosis Detection Reagent).

(c) Cells stained with Necrosis Detection Reagent (without Apoptosis Detection Reagent).

3.2 Detection of Apoptosis/Necrosis in Microscopy

Grow cells directly onto glass slides or polystyrene tissue culture plates until ~80 % confluent. Treat cells with compound of interest and negative control cells with vehicle, as described in Sect. 3.1.1 above. After the desired time of treatment, remove the medium and wash the cells with PBS, using enough volume to cover the cells. Carefully remove supernatant and dispense the detection reagent (described above in the flow cytometry section) in a volume sufficient for covering the cell monolayer. A third dye, such as Hoechst 33342 may be added (1 μM) to visualize all cells. Protect the cells from light and incubate at room temperature for 15 min.

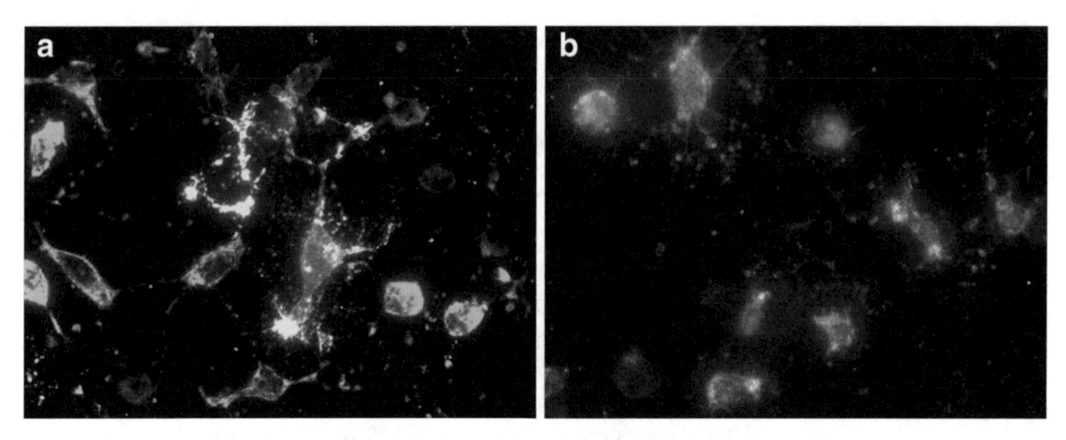

Fig. 2 Detection of apoptosis/necrosis by microscopy. HeLa cells were treated with 2 μM Staurosporine for 4 h (**b**) or with 2 μM Staurosporine for 2 hours (**a**) and stained as described in the text. The *green* is GFP, the *yellow* is Cy3-Annexin V staining of apoptotic cells and debris, and the *red* is 7-AAD staining of necrotic or broken cells

Remove the detection reagent by flicking onto paper towels. Add binding buffer to protect cells from drying out. Cover the cells, and observe under the microscope using the appropriate filter. It is recommended to have a positive control sample. Treat the cells with 2 μM Staurosporine for 4 h, prior to staining as described above. Some cell lines do not respond in the same way to the Staurosporine, so different conditions may have to be tried for each cell line. Figure 2 shows typical results of cells expressing GFP treated with Staurosporine for 2 (A) or 4 (B) hours. Long term treatment with the apoptosis inducing Staurosporine will cause more cells to become permeable and can be stained with the red necrosis detection reagent (7-AAD) and Cy3-Annexin V (Fig. 2b). In some cases, it might be necessary to set up a time course of treatment to distinguish apoptosis from necrosis.

If fixation is desired, it is important to remember the fixation process can disrupt cellular membranes and should be performed following incubation with the detection reagent (Annexin V/cell-impermeable DNA stain). 5 % formalin is the recommended fixative reagent. Standard immunofluorescence staining protocols using antibody conjugates should be administered post-fixation according to manufacturer instructions.

Untreated cell controls are vital in the apoptosis assay, because some cells, such as some platelets, have been shown to have phosphatidylserine on the surface. If this is the case, it is highly recommended to detect apoptosis with another assay, such as a chromosome condensation assay or caspase assay (see Sect. 2).

3.3 Detection of Autophagy by Flow Cytometry

The conventional way of monitoring autophagic activity is to measure the increased numbers of autophagosomes in cells in response to stimuli. However, the autophagosome formation is an intermediate stage in the whole dynamic autophagy process. The accumulation of autophagosomes can represent either the increased generation of autophagosomes or a block in fusion to lysosomes. Monitoring autophagic flux provides a meaningful way to distinguish these two sources of autophagosome accumulation. Autophagic flux is measured by inhibiting the fusion of the autophagosome to the lysosome, or inhibiting the degradation of the autophagosome in the lysosome, and measuring the increase in autophagosome formation in response to stimuli.

3.3.1 Treatment and Growth of Cells for Autophagy

The cells should be healthy and not overcrowded. Cell density should not exceed 1×10^6/ml. Cells should be maintained via standard tissue culture practice. Grow cells overnight to log phase in a humidified incubator at 37 °C, 5 % CO_2. Treat cells with the compound of interest for the desired length of time. Prepare negative control cells using vehicle treatment. It is highly recommended to set up positive and negative controls within the same experiment. To measure autophagic flux, include a lysosomal inhibitor, such as chloroquine. We have found that for long term treatment (greater than 5 h) starting with 10 μM chloroquine will hinder lysosome fusion without inducing apoptosis to a significant degree. For short term treatment, up to 300 μM chloroquine may be used without inducing apoptosis. These amounts can vary with different cell types, medium and length of treatment.

3.3.2 Staining and Analyzing Cells for Autophagy

At the end of the treatment, trypsinize (adherent cells), or collect cells by centrifugation (suspension cells). Samples should contain 1×10^5 to 1×10^6 cells per ml. Centrifuge at 1000 × g for 5 min to pellet the cells. Wash the cells by resuspending the cell pellet in cell culture medium, or other buffer of choice and again collect the cells by centrifugation.

Resuspend each live cell sample in 250 μl of 1× Assay Buffer, PBS or indicator free cell culture medium containing 5 % FBS. Add 250 μl of the diluted Cyto-ID® Autophagy Detection Reagent (see Section "Autophagy Specific Dyes", diluted 1 μl dye for every ml of medium or buffer) to each sample and mix well. Incubate for 30 min at room temperature or 37 °C in the dark. It is important to achieve a mono-disperse cell suspension at this step by gently pipetting up and down repeatedly.

After treatment, collect the cells by centrifugation and wash with 1× Assay Buffer from the Cyto-ID Autophagy kit or PBS. Resuspend the cell pellets in 500 μl of fresh 1× Assay Buffer or PBS.

Analyze the samples in green (FL1) or orange (FL2) channel of a flow cytometer.

Typical results are shown in Fig. 3, which shows HeLa cells starved in Earle's Balanced Salt Solution (EBSS) for three and a half hours. In Fig. 3, cells starved in EBSS without chloroquine do not show any change in intensity of the staining with Cyto-ID

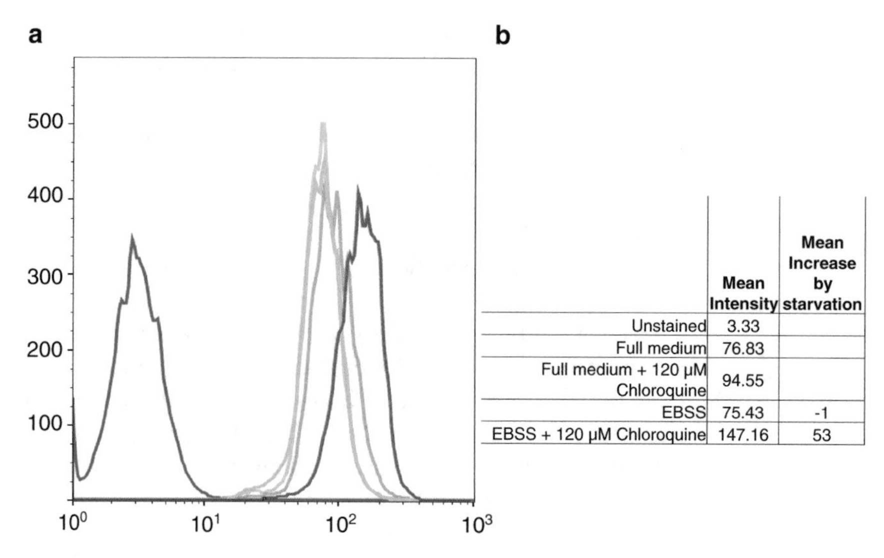

	Mean Intensity	Mean Increase by starvation
Unstained	3.33	
Full medium	76.83	
Full medium + 120 μM Chloroquine	94.55	
EBSS	75.43	-1
EBSS + 120 μM Chloroquine	147.16	53

Fig. 3 Detection of autophagy using flow cytometry. HeLa cells were plated at 4×10^5 per well in 6-well plates, and placed in a 37 °C incubator with 5 % CO_2 overnight. Cells were washed three times with EBSS (Sigma, St. Louis, MO) before starvation. After washing, cells were grown in full medium (DMEM, with 10 % FBS) or in EBSS with or without 120 μM Chloroquine for 4 h. Cells were stained as described above in PBS containing 5 % FBS. The *red line* is unstained cells, *aqua* is cells grown in full medium, *light green* is cells in EBSS, cells grown in full medium with chloroquine are depicted in *orange*, and *dark green* represents cells in EBSS with chloroquine

autophagy dye, but in the presence of 120 μM chloroquine, the mean fluorescent intensity increased 56 %. It should be noted that the chloroquine itself does increase the signal, but induction of autophagy significantly enhances the signal above this induction.

3.4 Detection of Autophagy by Microscopy

Grow cells on tissue culture treated slides or coverslips at a density to reach 50–70 % confluency after 16–48 h. Carefully remove medium and treat cells with the testing reagent (cell starvation, Rapamycin or other potential autophagy inducer) as required. As mentioned above, to properly see autophagic flux, it is best to inhibit lysosomal function with a drug such as chloroquine. It is important to have mock treated cells included as a control.

Post treatment, remove the medium and wash the cells twice with PBS, or medium without phenol red. Dispense 100 μl of the detection reagent (made by adding 1 μl of Cyto-ID Autophagy® Green Detection Reagent, Hoechst 33342 (~1.1 μg/ml) if desired, into 1 ml of PBS, assay buffer or medium without phenol red, supplemented with 5 % FBS). Make certain that all the cells are covered; add more dye for larger areas of cells. Stain at 37 °C for 15–30 min. Carefully remove staining reagent, and wash cells in PBS, assay buffer (supplied with the kit) or medium without phenol red. Analyze the stained cells by wide-field fluorescence or confocal microscopy (60× magnification is recommended). Use a

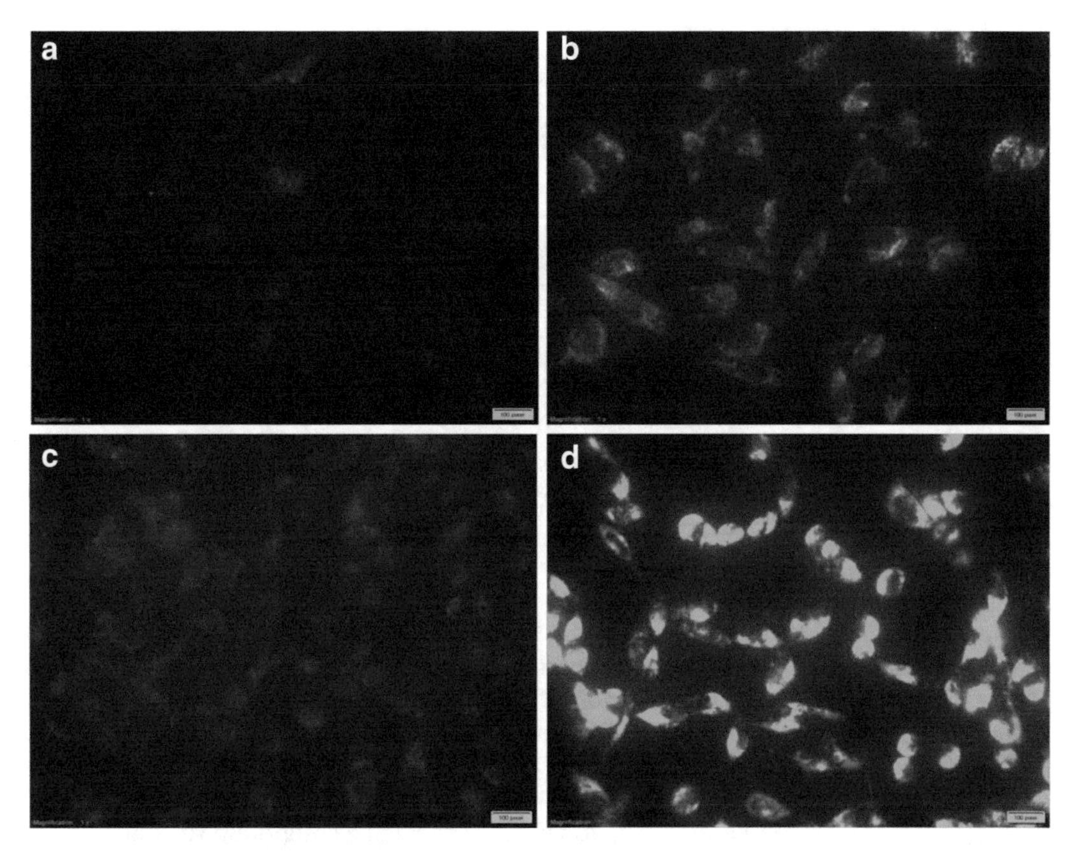

Fig. 4 Detection of starvation induced autophagy by microscopy. HeLa cells were grown as described on microscope slides overnight, then the cells were washed and overlaid with either fresh medium (DMEM with 10 % FBS, **a** and **b**) or EBSS (**c** and **d**) for 3.5 h. **b** and **d** also contained 40 μM chloroquine during the incubation. The incubation was followed by staining with Cyto-ID® Autophagy Green Detection Reagent as described in PBS with 5 % FBS

standard FITC filter set for imaging the autophagic signal. Optionally, image the nuclear signal using a DAPI filter set.

Figure 4 shows typical results of cells induced for autophagy by starvation in EBSS without added amino acids. In this case, it can be seen that autophagic flux can more clearly be seen in cells that have had the lysosomal activity inhibited with chloroquine. Other autophagy inducers, such as trehalose (Fig. 5) show autophagic induction without chloroquine. In this case, it could be that the trehalose is inhibiting lysosomal function as reported earlier [35].

3.5 Autophagy, Apoptosis and Necrosis in a Single Assay

These three assays can be run simultaneously, but one must be very careful interpreting the results. We prefer to run the autophagy assay separately from the apoptosis/necrosis assay.

If you plan to run the three assays simultaneously, the cells should be grown and treated as described above for the individual assays. Keep in mind that to see autophagic flux, lysosomal

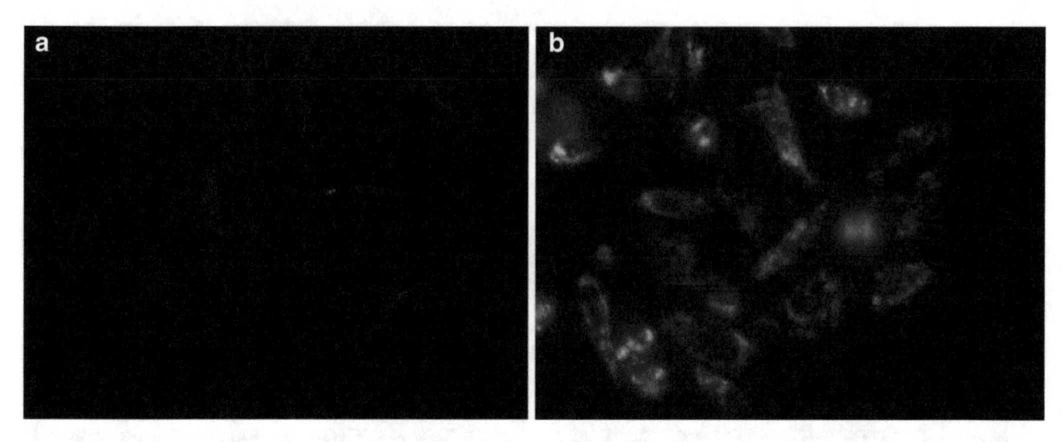

Fig. 5 Detection of trehalose induced autophagy. HeLa cells were induced with 50 mM Sucrose (**a**) or 50 mM trehalose (**b**) for 6 h, followed by staining with Cyto-ID® Autophagy Green Detection Reagent as described. In this case, no lysosomal inhibitor was added

degradation must be inhibited. The amount of chloroquine or other lysosomal inhibitor must be kept to a minimum to prevent induction of apoptosis. For long term treatment, a good starting concentration is 5–10 μM chloroquine. If the treatment is short term (up to 4 h), 100–300 μM chloroquine is a good starting point. Test these concentrations for each cell line and growth time. Alternatively, if you are interested in a compound that induces the accumulation of autophagosomes by itself, the assay can be run without a lysosomal inhibitor.

The dyes used must be distinguishable in the assay. We have used Annexin V-mFluor™ Violet 450 conjugate (AAT Bioquest, Sunnyvale, CA) for apoptosis, Necrosis Detection Reagent (ENZO Life Sciences, Farmingdale, NY) for necrosis, and Cyto-ID® Autophagy Green Detection Reagent (ENZO Life Sciences, Farmingdale, NY) for autophagy. The labeling and wash steps must be performed in the calcium containing binding buffer as described above. As mentioned above, each ml of the staining solution should have 1 μl of the Cyto-ID® Autophagy Green Detection Reagent stock, 5 μl Necrosis Detection Reagent, and 5 μl of the Annexin V-mFluor™ Violet 450.

After treatment of cells, wash one time with PBS or PBS with 5 % FBS. Add labeling solution to the cells. If the cells are suspension cells, resuspend them, and incubate at 37 °C in the dark for 15–30 min. Remove the staining solution, and wash the cells one time with binding buffer. The cells are ready for visualization by microscopy or flow cytometry.

Remember, there is a possibility the dyes might be detected in the wrong channel, so it is important to run positive samples stained with individual dyes to control for this or to calculate the compensation. If chloroquine was used in the assay, compounds that appear to induce apoptosis or necrosis should be retested in the absence of chloroquine to rule out any possible role of chloroquine in the results.

4 Notes

4.1 Health of Cells

One of the most critical variables in the study of apoptosis, necrosis, and autophagy is the health of the cells. Cells that are overgrown, or have depleted the medium of nutrients start to undergo apoptosis and autophagy on their own. For any assay, it is important to have cells that are not overcrowded, and have not been growing without fresh medium for a long time. As with most tests, cells that have undergone multiple passages may not react in the same way as fresh cells.

Overcrowded cells have been shown to have an attenuated apoptosis response to DNA damage in overcrowded cell densities [36]. This effect of cell crowding appears to be mediated by more rapid degradation of p53 in dense cultures.

References

1. Kroemer G, Levine B (2008) Autophagic cell death: the story of a misnomer. Nat Rev Mol Cell Biol 9:1004–1010
2. El-Khattouti A, Selimovic D, Haikel Y et al (2013) Crosstalk between apoptosis and autophagy: molecular mechanisms and therapeutic strategies in cancer. J Cell Death 6:19
3. Sperandio S, De Belle I, Bredesen DE (2000) An alternative, nonapoptotic form of programmed cell death. Proc Natl Acad Sci U S A 97:14376–14381
4. Hengartner MO (1997) Cell death. In: Riddle DL, Blumenthal T, Meyer BJ, Priess JR (eds) C. elegans II. Cold Spring Harbor, New York
5. Parone P, Priault M, James D et al (2003) Apoptosis: bombarding the mitochondria. Essays Biochem 39:41–51
6. Ferrer I, Planas AM (2003) Signaling of cell death and cell survival following focal cerebral ischemia: life and death struggle in the penumbra. J Neuropathol Exp Neurol 62:329–339
7. Gavrieli Y, Sherman Y, Ben-Sasson SA (1992) Identification of programmed cell death in situ via specific labeling of nuclear DNA fragmentation. J Cell Biol 119:493–501
8. Singh NP, Mccoy MT, Tice RR et al (1988) A simple technique for quantitation of low levels of DNA damage in individual cells. Exp Cell Res 175:184–191
9. Tone S, Sugimoto K, Tanda K et al (2007) Three distinct stages of apoptotic nuclear condensation revealed by time-lapse imaging, biochemical and electron microscopy analysis of cell-free apoptosis. Exp Cell Res 313:3635–3644
10. Siegel C, Mccullough LD (2011) NAD+ depletion or PAR polymer formation: which plays the role of executioner in ischaemic cell death? Acta Physiol 203:225–234
11. Shah GM, Poirier D, Duchaine C et al (1995) Methods for biochemical study of poly(ADP-ribose) metabolism in vitro and in vivo. Anal Biochem 227:1–13
12. Cuervo AM (2004) Autophagy: in sickness and in health. Trends Cell Biol 14:70–77
13. Dice JF (2007) Chaperone-mediated autophagy. Autophagy 3:295–299
14. Mizushima N (2005) The pleiotropic role of autophagy: from protein metabolism to bactericide. Cell Death Differ 12(Suppl 2):1535–1541
15. Yorimitsu T, Klionsky DJ (2005) Atg11 links cargo to the vesicle-forming machinery in the cytoplasm to vacuole targeting pathway. Mol Biol Cell 16:1593–1605
16. Mortimore GE, Lardeux BR, Adams CE (1988) Regulation of microautophagy and basal protein turnover in rat liver. Effects of

short-term starvation. J Biol Chem 263: 2506–2512

17. Cuervo AM, Knecht E, Terlecky SR et al (1995) Activation of a selective pathway of lysosomal proteolysis in rat liver by prolonged starvation. Am J Physiol 269:C1200–C1208
18. Iwata A, Christianson JC, Bucci M et al (2005) Increased susceptibility of cytoplasmic over nuclear polyglutamine aggregates to autophagic degradation. Proc Natl Acad Sci U S A 102:13135–13140
19. Kiffin R, Christian C, Knecht E et al (2004) Activation of chaperone-mediated autophagy during oxidative stress. Mol Biol Cell 15:4829–4840
20. Mizushima N, Yamamoto A, Matsui M et al (2004) In vivo analysis of autophagy in response to nutrient starvation using transgenic mice expressing a fluorescent autophagosome marker. Mol Biol Cell 15:1101–1111
21. Ravikumar B, Vacher C, Berger Z et al (2004) Inhibition of mTOR induces autophagy and reduces toxicity of polyglutamine expansions in fly and mouse models of Huntington disease. Nat Genet 36:585–595
22. Scherz-Shouval R, Shvets E, Fass E et al (2007) Reactive oxygen species are essential for autophagy and specifically regulate the activity of Atg4. EMBO J 26:1749–1760
23. Massey AC, Zhang C, Cuervo AM (2006) Chaperone-mediated autophagy in aging and disease. Curr Top Dev Biol 73:205–235
24. Yang Z, Klionsky DJ (2010) Eaten alive: a history of macroautophagy. Nat Cell Biol 12:814–822
25. Oeste CL, Seco E, Patton WF et al (2013) Interactions between autophagic and endolysosomal markers in endothelial cells. Histochem Cell Biol 139:659–670
26. Barth S, Glick D, Macleod KF (2010) Autophagy: assays and artifacts. J Pathol 221:117–124
27. Biederbick A, Kern HF, Elsasser HP (1995) Monodansylcadaverine (MDC) is a specific in vivo marker for autophagic vacuoles. Eur J Cell Biol 66:3–14
28. Niemann A, Takatsuki A, Elsasser HP (2000) The lysosomotropic agent monodansylcadaverine also acts as a solvent polarity probe. J Histochem Cytochem 48:251–258
29. Chan LL, Shen D, Wilkinson AR et al (2012) A novel image-based cytometry method for autophagy detection in living cells. Autophagy 8:1371–1382
30. Kobayashi S, Volden P, Timm D et al (2010) Transcription factor GATA4 inhibits doxorubicin-induced autophagy and cardiomyocyte death. J Biol Chem 285:793–804
31. Boya P, Gonzalez-Polo RA, Casares N et al (2005) Inhibition of macroautophagy triggers apoptosis. Mol Cell Biol 25:1025–1040
32. Peropadre A, Fernandez Freire P, Herrero O et al (2011) Cellular responses associated with dibucaine-induced phospholipidosis. Chem Res Toxicol 24:185–192
33. Anderson N, Borlak J (2006) Drug-induced phospholipidosis. FEBS Lett 580:5533–5540
34. Dulbecco R, Vogt M (1954) Plaque formation and isolation of pure lines with poliomyelitis viruses. J Exp Med 99:167–182
35. Porter K, Nallathambi J, Lin Y et al (2013) Lysosomal basification and decreased autophagic flux in oxidatively stressed trabecular meshwork cells: implications for glaucoma pathogenesis. Autophagy 9:581–594
36. Bar J, Cohen-Noyman E, Geiger B et al (2004) Attenuation of the p53 response to DNA damage by high cell density. Oncogene 23:2128–2137

Chapter 6

A Low-Cost Method for Tracking the Induction of Apoptosis Using FRET-Based Activity Sensors in Suspension Cells

Akamu J. Ewunkem, Carl D. Parson II, Perpetua M. Muganda, and Robert H. Newman

Abstract

Apoptosis, or programmed cell death, is a tightly regulated cellular event that plays an important role in both normal developmental processes and many pathological states. The induction of apoptosis is tightly regulated through the coordinated action of members of the caspase family of proteases. Here we discuss a relatively inexpensive protocol for monitoring the induction and progression of apoptosis using a genetically encoded fluorescence resonance energy transfer (FRET)-based biosensor of the executioner caspase, caspase-3, in living suspension cells.

Key words Genetically encoded biosensor, Apoptosis, Fluorescence resonance energy transfer, Caspase 3, Sensor for activated caspases based on FRET (SCAT3), Diepoxybutane

1 Introduction

Apoptosis is a form of cell death that plays an important role in both normal physiological processes and disease. Therefore, in order to maintain tissue homeostasis, apoptotic pathways must be properly regulated. Indeed, the dysregulation of apoptotic signaling mechanisms is associated with many pathological states [1–3]. For instance, the upregulation of apoptotic pathways leads to neuronal degeneration associated with dementia while their downregulation is a hallmark of many cancers. In the case of the latter, reduced apoptosis not only promotes tumor formation but it can also adversely affect chemotherapeutic and radiological interventions that induce apoptosis to eliminate cancerous cells [4].

Apoptosis can be divided into two related pathways, the intrinsic and extrinsic pathways (Fig. 1). In both cases, the induction of apoptosis relies on the coordinated activation of *c*ysteinyl, *asp*artate-specific prote*ase* (caspase) family members [5]. This coordination is dependent, in large part, on the specificity of the caspase family

Perpetua M. Muganda (ed.), *Apoptosis Methods in Toxicology*, Methods in Pharmacology and Toxicology,
DOI 10.1007/978-1-4939-3588-8_6, © Springer Science+Business Media New York 2016

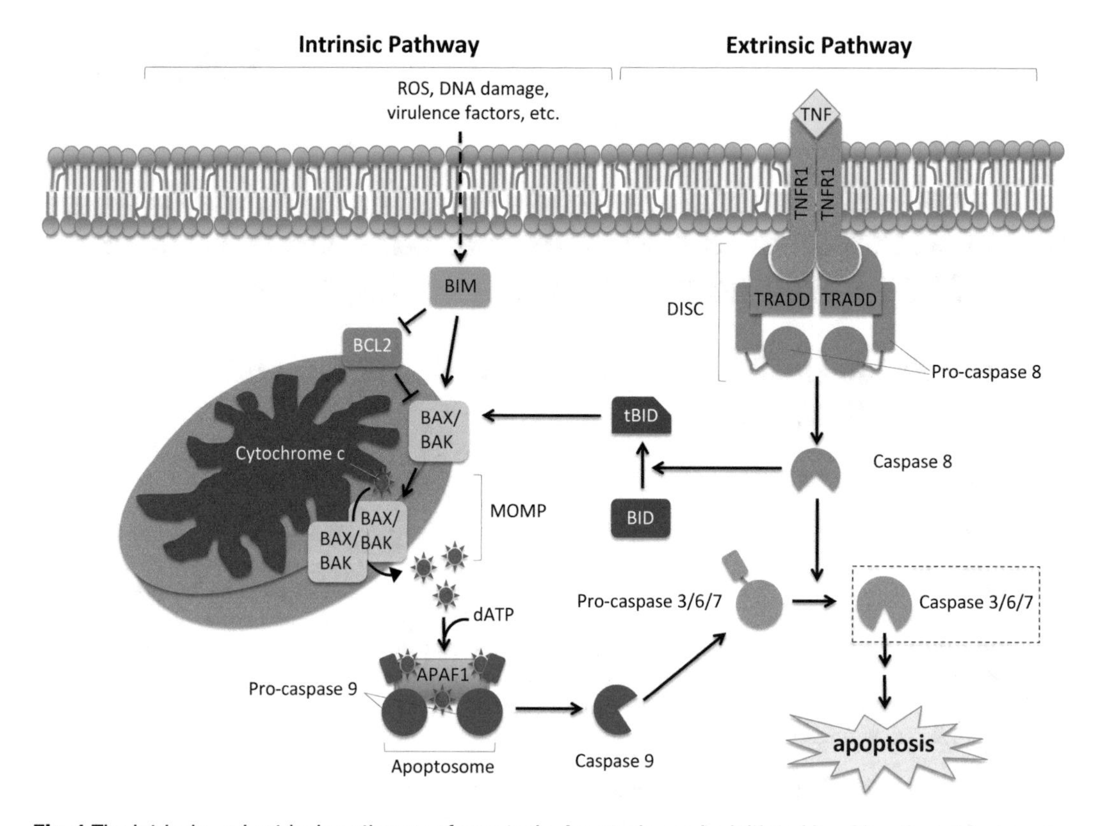

Fig. 1 The intrinsic and extrinsic pathways of apoptosis. Apoptosis can be initiated by either the intrinsic pathway (*left*) or the extrinsic pathway (*right*). The intrinsic, or mitochondrial, pathway is activated in response to a wide variety of cellular stresses, including DNA damage, viral virulence factors, and elevated levels of reactive oxygen species (ROS). This pathway is regulated by both pro- and anti-apoptotic members of the B-cell lymphoma-2 (BCL2) family, including the anti-apoptotic factor BCL2 and the pro-apoptotic factors Bcl2-interacting mediator of cell death (BIM), BCL2 antagonist/killer (BAK) and BCL2-associated X protein (BAX). Dimerization of BAK/BAX leads to mitochondrial outer membrane permeabilization (MOMP), resulting in the release of cytochrome c (*red starbursts*) from the mitochondrial intermembrane space into the cytosol. In the cytosol, cytochrome c associates with apoptosis activating factor-1 (Apaf-1) and pro-caspase-9 in a dATP-dependent manner to form a large, multimeric complex termed the apoptosome. At the apoptosome, pro-caspase-9 is converted to the active form, leading to cleavage and activation of executioner caspases, such as caspase-3, -6 and -7. Once activated, the executioner caspases process various downstream substrates, ultimately leading to apoptotic cell death. The extrinsic pathway is initiated by cytokines, such as tumor necrosis factor (TNF), which bind the TNF receptor-1 (TNFR1). Ligand binding promotes the activation of TNFR1, leading to its association with TNFR1-associated death domain protein (TRADD). TRADD helps sequester pro-caspase-8 in the death-inducing signaling complex (DISC), which promotes the cleavage and subsequent activation of caspase-8. Caspase-8 cleaves executioner caspases, leading to their activation and the induction of apoptosis. In some cellular contexts, caspase-8 can also cleave the BCL2-homology domain 3 (BH3)-interacting domain death agonist (BID). The truncated form of BID (tBID) then translocates from the cytosol to the mitochondria, where it helps initiate the intrinsic pathway via MOMP. In this way, the extrinsic and intrinsic pathways are linked. It is important to note that both the intrinsic and extrinsic pathways lead to the activation of executioner caspases (*dashed box*). The genetically encoded biosensor, SCAT-3, used in this protocol measures the activity of caspase-3

members for their substrates. Indeed, each caspase family member recognizes and cleaves a specific cleavage motif. For instance, the executioner caspase, caspase-3, recognizes the tetrapeptide sequence Asp-Glu-Val-Asp (DEVD/X) in its substrates, cleaving immediately after the second aspartate residue (represented by a "/").

In this chapter, we begin with a brief overview of the extrinsic and intrinsic apoptotic pathways before describing a relatively low-cost method to monitor the induction of apoptosis using fluorescence resonance energy transfer (FRET)-based biosensors in a microplate reader format. This method, which can offer insights about the timing of apoptotic induction in different cellular contexts, can also be used to screen a wide variety of pharmacological and toxicological agents for their effect on apoptosis.

1.1 The Intrinsic Pathway

The mitochondrial-dependent intrinsic pathway is activated in response to a wide variety of cellular stresses, including DNA damage, viral virulence factors, and elevated levels of reactive oxygen species (ROS) (Fig. 1a) [3, 6]. This pathway, which is regulated by both pro- and anti-apoptotic members of the Bcl-2 family, is initiated by mitochondrial outer membrane permeabilization (MOMP) [3, 7–9]. MOMP results in the diffusion of various proteins that typically reside in the mitochondrial intermembrane space into the cytosol. Chief among these is cytochrome c, which associates with apoptosis activating factor-1 (Apaf-1) and pro-caspase-9 in a dATP-dependent manner to form a large, multimeric complex termed the apoptosome [10–12]. At the apoptosome, the high local concentration of pro-caspase-9 leads to its autocatalytic conversion to the active form. Once activated, caspase-9 recognizes and cleaves executioner caspases, such as caspase-3, -6 and -7, which process various downstream substrates, leading to apoptotic cell death [5, 12].

1.2 The Extrinsic Pathway

Unlike the intrinsic pathway, which is triggered in response to changes in the intracellular environment, the extrinsic pathway relies upon receptor-mediated signaling to initiate apoptosis (Fig. 1b). In this context, cytokines, such as tumor necrosis factor (TNF), TNF-related apoptosis inducing ligand (TRAIL) and Fas ligand, bind so-called "death receptors" of the TNF receptor (TNFR) superfamily [13–15]. Ligand binding promotes the activation of the death receptors, leading to their association with death domain (DD)-containing adaptor proteins, such as Fas-associated death domain protein (FADD) and TNFR1-associated death domain protein (TRADD) [12, 15]. FADD and TRADD help sequester the zymogenic forms of the initiator caspases, pro-caspase-8 and pro-caspase-10, in the death-inducing signaling complex (DISC) via their death effector domains (DED) [12]. DISC formation leads to the cleavage and subsequent activation of the initiator caspases, which in turn cleave and activate executioner caspases, leading to the induction of apoptosis.

Recent evidence suggests that, in certain cell types (classified as type II cells), insufficient levels of caspase-8 are activated by the DISC to elicit apoptosis. If this is the case, cross talk between the extrinsic and intrinsic pathways via caspase-8-mediated processing of the Bcl-2 family member, BID, leads to apoptosis in a mitochondria-dependent manner (Fig. 1). Under these circumstances, cytosolic BID is cleaved by caspase-8 to generate a 15 kDa BID fragment (tBID) that is redistributed to the mitochondria via the N-myristoylation of an exposed glycine residue at the site of cleavage [16–18]. At the mitochondria, tBID initiates MOMP and the release of apoptogenic factors, such as cytochrome c, that promote the activation of the executioner caspase, caspase-3, and a full apoptotic response via the intrinsic pathway.

In contrast to type II cells, activation of the intrinsic pathway is not formally required for the induction of apoptosis in type I cells, which have been shown to undergo apoptosis despite the overexpression of anti-apoptotic regulators of the intrinsic pathway, such as BCL-2 and BCL-XL [3, 15]. Nonetheless, it is important to note that, even in type I cells, cross talk between the extrinsic and intrinsic pathways routinely occurs, serving to amplify and reinforce apoptotic initiation. Thus, in both type I and type II cells, caspase-3 activation is associated with induction of the apoptotic response. As a consequence, caspase-3 activity is an attractive marker of apoptosis induction [19]. Here, we describe a low-cost method to track changes in caspase-3 activity in real-time within the native cellular environment using the genetically encodable FRET-based biosensor, *s*ensor of activated *ca*spase-*3* using FRE*T* (SCAT-3) [20].

1.3 Genetically Encoded Biosensors for Measuring Induction of Apoptosis in Living Cells

Genetically encodable FRET-based reporters, which are able to track the dynamics of a variety of signaling molecules in the native cellular environment with high spatial and temporal resolution, have traditionally been used in conjunction with live cell imaging to better understand the regulation of complex cellular pathways [21, 22]. To date, FRET-based biosensors have been used to study the dynamic regulation of a wide range of cellular signaling molecules, including key second messengers (such as Ca^{2+} [23–26] and cAMP [27–31]), important signaling enzymes (such as protein kinases [32], phosphatases [33, 34] and small G-proteins [35–39]), and cell surface receptors (such as G-protein-coupled receptors [40] and receptor tyrosine kinases [41–43]) (reviewed in Refs. [21, 44]). However, due to the high cost of the fluorescence microscopy systems typically used for live cell imaging, these molecular tools are not readily accessible to many researchers. Recently, methods have been developed to monitor various cellular parameters utilizing FRET-based reporters using a microplate reader format [45–47]. For instance, Robinson et al. recently described a method to conduct live cell compound screens using

FRET-based biosensors of the cAMP-PKA signaling pathway [45]. Here, we describe a low-cost strategy to monitor the induction of apoptosis in non-adherent TK6 lymphoblasts using the caspase-3 activity reporter, SCAT-3, in conjunction with a fluorescent microplate reader.

1.4 Biosensor and Drugs Used in This Protocol

SCAT-3, which was constructed by sandwiching a short peptide containing the caspase-3-specific recognition sequence, DEVD, between enhanced cyan fluorescent protein (ECFP) and the yellow fluorescent protein, Venus, has been used to monitor the induction of apoptosis in many cellular contexts using live cell fluorescence microscopy (Fig. 2a) [19, 20, 48]. Due to the high degree of spectral overlap between the emission profile of ECFP (i.e., the donor) and the excitation profile of Venus (i.e., the acceptor), these fluorescent proteins (FPs) are able to undergo efficient FRET when they are in close proximity to one another (Fig. 2b, blue line). However, following caspase-3-mediated cleavage of the biosensor, the ECFP-Venus FP FRET pair is no longer linked together, resulting in a dramatic reduction in FRET as they diffuse away from one another (Fig. 2b, red line). In this way, SCAT-3 is able to track the activation of caspase-3 in real-time under a variety of cellular conditions. It is important to note that enzymatic amplification of the signal, coupled with the large change in FRET caused by biosensor cleavage, allows SCAT-3 to detect increases in caspase-3 activity associated with apoptosis at much earlier time points than traditional end-point assays, such as DNA fragmentation assays. On the other hand, unlike FRET-based biosensors of other cellular processes that typically rely on a signal-dependent conformational change in the biosensor to alter FRET, the irreversible nature of cleavage means that SCAT-3 and related FRET-based protease activity sensors are unable to report on the attenuation of the signal. Nonetheless, such FRET-based protease biosensors have proven to be powerful tools for monitoring changes in protease activity under a variety of cellular conditions, particularly during apoptosis.

Here, we use diepoxybutane (DEB)-induced apoptosis in SCAT-3 transfected TK6 human lymphoblasts as an example of how FRET-based biosensors can be used in a microplate reader format to monitor the induction of apoptosis (Fig. 3). DEB, which is the most toxic metabolite of the high production volume industrial chemical 1,3-butadiene, has been shown to induce cell death in TK6 lymphoblasts via the production of ROS and activation of the intrinsic apoptotic pathway [49].

We begin by discussing strategies for transfection of the biosensor DNA. We then outline the steps for performing the FRET assay and end with tips for data analysis. Though the details provided are specific for non-adherent TK6 human lymphoblasts, the protocol can be applied to numerous cell lines. Moreover, though

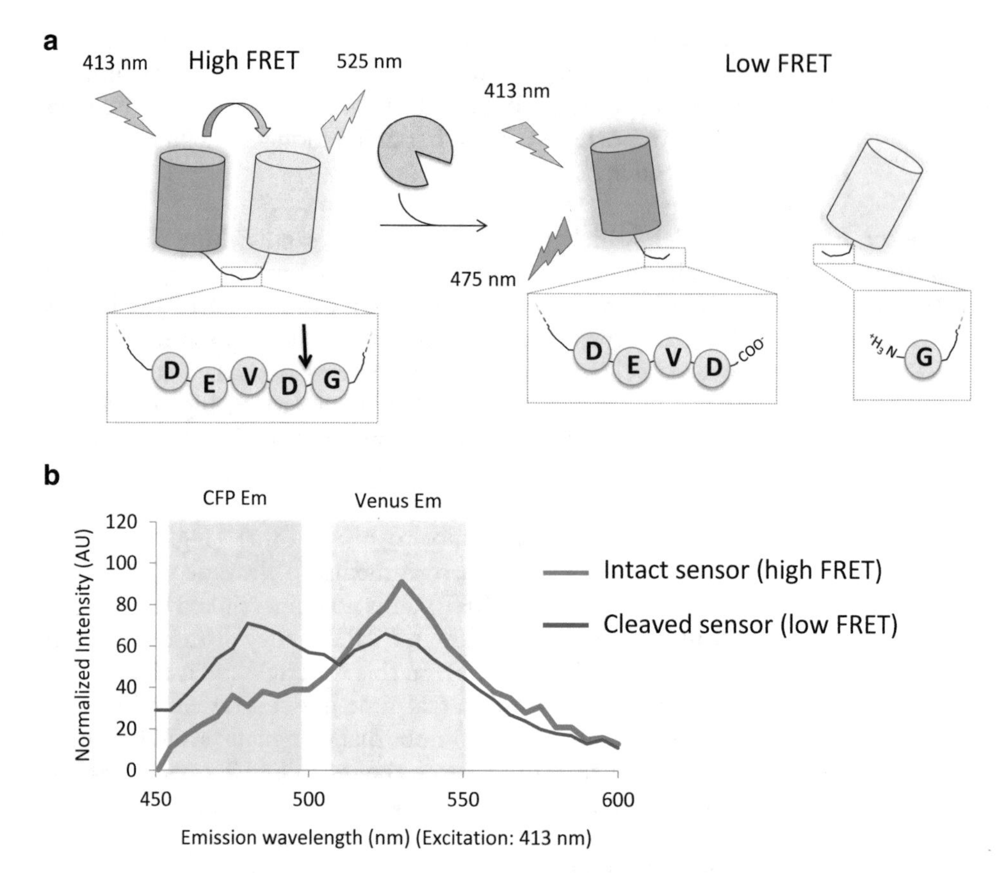

Fig. 2 Overview of the caspase-3 activity sensor, SCAT3. (**a**) The genetically encoded FRET-based biosensor, SCAT3, is composed of the caspase-3 recognition sequence (DEVD) sandwiched between the FP FRET pair, enhanced CFP (ECFP; *cyan cylinder*) and Venus (*yellow cylinder*). When the sensor is intact (*left side*), ECFP and Venus are in close proximity to one another. As a consequence, excitation of ECFP with 413 nm light (*light purple lightning bolt*) leads to non-radiative transfer of the excited state energy to Venus via FRET (*curved arrow*), causing emission at 525 nm (*yellow lightning bolt*). Caspase-3-mediated cleavage of the sensor causes an increase in the distance between the FP's, leading to reduced FRET between ECFP and Venus. As a consequence, excitation at 413 nm leads to emission by ECFP at 475 nm (*cyan lightning bolt*). (**b**) Representative SCAT3 emission scans from 450 to 600 nm following excitation at 413 nm before (*blue line*) and after (*red line*) caspase-3 activation. The regions of the spectrum used to quantitate ECFP emission (*cyan box*, 455–495 nm) and Venus emission (*yellow box*, 515–545 nm) are shown

we discuss only DEB as an apoptosis-inducing agent, the protocol can be used to measure the dynamics of apoptosis induction for a variety of xenobiotics (provided that their fluorescence spectra do not interfere with FRET measurements). Likewise, the protocol also can be easily adapted to screen compound libraries in order to assess the impact of a large number of compounds on the induction of apoptosis.

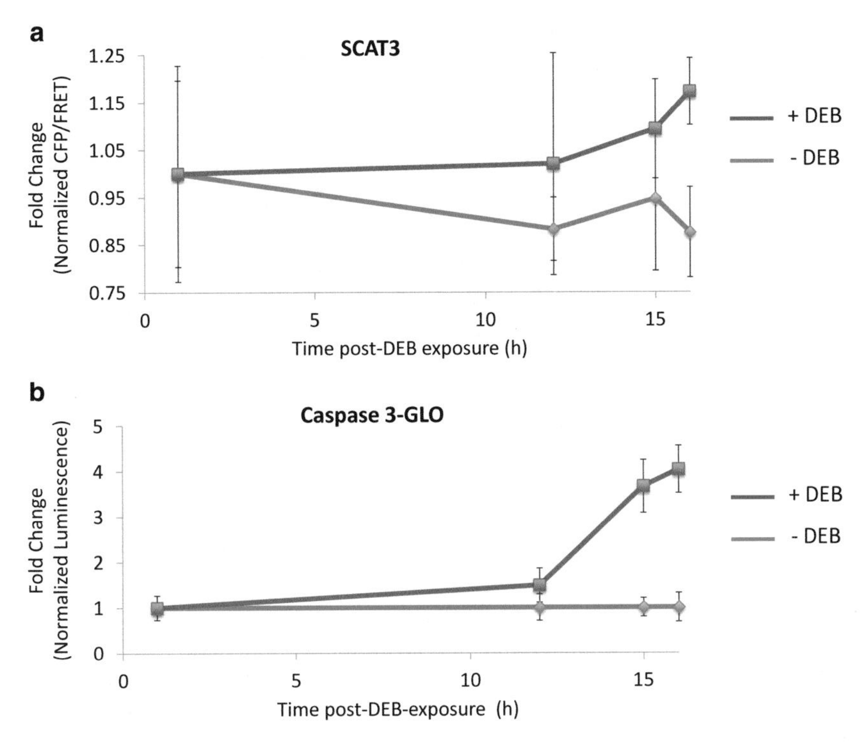

Fig. 3 Time course of DEB-induced apoptosis measured by orthogonal SCAT-3 and Caspase-3 GLO assays. Time course of DEB-induced apoptosis in TK6 lymphoblasts measured using SCAT3 (**a**) and Caspase-3 GLO (**b**). In both (**a**) and (**b**), the signal from cells treated with DEB are shown in *red* while that from untreated samples is shown in *blue*

2 Materials

2.1 Cell Culture

1. The TK6 human lymphoblastic cell line (The American Type Culture Collection Catalogue # CRL 8015).
2. Roswell Park Memorial Institute 1640 culture media (RPMI 1640, Gibco/BRL, Bethesda, MD) supplemented with 10 % fetal bovine serum (FBS, Atlanta Biologicals, Atlanta, GA).
3. T150 tissue culture flask.
4. 6-well tissue culture dishes.
5. 96-well assay plate, black walled, clear bottom with lid, polystyrene, sterile (Costar, No. 3603).

2.2 Transfection Materials

1. Biosensor DNA (e.g., SCAT3 and SCAT3-DEVG), kindly provided by Prof. Masayuki Miura (**Note 1**).
2. RPMI-1640 cell culture medium (Gibco/BRL, Bethesda, MD) supplemented with 10 % FBS.
3. Amaxa Nucleofector device with 96-well shuttle.
4. Nucleofector kit containing Nucleofection solution SF (Lonza, Walkersville, MD).

Table 1
The Infinite M200 PRO instrument settings for an emission scan following 413 nm excitation

Mode	Fluo top scan
Emission Wavelength Start	450 nm
Emission Wavelength End	600 nm
Emission Wavelength Step	2 nm
Emission Scan Number	76
Excitation Wavelength	413 nm
Bandwidth (Em)	280...850: 20 nm
Bandwidth (Ex) (Range 1)	230...315: 5 nm
Bandwidth (Ex) (Range 2)	316...850: 10 nm
Gain	Optimal
Number of Flashes	10
Integration Time	20 μs
Lag	0 μs
Settle Time	0 ms
Z-Position (Manual)	20,000 μm

2.3 Cell Counting

1. Hemocytometer.
2. 0.4 % trypan blue (Life Technologies, Carlsbad, CA).

2.4 FRET Assay Materials and Devices

1. Hank's Balanced Salt Solution (HBSS): 0.15 M NaCl, 4 mM KCl, 1.2 mM $MgCl_2 \cdot 6H_2O$, 55.5 mM glucose, 20 mM HEPES–KOH, pH 7.2; Store at 4 °C.
2. Tecan Infinite M200 PRO fluorescence microplate reader (Tecan, Inc.) set to conduct an emission scan following excitation at 413 nm (**Note 2**). The settings for the emission scan are given in Table 1.
3. Diepoxybutane (DEB; Sigma): 10 mM DEB stock solution (1000×). CAUTION: DEB is a potent carcinogen. Care must be taken when handling this reagent.

2.5 Data Analysis

1. Spreadsheet application (e.g., Microsoft Excel).

3 Methods

3.1 Cell Culture and Transfection

1. Propagation of cell cultures. TK6 cells are routinely propagated at a density of 2×10^5 cells/ml in RPMI-1640 medium containing 2 mM glutamine and 10 % FBS. Cells are passaged every 36 h when the density reaches 2×10^6 cells/ml, and are used for experiments at 12 h after they are passaged onto fresh medium.
2. Determine the distribution of experimental and control samples. Table 2 highlights suggested experimental and control samples, the biosensor to be transfected (if any) under each condition, the treatment condition, and the reason that each condition is included.
3. In a T150 tissue culture flask, grow 120 ml of TK6 human lymphoblasts to a density of ~1×10^6 cells/ml at 12 h prior to transfection (**Note 3**).
4. Transfect the cells via nucleofection (**Note 4**).
   (a) Determine the total number of cells to be nucleofected, as well as the number of nucleofection reactions to be conducted (**Note 5**).
   (b) Pellet the cells at $200 \times g$ for 10 min (**Note 6**).
   (c) While the cells are being pelleted:
   - Start the Nucleofector 96-well shuttle software and generate a parameter file from the predefined template for cell line optimization (for TK6 lymphoblasts, we use Program DS 137).
   - In the CO_2 incubator, pre-warm 6-well plates containing 2 ml of culture media per well.

Table 2
Sample experimental layout

Condition	Biosensor	Treatment	Parameter
1	None	None	Reference set for untreated cells
2	None	DEB	Reference set for treated cells
3	SCAT3	None	Untreated, SCAT3 control set
4	SCAT3	DEB	Experimental set
5	SCAT3-DEVG	None	Untreated, SCAT3-DEVG control set
6	SCAT3-DEVG	DEB	DEB-treated, SCAT3-DEVG control set

SCAT3: FRET-based caspase-3 sensor; SCAT3-DEVG: control construct in which the caspase-3 cleavage site, DEVD, has been mutated to DEVG; DEB: diepoxybutane

- In a 37 °C water bath, pre-warm a separate aliquot of culture medium (use 200 μl of media per nucleofection reaction).
- Pre-warm Nucleofector solution SF to room temperature.
- For each nucleofection reaction, add 400 ng of the appropriate biosensor DNA to a sterile PCR tube in a strip format (this will allow a multichannel pipette to be used in subsequent transfection steps).

(d) Following centrifugation, carefully decant the supernatant. Remove supernatant completely (**Note 7**).

(e) Gently resuspend the cell pellet in Nucleofector solution SF to a final concentration of 1×10^6 cells per 21 μl (e.g., for 120×10^6 cells, resuspend cells in 2.52 ml of Nucleofector solution SF). Add the cell suspension to a sterile small volume multichannel pipette reservoir. Avoid creating bubbles at all times when working with cells in nucleofection solution.

(f) Using a multichannel pipette, add 21 μl of the resuspended cells (i.e., 1×10^6 cells) to each PCR tube containing DNA and mix by pipetting up and down several times; avoid creating bubbles. Using the same pipet and tips, transfer 20.5 μl of the DNA/cell suspension to the appropriate nucleofection cuvette wells, taking care to deliver the DNA/cell suspension to the bottom of the cuvettes.

(g) Put the lid onto the 96-well shuttle plate, and gently tap the cuvette to ensure that the sample covers the bottom of the cuvettes.

(h) Put the 96-well shuttle plate into the nucleofector device and start the nucleofection process by pressing "upload and start" in the shuttle software.

(i) Following nucleofection, open the retainer and carefully remove the 96-well cuvette strips from the retainer.

(j) Add 100 μl of pre-warmed culture media to each cuvette and transfer the cells into the corresponding well of the 6-well plates. To increase recovery of cells, repeat once more.

(k) Incubate the nucleofected cells at 37 °C for 12 h in a CO_2 incubator.

5. Measure the transfection efficiency 12 h after transfection. Proceed only if greater than 50 % transfection efficiency is achieved (**Note 8**).

3.2 Induction of Apoptosis and Fluorescence Data Acquisition

On the day of the experiment, pool all cells containing the same biosensor (e.g., all cells expressing SCAT3) and measure the density of the pooled cells using a hemocytometer or automated cell counting device. If using a hemocytometer:

- Add 10 μl of cells to 10 μl of 0.4 % trypan blue.
- Add the solution to the hemocytometer and count all of the unstained cells in one quadrant (large square).
- Determine the number of live cells per ml according to the following equation (**Notes 9** and **10**):

$$\text{Cells / ml} = (\#\,\text{cells}) \times (\text{dilution factor}\,(\text{e.g.}, 2) \times 10^4\ /\ \text{ml}$$

1. Prior to DEB addition, dilute each set of pooled cells to a density of 2.0×10^5 cells/ml using pre-warmed RPMI, 10 % FBS culture media.
2. Split each set into two groups. The first group, which will be an untreated control, will contain 40 % of the pooled cells. The second group, which will be treated with DEB, will contain 60 % of the pooled cells. This is done in order to account for the fact that DEB inhibits cell proliferation.
3. Add DEB to the cells according to the experimental scheme outlined in Table 1. Add DEB to a final concentration of 10 μM with gentle agitation to ensure even distribution of the xenobiotic (**Note 11**). Unexposed control cells should receive vehicle only (culture medium, in this case).
4. Measure the baseline fluorescence immediately after DEB addition.
   (a) Remove 2.4×10^6 cells from each group and pellet the cells at $800 \times g$ for 3 min. While the cells are being pelleted, place the remaining cells back in the CO_2 incubator.
   (b) Following centrifugation, remove the media taking care not to disturb the cell pellet (**Note 11**).
   (c) Wash each pellet once with 1/5 volume of HBSS imaging buffer (e.g., if 6 ml of cells in culture were initially pelleted, the cell pellet should be washed with 1.2 ml of HBSS).
   (d) Pellet the cells again at $800 \times g$ for 3 min.
   (e) Carefully remove the supernatant and gently resuspend each cell pellet in 330 μl of HBSS imaging buffer.
   (f) Transfer 3×100 μl of the cell suspension to the appropriate well in the 96-well black-walled assay plate.

(g) Insert the 96-well plate into the microplate reader and conduct an emission scan in each well using 413 nm excitation (please refer to Table 1 in Sect. 2.4, item 2 for microplate reader settings) (**Note 12**).

5. At each time point, remove 2.4×10^6 cells and conduct an emission scan using an excitation wavelength of 413 nm, as outlined in step 4 (**Note 13**).

3.3 Data Analysis

1. The output of an emission scan will be a series of data points corresponding to the fluorescence emission at each wavelength across the designated scan range. For example, using the settings shown in Table 1, the fluorescence emission following excitation at 413 nm will be measured at 2 nm intervals from 450 to 600 nm (for a total of 76 data points per well).
2. Using a spreadsheet application (e.g., Microsoft Excel), calculate the average fluorescence at each wavelength in the wells that contained non-transfected cells. For example, calculate the average fluorescence at each wavelength for the non-transfected, untreated cells. Do the same for the non-transfected, DEB-treated cells. These values represent the background fluorescence with and without DEB treatment.
3. Subtract the background fluorescence at each wavelength from the analogous value for each condition. Be sure to use the appropriate reference set for background correction (e.g., pair "untreated with untreated" and "treated with treated"). These values serve as the background corrected values.
4. To determine the emission intensity directly from CFP (i.e., CFP direct), add all of the background corrected values from 455 to 495 nm (this corresponds to a 475 nm emission peak with a 20 nm bandwidth). To determine the emission intensity from YFP FRET (i.e., YFP FRET), add all of the background corrected values from 505 to 545 nm (this corresponds to a 525 nm emission peak with a 20 nm bandwidth).
5. For each time point, take the ratio of the CFP direct (i.e., donor) to YFP FRET (i.e., acceptor) according to the following equation:

$$\text{Emission Ratio} = \frac{\text{Background corrected CFP direct}}{\text{Background corrected YFP FRET}}$$

6. Normalize the calculated ratio at each time point to the average ratio of the corresponding set immediately following DEB addition (i.e., t_{1h}). This represents the fold change at each point.
7. Calculate the average fold change and standard error at each point.
8. To obtain a time course of DEB-induced apoptosis, plot the normalized ratios over time (Fig. 3).

4 Notes

1. Early versions of FRET-based caspase sensors employed enhanced yellow fluorescent protein (EYFP) as the FRET acceptor rather than the next generation YFP, Venus [50–52]. Because EYFP is acutely sensitive to changes in pH and the Cl^- concentration, changes in the cellular environment not directly related to caspase-3 activity could quench EYFP fluorescence, leading to an apparent reduction in FRET in the absence of sensor cleavage. Though Venus fluorescence is not affected by either pH or Cl^- at physiologically relevant levels, SCAT3-DEVG may be employed as a negative control in these experiments. SCAT3-DEVG contains a modified tetrapeptide sequence in which the caspase-3-specific cleavage site, DEVD, has been mutated to DEVG. As a consequence, SCAT3-DEVG is not cleaved by caspase-3. Likewise, SCAT9, which contains the caspase-9-specific cleavage motif, LEHD, in place of DEVD, is also available. SCAT9 specifically monitors caspase-9 activity.
2. The Infinite M200 PRO is a monochromator-based "filterless" microplate reader. If a filter-based system is used, then the following filters should be used: one 420DF20 excitation filter and two emission filters (470DF40 for donor CFP and 535DF25 for acceptor Venus).
3. Here, we describe a protocol using TK6 lymphoblasts as an example. Optimization of cell culture and transfection techniques, as well as plating density if adherent cells are used, may be necessary for other cell types. For TK6 lymphoblasts, this cell density should be sufficient for five time points at 8.0×10^5 cells per well measured in triplicate. If additional time points are desired, the number of cells should be scaled accordingly. The total DNA should also be scaled accordingly.
4. All transfection steps should be carried out in a sterile biosafety cabinet.
5. For each condition (e.g., SCAT3 + DEB), approximately 12.0×10^6 cells will be required. To maintain consistency, treated and non-treated cells containing the same biosensor should be derived from the same population of transfected cells. Therefore, at the time of transfection, approximately 24×10^6 cells should be transfected with each biosensor (12×10^6 cells per condition $\times$ 2 conditions). Since 1×10^6 cells are used per nucleofection reaction, this would require a total of 24 nucleofection reactions.
6. TK6 lymphoblasts are non-adherent suspension cells. If using adherent cells for nucleofection, the cells must first be trypsinized prior to centrifugation. Other transfection procedures that are effective for the cell type under study can also be utilized.

7. Be sure not to aspirate off the pellet. It is best to turn the centrifuge tube horizontally to aspirate the media off of the side of the tube, keeping the tip away from the pellet.
8. If the transfection efficiency is <50 %, then the signal-to-noise ratio may be too low to achieve high sensitivity. If transfection efficiencies >50 % cannot be achieved, a cell line stably expressing the biosensor-of-interest can be generated and used as an alternative. Likewise, if using a fluorescent biosensor with a signal-to-noise ratio lower than that of SCAT3, higher transfection efficiencies may be necessary.
9. Live TK6 lymphoblasts will appear round and translucent while dead cells will appear blue and may be irregularly shaped. Count only the live cells.
10. The volume of one square on the hemocytometer is 0.1 mm^3 (or 1.0×10^{-4} ml) so the quick calculation is: cells/ml = # cells × Dilution Factor × 10^4/ml. For more accurate cell counts, count the number of cells in two to four quadrants (large squares) and use the average to calculate the cell density
11. DEB is a potent carcinogen. Care should be taken when working with DEB-treated cells and any by-products (e.g., supernatants) from the experiments should be disposed of in accordance with institutional, state and federal regulations.
12. If a need to verify assay performance arises, an orthogonal detection assay, such as the bioluminescence-based Caspase 3-GLO assay (Promega, Madison, WI), may be conducted according to the manufacturer's instructions once the emission scan has been completed.
13. If the microplate reader being used for the experiments is equipped with temperature and CO_2 control modules, continuous measurements can be obtained (e.g., at 1 h intervals for 36 h) by substituting RPMI-1640 media lacking phenol red for the standard RPMI-1640 media.

References

1. Ashkenazi A, Dixit VM (1998) Death receptors: signaling and modulation. Science 281: 1305–1308
2. Nagata S (1997) Apoptosis by death factor. Cell 88:355–365
3. Plati J, Bucur O, Khosravi-Far R (2011) Apoptotic cell signaling in cancer progression and therapy. Integr Biol 3:279–296
4. Cotter TG (2009) Apoptosis and cancer: the genesis of a research field. Nat Rev Cancer 9:501–507
5. Shalini S, Dorstyn L, Dawar S, Kumar S (2015) Old, new and emerging functions of caspases. Cell Death Differ 22:526–539
6. Mohamad N, Gutierrez A, Nunez M, Cocca C, Martin G, Cricco G, Medina V, Rivera E, Bergoc R (2005) Mitochondrial apoptotic pathways. Biocell 29:149–161
7. Khosravi-Far R, Esposti MD (2004) Death receptor signals to mitochondria. Cancer Biol Ther 3:1051–1057
8. Green DR, Kroemer G (2004) The pathophysiology of mitochondrial cell death. Science 305:626–629
9. Hail N Jr, Carter BZ, Konopleva M, Andreeff M (2006) Apoptosis effector mechanisms: a requiem performed in different keys. Apoptosis 11:889–904

10. Czabotar PE, Lessene G, Strasser A, Adams JM (2014) Control of apoptosis by the BCL-2 protein family: implications for physiology and therapy. Nat Rev Mol Cell Biol 15:49–63
11. Wang ZB, Liu YQ, Cui YF (2005) Pathways to caspase activation. Cell Biol Int 29:489–496
12. Jin Z, El-Deiry WS (2005) Overview of cell death signaling pathways. Cancer Biol Ther 4:139–163
13. Fas SC, Fritzsching B, Suri-Payer E, Krammer PH (2006) Death receptor signaling and its function in the immune system. Curr Dir Autoimmun 9:1–17
14. Gaur U, Aggarwal BB (2003) Regulation of proliferation, survival and apoptosis by members of the TNF superfamily. Biochem Pharmacol 66:1403–1408
15. Guicciardi ME, Gores GJ (2009) Life and death by death receptors. FASEB J 23: 1625–1637
16. Li H, Zhu H, Xu CJ, Yuan J (1998) Cleavage of BID by caspase 8 mediates the mitochondrial damage in the Fas pathway of apoptosis. Cell 94:491–501
17. Lovell JF, Billen LP, Bindner S, Shamas-Din A, Fradin C, Leber B, Andrews DW (2008) Membrane binding by tBid initiates an ordered series of events culminating in membrane permeabilization by Bax. Cell 135:1074–1084
18. Luo X, Budihardjo I, Zou H, Slaughter C, Wang X (1998) Bid, a Bcl2 interacting protein, mediates cytochrome c release from mitochondria in response to activation of cell surface death receptors. Cell 94:481–490
19. Wu Y, Xing D, Chen WR (2006) Single cell FRET imaging for determination of pathway of tumor cell apoptosis induced by photofrin-PDT. Cell Cycle 5:729–734
20. Takemoto K, Nagai T, Miyawaki A, Miura M (2003) Spatio-temporal activation of caspase revealed by indicator that is insensitive to environmental effects. J Cell Biol 160:235–243
21. Newman RH, Fosbrink MD, Zhang J (2011) Genetically encodable fluorescent biosensors for tracking signaling dynamics in living cells. Chem Rev 111:3614–3666
22. Newman RH, Zhang J (2014) The design and application of genetically encodable biosensors based on fluorescent proteins. Methods Mol Biol 1071:1–16
23. Mank M, Griesbeck O (2008) Genetically encoded calcium indicators. Chem Rev 108:1550–1564
24. Miyawaki A, Llopis J, Heim R, McCaffery JM, Adams JA, Ikura M, Tsien RY (1997) Fluorescent indicators for Ca^{2+} based on green fluorescent proteins and calmodulin. Nature 388:882–887
25. Palmer AE, Giacomello M, Kortemme T, Hires SA, Lev-Ram V, Baker D, Tsien RY (2006) Ca^{2+} indicators based on computationally redesigned calmodulin-peptide pairs. Chem Biol 13:521–530
26. Truong K, Sawano A, Mizuno H, Hama H, Tong KI, Mal TK, Miyawaki A, Ikura M (2001) FRET-based in vivo Ca^{2+} imaging by a new calmodulin-GFP fusion molecule. Nat Struct Biol 8:1069–1073
27. Zaccolo M, Pozzan T (2002) Discrete microdomains with high concentration of cAMP in stimulated rat neonatal cardiac myocytes. Science 295:1711–1715
28. Ponsioen B, Zhao J, Riedl J, Zwartkruis F, van der Krogt G, Zaccolo M, Moolenaar WH, Bos JL, Jalink K (2004) Detecting cAMP-induced Epac activation by fluorescence resonance energy transfer: Epac as a novel cAMP indicator. EMBO Rep 5:1176–1180
29. Nikolaev VO, Bunemann M, Schmitteckert E, Lohse MJ, Engelhardt S (2006) Cyclic AMP imaging in adult cardiac myocytes reveals far-reaching beta1-adrenergic but locally confined beta2-adrenergic receptor-mediated signaling. Circ Res 99:1084–1091
30. Nikolaev VO, Bunemann M, Hein L, Hannawacker A, Lohse MJ (2004) Novel single chain cAMP sensors for receptor-induced signal propagation. J Biol Chem 279:37215–37218
31. Allen MD, DiPilato LM, Rahdar M, Ren YR, Chong C, Liu JO, Zhang J (2006) Reading dynamic kinase activity in living cells for high-throughput screening. ACS Chem Biol 1: 371–376
32. Zhang J, Allen MD (2007) FRET-based biosensors for protein kinases: illuminating the kinome. Mol Biosyst 3:759–765
33. Newman RH, Zhang J (2008) Visualization of phosphatase activity in living cells with a FRET-based calcineurin activity sensor. Mol Biosyst 4:496–501
34. Mehta S, Aye-Han NN, Ganesan A, Oldach L, Gorshkov K, Zhang J (2014) Calmodulin-controlled spatial decoding of oscillatory Ca^{2+} signals by calcineurin. eLife 3:e03765
35. Mochizuki N, Yamashita S, Kurokawa K, Ohba Y, Nagai T, Miyawaki A, Matsuda M (2001) Spatio-temporal images of growth-factor-induced activation of Ras and Rap1. Nature 411: 1065–1068
36. Itoh RE, Kurokawa K, Ohba Y, Yoshizaki H, Mochizuki N, Matsuda M (2002) Activation of rac and cdc42 video imaged by fluorescent resonance energy transfer-based single-molecule

probes in the membrane of living cells. Mol Cell Biol 22:6582–6591

37. Yoshizaki H, Ohba Y, Kurokawa K, Itoh RE, Nakamura T, Mochizuki N, Nagashima K, Matsuda M (2003) Activity of Rho-family GTPases during cell division as visualized with FRET-based probes. J Cell Biol 162:223–232
38. Yoshizaki H, Aoki K, Nakamura T, Matsuda M (2006) Regulation of RalA GTPase by phosphatidylinositol 3-kinase as visualized by FRET probes. Biochem Soc Trans 34:851–854
39. Pertz O, Hodgson L, Klemke RL, Hahn KM (2006) Spatiotemporal dynamics of RhoA activity in migrating cells. Nature 440:1069–1072
40. Lohse MJ, Nikolaev VO, Hein P, Hoffmann C, Vilardaga JP, Bunemann M (2008) Optical techniques to analyze real-time activation and signaling of G-protein-coupled receptors. Trends Pharmacol Sci 29:159–165
41. Itoh RE, Kurokawa K, Fujioka A, Sharma A, Mayer BJ, Matsuda M (2005) A FRET-based probe for epidermal growth factor receptor bound non-covalently to a pair of synthetic amphipathic helixes. Exp Cell Res 307: 142–152
42. Kurokawa K, Mochizuki N, Ohba Y, Mizuno H, Miyawaki A, Matsuda M (2001) A pair of fluorescent resonance energy transfer-based probes for tyrosine phosphorylation of the CrkII adaptor protein in vivo. J Biol Chem 276:31305–31310
43. Sato M, Umezawa Y (2004) Imaging protein phosphorylation by fluorescence in single living cells. Methods 32:451–455
44. Mehta S, Zhang J (2011) Reporting from the field: genetically encoded fluorescent reporters uncover signaling dynamics in living biological systems. Annu Rev Biochem 80:375–401
45. Robinson KH, Yang JR, Zhang J (2014) FRET and BRET-based biosensors in live cell compound screens. Methods Mol Biol 1071: 217–225
46. Tian H, Ip L, Luo H, Chang DC, Luo KQ (2007) A high throughput drug screen based on fluorescence resonance energy transfer (FRET) for anticancer activity of compounds from herbal medicine. Br J Pharmacol 150: 321–334
47. Zhu X, Fu A, Luo KQ (2012) A high-throughput fluorescence resonance energy transfer (FRET)-based endothelial cell apoptosis assay and its application for screening vascular disrupting agents. Biochem Biophys Res Commun 418:641–646
48. Koike-Kuroda Y, Kakeyama M, Fujimaki H, Tsukahara S (2010) Use of live imaging analysis for evaluation of cytotoxic chemicals that induce apoptotic cell death. Toxicol In Vitro 24:2012–2020
49. Yadavilli S, Martinez-Ceballos E, Snowden-Aikens J, Hurst A, Joseph T, Albrecht T, Muganda PM (2007) Diepoxybutane activates the mitochondrial apoptotic pathway and mediates apoptosis in human lymphoblasts through oxidative stress. Toxicol In Vitro 21:1429–1441
50. Luo KQ, Yu VC, Pu Y, Chang DC (2001) Application of the fluorescence resonance energy transfer method for studying the dynamics of caspase-3 activation during UV-induced apoptosis in living HeLa cells. Biochem Biophys Res Commun 283: 1054–1060
51. Rehm M, Dussmann H, Janicke RU, Tavare JM, Kogel D, Prehn JH (2002) Single-cell fluorescence resonance energy transfer analysis demonstrates that caspase activation during apoptosis is a rapid process. Role of caspase-3. J Biol Chem 277:24506–24514
52. Tyas L, Brophy VA, Pope A, Rivett AJ, Tavare JM (2000) Rapid caspase-3 activation during apoptosis revealed using fluorescence-resonance energy transfer. EMBO Rep 1: 266–270

Chapter 7

FRET-Based Measurement of Apoptotic Caspase Activities by High-Throughput Screening Flow Cytometry

Christian T. Hellwig, Agnieszka H. Ludwig-Galezowska, and Markus Rehm

Abstract

Unwanted and excessive apoptosis contributes to various degenerative diseases, and apoptosis-inducing drugs are a mainstay of anticancer treatment regimens. The fields of pharmacology and toxicology consequently have a long history of investigating apoptotic cell death in the context of drug safety and efficacy studies. Canonical apoptotic cell death is crucially dependent on type II cysteinyl aspartate-specific proteases (caspases), and their activation is therefore widely used as a marker for this cell death modality. Here we describe a flow cytometric method for noninvasive, highly sensitive and reproducible FRET-based measurements of caspase activation. Compared to other flow cytometric techniques for apoptosis detection, this approach requires only minimal sample handling steps and provides a highly cost efficient option for large scale drug interaction studies and screens of compound libraries.

Key words Apoptosis, Cancer, Caspases, Flow cytometry, Förster resonance energy transfer (FRET), High-throughput screening (HTS)

1 Introduction

The major drivers of apoptosis execution are effector caspases, in particular caspases-3 and -7 [1, 2]. Both are expressed as inactive zymogens, proteolytically activated by upstream initiator caspases, and preferably cleave after exposed tetrapeptide aspartic acid-glutamic acid-valine-aspartic acid (DEVD) sequences [3]. Their high catalytic rates allow for the efficient cleavage of hundreds of cellular proteins; these include numerous key regulators of cell survival, repair, and proliferation, as well as crucial components of cellular structures, such as nuclear envelope proteins and cytoskeletal proteins [4]. Measurements of effector caspase activation and the resulting consequences have therefore served as markers for the induction of apoptotic cell death for many years. Dye-based fluorescent staining techniques are widely used to measure apoptosis by flow cytometry. These will be introduced briefly, before

Perpetua M. Muganda (ed.), *Apoptosis Methods in Toxicology*, Methods in Pharmacology and Toxicology,
DOI 10.1007/978-1-4939-3588-8_7, © Springer Science+Business Media New York 2016

describing Förster resonance energy transfer (FRET)-based measurements of caspase activity. This will allow us to illustrate important conceptual differences, advantages and disadvantages of FRET-based caspase activity measurements when compared to more conventional assays.

Since effector caspases are activated by proteolysis, the detection of processed effector caspase subunits by target-specific antibodies, either directly labeled with fluorophores or through indirect detection by fluorophore-labeled secondary antibodies, is widely used. This approach requires a number of laborious and time consuming sample processing steps prior to sample reading, including cell fixation and permeabilization, as well as a number of antibody incubation and wash steps. A considerable drawback of this approach is that the strength of the fluorescence signal measures caspase processing rather than the resulting caspase activity. The latter crucially depends on the amounts of various members of the inhibitor of apoptosis protein (IAP) family, which directly bind and inhibit effector caspases [5]. This means that even if a considerable amount of effector procaspases was converted to the processed active form, only a fraction of this may be active. Secondly, active caspases are short-lived proteins, so that the amount of processed caspases at any one time may not reflect the overall amount of processed caspases or their activity. This problem can be circumvented by measuring caspase activity by the cleavage of hallmark effector caspase substrates. Most widely used is the staining of caspase-cleaved poly ADP ribose polymerase (PARP) [6]. Nevertheless, sample processing remains complex, and the measured signals are influenced by cell-to-cell differences in PARP abundance.

Another well-established approach to measure apoptosis by flow cytometry is the quantification of phosphatidylserine (PS) exposure on the cell surface. Due to the asymmetry in the phospholipid composition of the inner and outer layer of the plasma membrane bilayer, PS normally is not found in the outer lipid layer. However, once effector caspases are activated and cleave the flippase ATP11C, PS is exposed on the cell surface and can be stained by fluorophore-labeled Annexin-V [7, 8]. Maintenance of membrane asymmetry is an energy-dependent process, and changes in PS exposure may also be observed in other, non-apoptotic stress scenarios [9–11]. Annexin V-based staining of PS requires expensive reagents, as well as time to conduct the staining steps, rendering this approach less attractive for large scale studies. However, protocols to generate recombinant Annexin V for apoptosis detection are available, and can contribute to reducing costs [12]. Importantly, also necrotic cells can stain positive due to fluorophore-labeled Annexin-V penetrating the ruptured plasma membrane and staining PS on the intracellular leaflet of the membrane bilayer.

As an alternative for measuring caspase activation, the cleavage of recombinantly expressed caspase substrates which exhibit Förster Resonance Energy Transfer (FRET) can be considered. FRET is a radiation-less process by which a short wavelength-absorbing donor fluorophore transmits its excited state energy to an acceptor fluorophore with a longer wavelength excitation spectrum [13, 14]. Recombinant caspase FRET substrates typically comprise variants of green fluorescent protein (GFP) as donors and acceptors. The efficiency of FRET strongly depends on the overlap of donor emission and acceptor excitation spectra, their extinction coefficients and quantum yields, spatial orientation of donor and acceptor and also strongly on donor-acceptor distance [13]. Half maximal energy transfer efficiency (Förster radius) between GFP variants are typically measured in the range for 5–10 nm, with transfer efficiency dropping with the inverse sixth power of the distance. The most efficient GFP-based FRET pairs are composed of cyan and yellow fluorescent protein (CFP and YFP, respectively) and their variants, which have been optimized for brightness, quantum yield, pH insensitivity and maturation times [15]. For the measurement of caspase activities, these fluorophores are linked by short amino acid sequences which contain optimal substrate recognition motifs for caspases. Cleavage of these linker sequences and dissociation of the fluorophores disrupts energy transfer to the acceptor, and thereby results in a higher quantum yield of donor fluorescence emission (Fig. 1).

Using FRET substrates to detect caspase activation has a number of advantages. Since the probes are expressed as intracellular fluorescent proteins, measurements can be conducted without further cell fixation, permeabilization or staining, resulting in significant time and cost savings. In contrast to technically very challenging FRET studies on protein-protein interactions, the use of caspase substrates in which donor and acceptor fluorophores are present in equimolar amounts simplifies the analysis considerably and allows generating ratiometric readouts which correct for cell-to-cell heterogeneities in probe expression amounts and cell volumes (see protocols below). FRET-based caspase activity measurements therefore provide very high signal to noise levels, allowing for the detection of small, submaximal amounts of substrate cleavage. FRET substrates with recognition motifs optimized for effector caspases-3, -7 (DEVD), initiator caspases-8/-10 (IETD), and initiator caspase-2 (VDVAD) have been described by us and others before [16–22]. However, it needs to be noted that substrate specificities in the caspase family strongly overlap [23] and that, unless well characterized reporter cell lines are used in which initiator caspase activities can be uncoupled from effector caspase activation [18, 21, 24–27], readouts for any of the currently available FRET probes are strongly dominated by the activity of effector caspases. An overview of probes readily available,

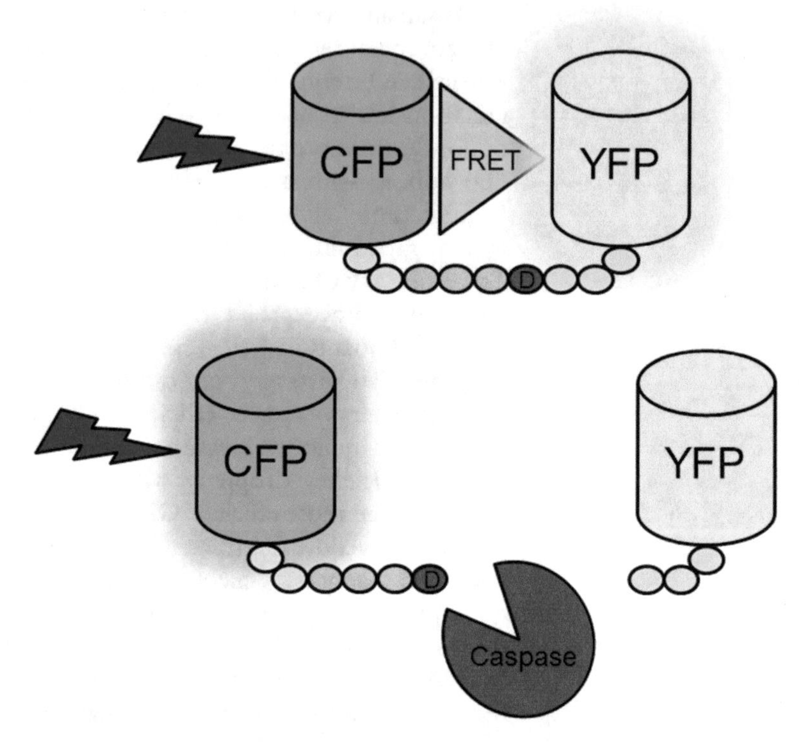

Fig. 1 Förster Resonance Energy Transfer (FRET) used in a marker protein to detect caspase activity. A recombinantly expressed FRET pair (here: CFP and YFP) is linked by a short amino acid sequence containing a motif recognized and cleaved by a caspase (e.g., DEVD for caspase-3, see Table 1). FRET occurs when the donor CFP is excited by light within the excitation spectrum (e.g., 405 nm, *violet laser*) and its emission overlaps with the excitation spectrum of the acceptor YFP. YFP then emits light with a longer wavelength than CFP (*upper sketch*). Upon cleavage after the aspartic acid (D) in the linker by a caspase and dissociation of the two fluorophores, FRET is disrupted and excited CFP then emits light with a lower wavelength. See Table 2 for the filter setup to detect the two emission spectra

their recognition sites, sources from which these probes are available, and notes on precautions to be taken in the context of readout specificity are listed in Table 1. For the remainder of this protocol, we will mostly restrict ourselves to examples based on DEVD FRET probes, but probes with other cleavage sequences can be used accordingly.

2 Materials

For the experiments described in the Sect. 3, the following materials are required:

2.1 Cell Cultivation

To establish the method, we recommend using the human cervical carcinoma cell line HeLa, since it is a widely used model, fast

Table 1
FRET probes optimized for caspase activity detection

Cleavage motif	Optimized for	Comment	References
DEVG	Non-cleavable	Negative control	[16, 17]
DEVD	Caspase-3,-7	Mild contribution of Casp-8/-10 to cleavage	[16, 21, 22], 17]
DEVDR	Caspase-3,-7	Contribution by Casp-8/-10 minimized	[21]
IETD	Caspase-8,-10	Efficiently cleaved by Casp-3/-7; Apoptosis execution phase needs to be blocked	[18]
IETD-IETD	Caspase-8,-10	Cleavage rate increased. Efficiently cleaved by Casp-3/-7; Apoptosis execution phase needs to be blocked	[21]
VDVAD	Caspase-2	Efficiently cleaved by Casp-3/-7; Apoptosis execution phase needs to be blocked	[20]

growing, and easily transfectable. HeLa cells can be cultured in RPMI1640 medium (Sigma), supplemented with 10 % heat-inactivated fetal bovine serum (Sigma), 4 mM L-glutamine, penicillin (100 U/ml) and streptomycin (100 μg/ml). Cells are then grown in a humidified 37 °C incubator with 5 % CO_2 atmosphere. Cell harvest before measurement requires an 8- or 12-channel pipette and a multi-pipette, as well as a Trypsin/EDTA solution (Sigma) to detach adherent cells. The FRET method described below is also suitable for other cell lines, as well as most primary cells.

2.2 Cell Transfection

Cells need to be transfected transiently or stably with vectors coding for CFP-YFP FRET-probes of choice (see Table 1 for a selection of probes, including a negative control). In the following, we mostly refer to examples using DEVD motif-containing FRET probes. HeLa cells can be efficiently transfected with plasmids using cationic lipid transfection agents, such as Lipofectamine® 2000 (Invitrogen by ThermoFisher Scientific) in Opti-MEM® I reduced-serum medium (Gibco by Life Technologies).

2.3 Drugs and Inhibitors

Any drug or compound that induces apoptosis can be used as stimulus. In our examples, we predominantly refer to treatments with the chemotherapeutic cisplatin (Sigma). Co-treatments with pan-caspase inhibitors, such as z-Val-Ala-Asp-fluoromethylketone (zVAD-fmk) (Alexis) can serve as controls to demonstrate that FRET probe cleavage is caspase-dependent.

2.4 Flow Cytometer and Configuration

We routinely use a BD™ LSR II flow cytometer (Becton, Dickinson and Company (BD) Biosciences), equipped with the BD High

Table 2
Laser and filter setup for combined CFP-YFP FRET and PI measurements

Fluorophore	Laser (nm)	Long pass filter	Bandpass filter (nm)
CFP	405		450/50
FRET (CFP→YFP)	405	505 nm LP	585/42
YFP	488	505 nm LP	525/50
PI	561	750 nm LP	605/40

Throughput Sampler (HTS) module, and controlled by the BD FACSDiva™ software. In a common configuration, the instrument features three continuous wave (CW) solid-state lasers (Coherent® Radius™ 405 with a wavelength of 405 nm (violet), Coherent® Sapphire™ with a wavelength of 488 nm (blue), and LSR 561-50 RTR LSR II with a wavelength of 561 nm (yellow)). The blue laser (Coherent® Sapphire™) is commonly used to measure forward scatter (cell size) and side scatter (cell granularity) signals.

For CFP and FRET measurements, CFP is excited by the violet laser with a wavelength of 405 nm. CFP emission is detected after transmission through a 450 ± 25 nm band pass filter, whereas FRET (YFP) emission is detected after transmission through a 505 nm long pass filter, followed by a 585 ± 21 nm band pass filter. YFP is excited by the blue laser with a wavelength of 488 nm, and the YFP emission is detected after transmission through a 505 nm long pass filter and a 525 ± 25 nm band pass filter (Table 2).

2.5 Software for Data Analysis

The resulting flow cytometry data can be analyzed with free flow cytometry analysis software, such as Cyflogic (Perttu Terho & CyFlo Ltd, http://www.cyflogic.com). Alternatively, any other commercial software can be used that features scatter plots, ratiometry, and histogram visualizations of FCS data files.

3 Methods

In this section, we describe the necessary steps for carrying out FRET-based flow cytometric measurements to analyze caspase activities in single cells. For easy navigation, we divided the methods into subsections which deal with the cell culture and transfection protocol to generate FRET probe-containing cells (Sect. 3.1), the experimental setup and cell treatment to induce caspase activation (Sect. 3.2), cell preparation and FRET-based flow cytometry measurements (Sect. 3.3), and software-based data analysis (Sect. 3.4).

3.1 Cell Transfection

This section briefly describes how cultured cell lines can be transfected to generate cells that express fluorescent probes for FRET-based flow cytometry measurements of caspase activities. The below protocol is optimized for HeLa cervical cancer cells, and may need to be adjusted for other cell lines or primary cells. All volumes and amounts refer to requirements for the transfection of cells in the well of a typical 6-well plate (approx. 9.5 cm^2). Amounts can be scaled up or down as needed.

1. Seed 2×10^5 cells into a well of a 6-well-plate and cultivate overnight in 2 ml growth medium to let the cells attach.
2. On the next day, dilute 4 μl of Lipofectamine transfection reagent in 250 μl Opti-MEM; incubate this solution at room temperature for 5 min.
3. Dilute 1 μg plasmid DNA in 250 μl Opti-MEM, then slowly pipette the DNA-containing solution onto the transfection reagent-containing solution and incubate this mix (4:1 volume to DNA mass ratio [μl/μg]) for 30 min at room temperature.
4. Aspirate the growth medium off the cells and wash the cells with phosphate-buffered saline. Then add 2 ml Opti-MEM to the cells.
5. Slowly drip the 500 μl DNA/transfection agent mix onto the cells. Incubate the cells for 4–6 h at 37 °C and 5 % CO_2 in the incubator.
6. After the incubation, aspirate the supernatant off the cells and add 2 ml RPMI growth medium. Do not wash the cells with PBS prior to addition of RPMI.

Typically, the cells can be used for experiments 24–48 h after transfection, depending on transfection efficiency and expression level of the FRET probe. The success rate of the transfection can be determined after 24 h by assessing the rate of fluorescent versus nonfluorescent cells in one or more fields of view using a fluorescence microscope suitable for YFP detection. For the generation of a cell line stably expressing the FRET probe, single colonies of fluorescent cells can be selected manually following antibiotic resistance selection (antibiotic resistances are coded by the plasmid vectors) or by fluorescence-assisted cell sorting (FACS).

3.2 Experimental Design for High-Throughput Screening of Caspase Activation in 96-Well Format

A high-throughput screening approach is typically chosen to address research questions that require comparisons of numerous treatments or treatment combinations. Possible applications are for example dose-response measurements of effector caspase activation as well as complex dose combinations of two or more stimuli which may act antagonistically or agonistically (e.g., additivity, synergy, or potentiation). In this section, we describe the protocol for a typical HTS experiment to measure caspase-3/-7 activities during apoptosis in Hela cells.

The experimental design for such an experiment should include triplicates for each condition. The conditions must include negative controls (untreated and vehicle-treated cells), a positive control for maximal caspase activation (if such a stimulus or condition is known), and the treatment conditions to be investigated. For the treatment for which the strongest responses are expected, a co-treatment condition in presence of caspase inhibitor zVAD-fmk can be included to verify that the treatment-induced FRET disruption is caspase-dependent. With high-throughput sampling of an entire 96-well plate, cells treated with as many as 30 different drug combinations and concentrations can be measured, with the remaining six wells left for the controls.

The following is a typical workflow for combination treatments with two drugs (drug A and drug B) in the 96-well plate format. This design can be scaled down or up (across multiple plates) as needed:

1. Seed 5000 FRET probe-expressing Hela cells in 200 μl growth medium per well into a 96-well plate. To fill an entire plate, suspend 500,000 cells in 20 ml medium and transfer 200 μl of this suspension into each well with either a multichannel or a multi-pipette.
2. Place the plate in the incubator and allow the cells to attach overnight. On the day after seeding, check if the cells are attached, healthy, and fluorescent.
3. In the meantime, prepare a treatment scheme for the 96-well plate which systematically arranges the different conditions across the plate. Figure 2 provides a template for an experiment with two drugs in 5 × 5 concentration combinations that can be used as a map. It is advisable to prepare a general template using Microsoft EXCEL or similar programs to allow reuse and easy amendments. Choose the desired drug concentrations. For a 5 × 5 matrix with two drugs, choose five steps for factorial dilutions from the highest concentration to the lowest concentration/negative control. For example, for a dilution factor of 10 and a highest concentration of 1 μM, these steps will be 1000, 100, 10, 1, and 0 nM (growth medium only, negative control).
4. Prepare one master solution per drug in growth medium, and from this prepare drug solutions required for the dilution series. Each master solution must have a final drug concentration of twice the highest treatment concentration (e.g., 2 μM in our example; mixing of the drug solutions in the plates will dilute the drugs to the desired final concentrations (see later steps)).
   (a) Calculate the volumes needed for each master drug solution according to the following guidelines. In our 5 × 5

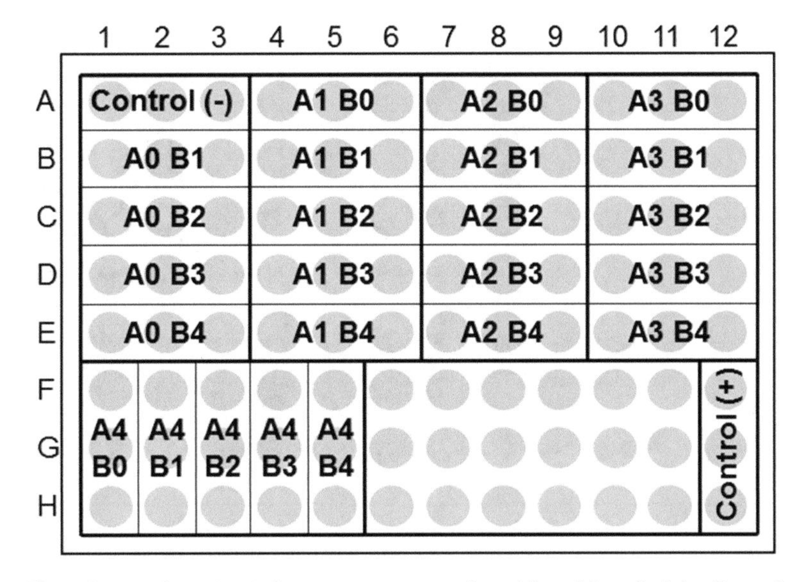

Fig. 2 Layout for a 5 × 5 dose-response experiment in a 96-well plate. Drugs A and B are added in five different concentrations, A0 and B0 have the concentration 0. Control (+) (wells F12, G12, H12) can be a treatment with a potent apoptosis inducer

example, the highest concentration for each drug will be required for five conditions each, and each of these is represented in triplicate (see template in Fig. 2) (5 × 3 = 15 wells). 50 μl of the master solution will be required for each of the 15 wells (50 μl × 15 = 750 μl). Adding an additional 50 μl will compensate for potential volume losses, resulting in 800 μl master solution for drug A and drug B, respectively. Next, consider the additional volume required for the dilution series in the 5 × 5 matrix. For our example of a 10× dilution series, an additional 100 μl should be prepared (total final master solutions per drug = 900 μl). 90 μl of this added volume can then be diluted 1:10 in 810 μl growth medium to obtain 900 μl of drug solution for achieving the next lower concentration. The further dilution steps are conducted accordingly. We have provided an overview of this in Fig. 3.

(b) The serial dilutions described above in our 5 × 5 example will result in 4 × 900 μl volumes of drugs A and B at concentrations of 2000, 200, 20 and 2 nM.

(c) Prepare a master solution of a known potent apoptosis inducer in growth medium which can be used as a positive control in spare wells. For example, for HeLa cells prepare a 350 μl master solution of 1 μM staurosporine or 40 μM cisplatin of which 100 μl can be added per well.

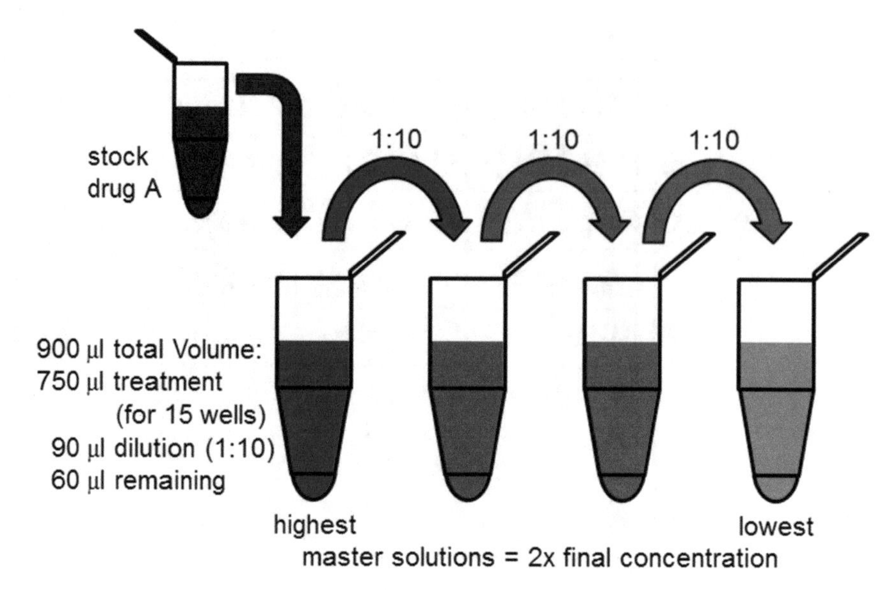

Fig. 3 Example for the preparation of master solutions for treatments with multiple drug concentrations

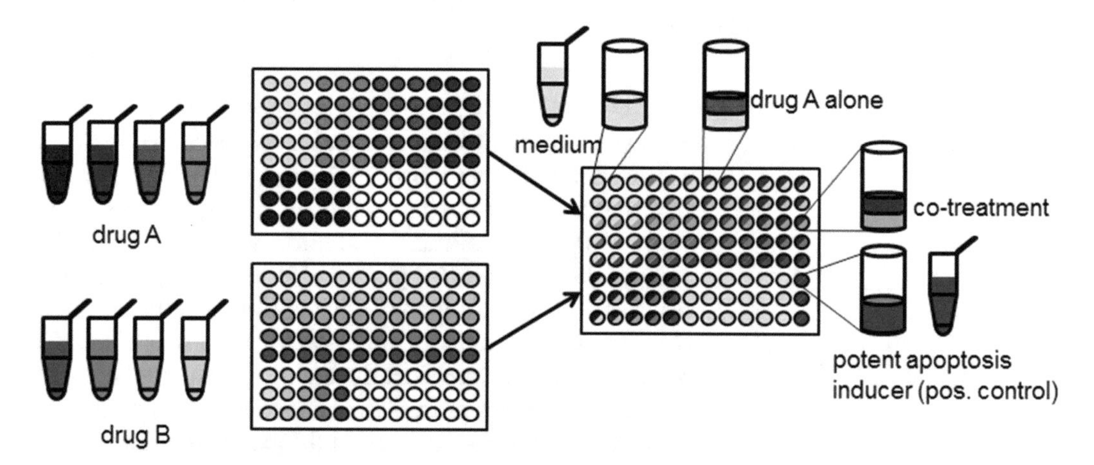

Fig. 4 Pipetting scheme for a 5 × 5 dose-response experiment in a 96-well plate, based on the layout shown in Fig. 2 and using the master solutions described in Fig. 3. Pre-pipetting into two separate 96-well plates or pre-mixing of both drugs in another plate is optional; the drugs can be combined in the plate with the cells directly

5. Remove the growth medium from the cells. Then add 50 µl of the master drug solutions or the respective drug dilutions to each well according to the design of the treatment matrix (see Figs. 2 and 4). Add 50 µl of growth medium to wells that receive only drug A or drug B. Untreated control wells receive 100 µl fresh growth medium. This will result in a final volume of 100 µl in all wells.
6. Add 100 µl of the potent apoptosis inducer solution into positive control wells after removing the growth medium.

7. Incubate the 96-well plate in the incubator for the chosen treatment duration.

In the 5×5 example in Fig. 4 drug A is arranged in blocks of the same concentration, whereas drug B is added in lines of the same concentration to facilitate pipetting. 18 wells remain vacant (F6–11, G6–11, H6–11) and can be used for additional controls, e.g., treatments with the pan-caspase inhibitor zVAD-fmk to test for caspase-dependent FRET-probe cleavage. Optionally, for the generation of the treatment combination matrix, drugs A and B can also be diluted and premixed in additional 96-well plates, so that the final treatment volumes can be conveniently and rapidly transferred with a multichannel pipette.

3.3 Flow Cytometer Adjustments

The BD LSR II flow cytometer needs to be equipped with a high-throughput sampling system suitable for 96-well plate format. A software-controlled arm automatically takes samples from each of the 96-wells in user-defined order, speed and volume. The software FACSDiva which controls the HTS module and the flow cytometer, stores the results in exportable FCS files. Before harvesting and measuring the cells, the sampling scheme needs to be set up on the instrument:

1. Check the supply of sheath solution and switch on the sheath tank-filling pump. The liquid waste container should be emptied prior to measurement.
2. Switch on the flow cytometer to warm up the lasers for at least 30 min.
3. Attach the HTS module and switch it on.
4. Start the FACSDiva software and wait until it connects with both instruments.
5. In the FACSDiva software, add a new folder in the Browser window and name it after the experimenter or project. All related future experiments can be placed in this folder.
6. In the opened folder, add a new experiment by clicking on the "New Experiment" icon in the Browser toolbar. The experiment can contain a number of individual experimental layouts (96-well plates and single tubes) and should be labeled to describe the type of experiments that will be collected here, e.g., "Caspase FRET measurements."
7. Double click on the experiment to open it. Click on the "New Plate" icon in the Browser toolbar to add a new 96-well plate (U-bottom) to the experiment. Choose a name for the plate that describes the experiment, including the date of the experiment, the cell line, the drugs and the incubation time, e.g., "20140511 Hela 5×5 drugA drugB 24 h" for a 5×5 dose response experiment on the 11th of May 2014 with Hela cells challenged for 24 h with drugs A and B.

8. Create the treatment layout of the experiment in the plate window by selecting each triplicate of wells with the same condition separately and clicking on the "Add Specimen Wells" icon in the plate window toolbar, i.e., grouping wells A1 to A3, wells B1 to B3, and so forth. Label the groups of wells according to the treatments layout on the plate (Fig. 2). For a general 5 × 5 experiment with two drugs A and B with the concentrations A0 (no drug A), A1, A2, A3, A4, B0 (no drug B), B1, B2, B3, and B4, the names of the triplicates are combinations of the two drugs, e.g., the group of the wells B1 to B3 is named "A0 B1" (= treatment with concentration 1 of drug B but not drug A), and the group of the wells C4 to C6 is labeled "A1 B2" (= treatment with concentration 1 of drug A combined with concentration 2 of drug B), whereas the group name of the wells A1 to A3 becomes "negative control" (= no treatment with either drug) and the wells F12, G12, and H12 become the specimen "positive control." Precise labels will facilitate the later analysis. The FACSDiva software automatically saves these labels and the plate layout and will save the data of each well under these names during measurement.
9. Select channels and their settings for subsequent measurements in the Cytometer FACSControl window on the Parameters sheet. Select the FSC and SSC channels (selected by default, excited by the 488 nm laser), the channels for CFP and FRET (405 nm laser excitation), as well as the YFP channel (488 nm laser). Unselect all other possible channels listed. Furthermore, select the CFP/FRET ratio as a data output in the Ratio sheet of the same window.
10. Create a scatter plot in the Global Worksheet window, with FSC values on the x-axis and SSC values on the y-axis. The cell population should later form a cloud in the center, whereas dead cells will shift towards the left side (smaller size, lower FSC values). Draw a polygonal region ("P1") in the x=FSC/y=SSC scatter plot (Fig. 6a), which excludes events that are too small and lack granularity (considered to be debris) and events that are too big (multiple cells clustered together). This polygonal region will have to be adjusted to the cells at the start of the measurements or by using a separate batch of cells. Define all events within P1 to be cells so that only these are counted towards the targeted number of measured events.
11. In the Acquisition Dashboard window, set the stopping gate to "events in P1" and the events to record to 10,000 events. Once 10,000 events are counted or when the entire sample volume has passed through the chamber, the measurement will be stopped. A stopping time does not need to be defined and is calculated from the sample volume and the sample flow rate.

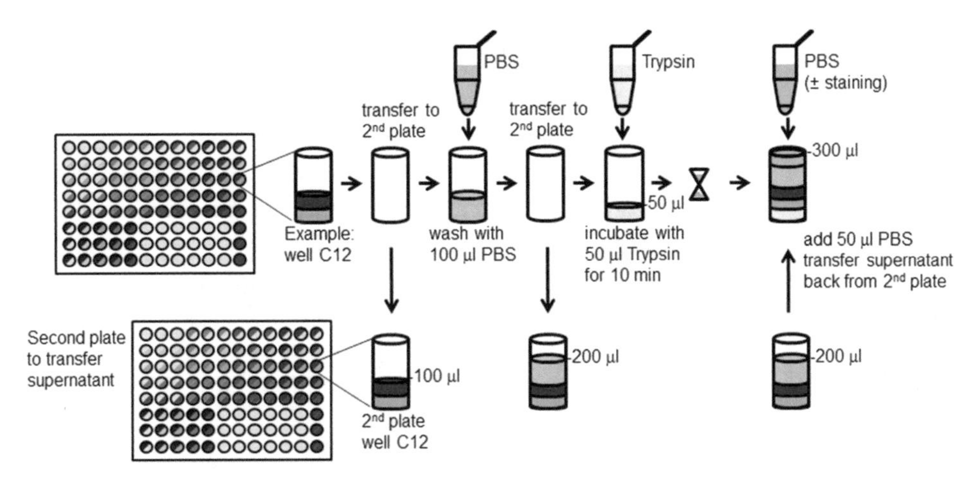

Fig. 5 Schematic describing cell harvest for HTS flow cytometry measurement

12. In the plate window, set the sampling volume to 100 μl, which will be a third of the total well volume (300 μl) at the time of measurement. Select the 'standard' mode. The sample flow rate can be set to 3 μl/s so that a sample of 100 μl is measured within 33 s.
13. Other settings in the Plate window like mixing volume, mixing speed, and number of mixes, as well as the wash volume, must be defined. For Hela cells, two mixes of 50 μl at a speed of 100 μl/s before acquiring a 100 μl sample are favored, as it guarantees a good mixing of the detached cells and does not cause shearing to kill them.
14. In the Global Worksheet window, create histograms for CFP, YFP, FRET, and the CFP/FRET ratio. The measurements will be displayed here in real-time.

3.4 Cell Harvesting and Preparation for HTS FRET Measurements

At the end of the treatment, the cells need to be detached and analyzed by FRET flow cytometry (Fig. 5).

1. Transfer the supernatant of each well with a multichannel pipette into a new 96-well plate with the same layout. These supernatants may contain detached and floating cells which will be added back to the original plate later on.
2. Wash each well in the original plate with 100 μl phosphate-buffered saline using a multichannel pipette. Transfer these wash volumes into the corresponding wells of the 96-well plate which already contained the supernatants (see previous step).
3. Add 50 μl Trypsin/EDTA solution into each well of the original 96-well plate. Gently shake the plate. Cells should detach after 10 min at room temperature. Short incubation at 37 °C can accelerate this process.

4. Add 50 μl phosphate-buffered saline solution into each well of the original plate. Gently pipette up and down to detach and separate the cells. If needed, additional dyes to stain the cells can be added with the phosphate-buffered saline solution. For example, we commonly add propidium iodide at a final concentration of 1.33 μM to measure cell death.
5. Transfer the volumes collected in the second 96-well plate back into the wells of the original plate. The final volume per well will now be 300 μl.

3.5 FRET-Based HTS Measurement of Caspase Activation

1. Place the plate with the detached cells into the HTS module of the flow cytometer.
2. If needed, spare wells with cells can be used for final adjustments of the photomultiplier tube (PMT) voltages for FSC and SSC (use known values from cell lines previously measured). Further amend the polygonal P1 in the x = FSC/y = SSC scatter plot (Fig. 6a) to exclude the debris (events in the lower left corner, low values of both FSC and SSC), if needed.
3. Adjust the voltage for the detection of YFP-, CFP-, and FRET emission by measuring non-apoptotic FRET-probe expressing cells from spare wells (negative control peak) and non-transfected controls (if seeded in additional wells, optional). The values are increased or decreased in the Cytometer FACSControl window on the Parameter sheet to place the population of events in the YFP-, CFP-, and FRET-histograms within the logarithmic detector range (the CFP peak to the left to account for an increase, the FRET peak to the right to be able to detect a decrease). These PMT voltage values depend on the fluorescent protein expression levels (CFP-YFP FRET-probe). Measuring cells that do not express the fluorescent FRET-probe can be used to identify the intensities of non-transfected cells (autofluorescence and background noise), and serves to distinguish transfected from non-transfected cells.
4. After the settings of the flow cytometer have been adjusted, the plate is automatically measured using the command "Run Plate" or "Run Well(s)", if just a selection of wells has been chosen.
5. The data of the experiment is automatically saved by the FACSDiva software during measurement. After the run is completed, the data can be exported and saved as FCS files into one folder.

3.6 Data Analysis

Free software are available for the analysis of flow cytometry data (e.g., Cyflogic (Perttu Terho & CyFlo Ltd, http://www.cyflogic.com)). The following steps describe the data analysis routine in the software environment of Cyflogic v. 1.2.1.

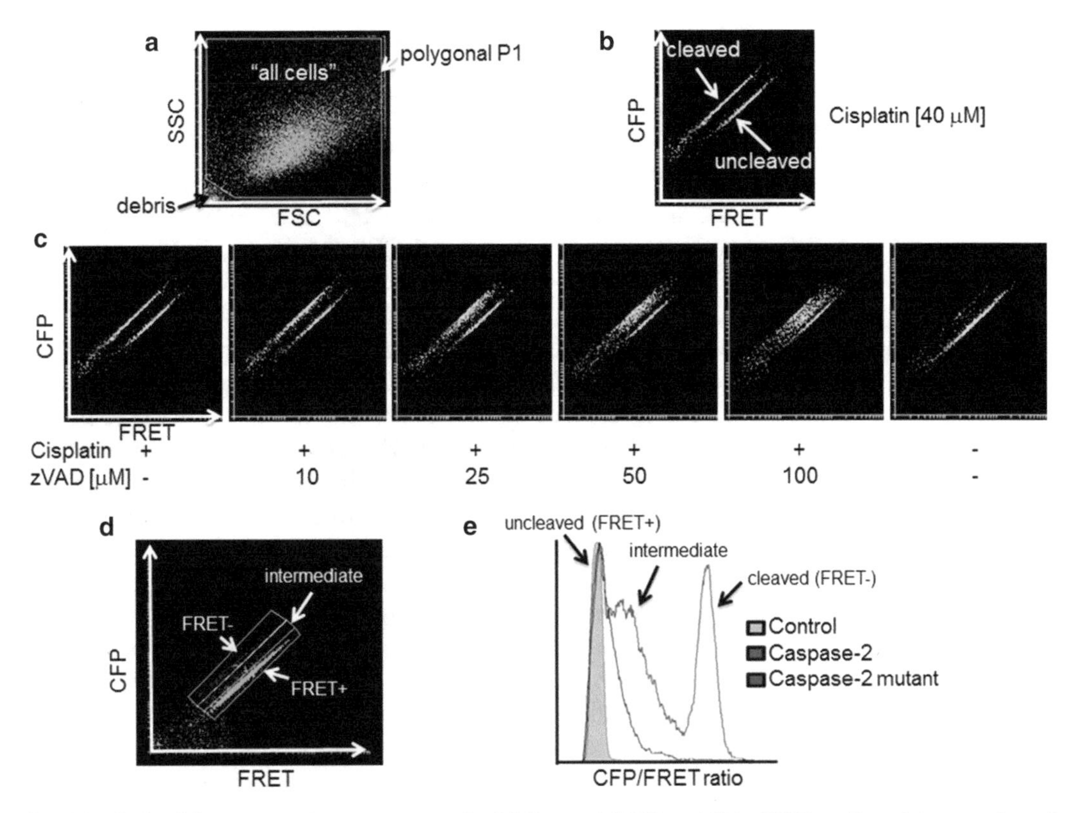

Fig. 6 Analysis of flow cytometry measurements. (**a**) Forward (FSC) and Side (SSC) scatter plot presenting all measured events. The polygonal P1 (during analysis renamed into "all cells") defines which events are cells and excludes the cell debris. (**b**) FRET and CFP scatter plot of FRET probe-expressing cells forming two lines after treatment with the apoptosis inducer cisplatin. Normal cells retain an intact FRET probe (uncleaved), whereas in apoptotic cells FRET is disrupted (cleaved). (**c**) Analysis of an experiment with cisplatin and increasing concentrations of a pan-caspase inhibitor (zVAD-fmk). The higher the inhibitor concentration, the more cells shift from "cleaved" to "uncleaved", forming a group with intermediate FRET probe cleavage. (**d**) Analysis of the three subpopulations in a FRET and CFP scatter plot by drawing regions around the two clusters of cells (FRET+ and FRET−) and the area between (intermediate). (**e**) CFP/FRET ratio histogram to analyze the three subpopulations. The cells express a VDVAD FRET probe to measure caspase-2 activity and are either control MCF-7 cells, caspase-2 overexpressing cells, or cells expressing a caspase-2 mutant. Reproduced from Delgado 2013 [20] with kind permission from Elsevier

1. Create an FSC and SSC scatter plot by clicking on "Dot plot" in the menu tab "Create". Open an FCS file from the experiment by clicking on "Open FCS…" in the menu tab "File". Choose a negative control FCS file.
2. Gate for events that have the correct size to be defined as cells. Move the mouse to the right side until the region panel appears, then chose the first population and click on the create button. Now click into the scatter plot to draw the region as a polygon (Fig. 6a). This gate should match the P1 polygonal used during the measurement on the flow cytometer. Rename the polygonal

on the region panel "all cells". All events within this gate are colored in red by default, the events outside stay white.

3. Create a second scatter plot with cellular FRET intensities on the *x* axis and CFP intensities on the *y* axis (both in logarithmic scale, Fig. 6b) by creating a new dot plot and right-clicking on the axis labeling (the default is FSC and SSC) to change them to FRET for the *x* axis and CFP for the *y* axis. The two variables positively correlate and form a line of dots from lower left to upper right in cells with an intact FRET probe. If a cell presents with cleaved probe (apoptotic), a shift from low CFP intensities and high FRET intensities to high CFP- and low FRET intensities can be observed, creating a parallel line of dots which represent cells with disrupted resonance energy transfer (see Fig. 6b for two distinct populations of apoptotic and non-apoptotic cells after treatment and Fig. 6c for control (last graph) versus treatment (first graph)). Cells located between intact and cleaved probe represent subpopulations with intermediate FRET-probe cleavage (dots between both lines, see Fig. 6c for cells treated with increasing concentrations of zVAD-fmk). All three populations can be quantified by drawing regions around them.
4. Right-click on the x=FRET/y=CFP scatter plot and choose "all cells" in the menu tab "Show populations" to display only the events in the x=FSC/y=SSC scatter plot that were defined to be cells. All events still displayed are now shown in red. Optionally, to change the color, double-click on the red color square in the region panel and choose a color in the color window. For the scatter plots in Fig. 6a, we chose the color yellow for the "all cells" region. Note that the different colors in Fig. 6a are due to subpopulations defined in the next step.
5. Create three regions in the x=FRET/y=CFP scatter plot for populations of FRET-positive cells ("FRET+"), cells with intermediate cleavage of the FRET-probe ("intermediate"), and FRET-negative cells (FRET−). FRET+ is the population furthest to the lower right, FRET− the parallel line of dots furthest to the upper left, and intermediate is the region between the other two, containing all cells that are neither FRET+ nor FRET− (Fig. 6d). The regions can be renamed and recolored. In Fig. 6d we chose the colors green (FRET+), orange (intermediate), and red (FRET−). Since most cells in Fig. 6d (untreated control) are non-apoptotic and FRET+, the majority of cells in the corresponding x=FSC/y=SSC scatter plot (Fig. 6a) are green.
6. Create a histogram with YFP intensities, choose to display "all cells" only, and gate for YFP-positive cells.
7. Create a histogram with the CFP/FRET ratio and choose to display "all cells" that are YFP-positive only (use a definition).

FRET+ cells form a peak at the left side, whereas FRET− cells form a second, distinct peak further to the right. Cells with intermediate FRET probe cleavage are accumulating in between (Fig. 6e, from Delgado 2013 with kind permission of Elsevier) [20]. Create regions in the histogram for FRET+, intermediate, and FRET− cell populations.

8. Right-click on the x = FRET/y = CFP scatter plot as well as the CFP/FRET ratio histogram and choose open statistics. Right-click on the statistics window and choose "Number" (total number of cells visible in the plot/region) and "% of vis" (percentage of cells in the region in relation to all cells visible in the plot) to be displayed.
9. Analyze the experiment. Click on "Export Statistics to File" in the Tools menu and name the file "[experiment name] data". The software automatically reads out all "Number" and "% of vis" data from all FCS files in the folder of the opened FCS file.
10. Import the data file into EXCEL or similar programs to appear as a list. The data of each well is labeled with the well position on the plate (A to H and 1–12), as well as the name given to the respective groups previously (see Sect. 3.3, step 8 above).
11. Calculate the mean and standard deviation for each triplicate. Then create a bar graph displaying the mean percentages ± s.d. of FRET-negative cells for the different treatment conditions (Fig. 7a). A 5 × 5 dose-response experiment can also be presented as a 3D column diagram, where the x axis shows increasing amounts of drug A, the y axis shows increasing amounts of drug B, and the z axis the percentage of apoptotic cells (% of FRET− cells) (Fig. 7b). Similar display items can be generated for FRET+ cells and cells with intermediate probe cleavage.
12. Finally, the characteristics of the drug interactions can be studied by synergy analysis. This is commonly performed by calculating the combination index (CI) using Webb's fractional product method [28]. The following formula applies:

$$\mathbf{CI} = \frac{\%\ \text{FRET+ co-treatment}}{(\%\ \text{FRET+ drug A}) \times (\%\ \text{FRET+ drug B}) / 100}$$

The drug combination effect is additive when CI = 1.0, synergistic when CI < 1.0 and antagonistic when CI > 1.0. For example: After a treatment with 1 nM drug A, 81 % of cells present with intact FRET-probe (FRET+). Treatment with 10 nM drug B leaves 89 % of cells in the group of FRET+ cells. However, when combined, treatment with 1 nM drug A and 10 nM drug B reduces FRET+ cells to 43 %. CI = 43/(81 × 89/100) = 0.596; 0.596 < 1.0, therefore a synergistic effect is observed for these concentrations. Figure 7c presents the synergy analysis of the data presented in Fig. 7b.

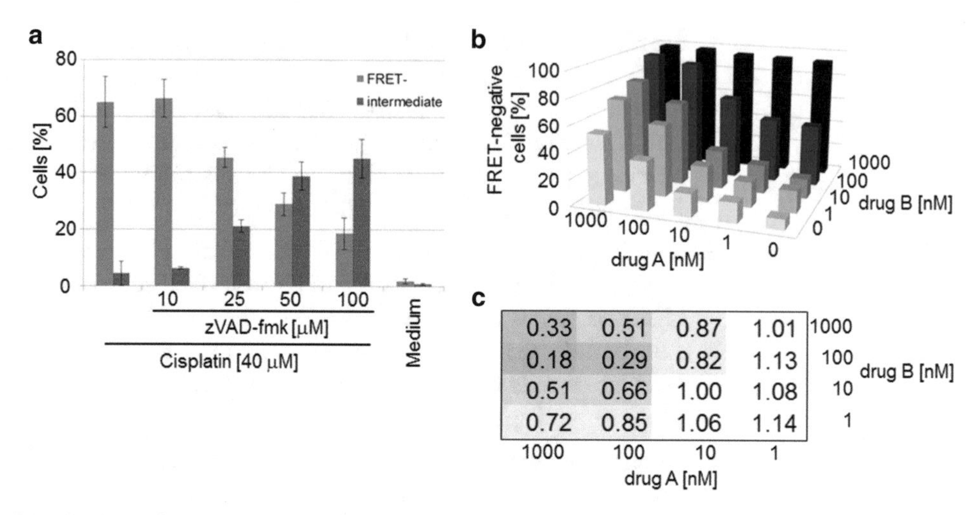

0.33	0.51	0.87	1.01	1000
0.18	0.29	0.82	1.13	100
0.51	0.66	1.00	1.08	10
0.72	0.85	1.06	1.14	1
1000	100	10	1	

Fig. 7 (**a**) Analysis of the experimental data shown in Fig. 6c, data are means of triplicates, error bars are s.d. (**b**) Example 3D bar graph of an entire 5 × 5 dose-response experiment with drug A and drug B. (**c**) Analysis of synergy by Webb's fraction product method. Values below 1.0 are defined as synergistic

4 Notes

In this section, we extend the subchapters of the methods chapter by describing alternatives, discussing possible obstacles, and offering troubleshooting advice.

4.1 Cell Transfection

Stable cell lines expressing the FRET probe greatly reduce the experiment duration and increase the number of observable events. Especially for high-throughput sampling and repeated experiments, stable cell lines save time and increase the reproducibility. It also reduces potential cell death contributions and sensitization caused by transfection stress. Fluorescent cells can most easily be selected by microscopy after re-seeding transfected cells in low numbers and letting them form single colonies or by fluorescence-assisted cell sorting. When picked from single colonies, clonal populations with different levels of brightness can be established, which can contribute to minimize fluorescence crosstalk into detection channels used for other dyes.

4.2 Experimental Design for High-Throughput Screening of Caspases Activation

It is desirable to have a few spare wells available as additional positive and negative controls to adjust PMT Voltage gain without wasting samples on the day.

The small volume of medium per well may become limiting for long treatment durations (>48 h), especially in wells with untreated and continuously growing control cells. This can be compensated by plating lower amounts of cells.

For combinations of more than two drugs or treatments per well, the concentrations of the drug master solutions and the dilutions generated thereof can be increased accordingly (see Sect. 3.2, steps 3–5), so that a final total volume of 100 μl during treatment is maintained.

FRET-based flow cytometry can also be performed with sampling tubes instead of the HTS module, i.e., if higher cell numbers are needed. On a 6-well plate, 2×10^5 FRET probe-expressing cells are seeded in 2 ml medium per well and treated on the next day. In contrast to the HTS approach, cells treated in 6-well plates have to be collected by centrifugation prior to measurement to reduce the volume of the cell suspension, which usually consists of 2 ml treatment medium, 1 ml phosphate-buffered saline solution to wash the cells after incubation, 0.5 ml trypsin solution, and an additional 1 ml phosphate-buffered saline solution to take up the cells from the well. Since the centrifugation step reduces the volume in the tubes, the samples can be transferred into a 96-well plate and measured with the HTS module, too. To rule out pipetting mistakes that could cause variations in the results, a test run with a common laboratory dye can be performed. After creating different master solutions with stepwise diluted concentrations of the dye, pipette the same volumes as in the protocol (50 or 100 μl, see Sect. 3.2, step 5) into a 96-well plate, and then measure the absorbance with a plate reader.

4.3 Flow Cytometer Adjustments

YFP measurements can be used as internal controls for intact cells that express the FRET-probe, since the intensities do not change upon probe cleavage. Lack of YFP signals mark non-transfected cells in untreated controls. YFP negative cells following treatment indicate cell membrane rupture, e.g., as a consequence of secondary necrosis at a late apoptosis stage, and can be used as a surrogate cell death marker.

If the flow cytometer is equipped with a third laser with a wavelength of 561 nm (yellow), an additional dye in the red color range can be measured in parallel without interfering with the FRET measurements and the long emission shoulder of YFP. For cells undergoing apoptosis, the red cell death stain propidium iodide (PI) can provide extra information about cells with disrupted plasma membrane, which may have lost the cytosolic FRET probe. PI can be excited with the yellow laser (561 nm) and its emission is detected after transmission of a 570 nm long pass filter and a 605 nm ± 20 nm band pass filter.

4.4 Cell Preparation and FRET Measurement

Measurement of an entire 96-well takes approximately 1 h (96 measurements of 100 μl cell suspension at a rate of 3 μl/s, plus a few seconds each well for moving the sampling arm, mixing, and taking up the sample), in which the detached cells are in the HTS module at room temperature. To rule out any effects or biases that affect FRET probe cleavage during the time of measurements, wells with positive and negative controls should be run both in the beginning and at the end of the experiment to compare their readouts. Alternatively, cooled HTS systems can be used. To test the accuracy and reproducibility of cell handling, cells can be seeded into a 96-well plate, incubated without treatment, and then harvested and measured as described in Sects. 3.4 and 3.5. Compare all wells in regard to cell amount per measurement duration (e.g., number of cells detected in 33 s, as described in Sect. 3.3, step 12) to validate that the cell concentration is similar in all wells. Also, analyze the percentage of apoptotic cells (baseline cell death).

4.5 Data Analysis

Since experiments need to be repeated and designs are frequently reused with only slight changes, it is advisable to label the wells in the FACSDiva software precisely. Separate layouts can be stored for experiments with different drugs and different cell lines. Keeping the format not only minimizes errors during the experiment preparation, but facilitates a quicker analysis. The analysis software Cyflogic allows saving the analysis view, the scatter plots and histograms including tables and gating regions. These analysis views can be reused and amended for the new experiment instead of starting all over. The output data can then be imported into an existing Microsoft EXCEL file with preprepared tables, formulas and diagrams and the further data analysis and visualization is thus automated.

Acknowledgements

This work was supported by a postdoctoral fellowship grant from the Irish Research Council (GOIPD/2013/102) awarded to Christian Hellwig, an NBIP Career Enhancement and Mobility Fellowship cofunded by Marie Curie Actions (EU FP7), the Irish HEA PRTLI cycle 4 and the Italian National Research Council, awarded to Agnieszka Ludwig-Galezowska, and by funding from the European Union (Horizon 2020 Marie S. Curie ETN MEL-PLEX) awarded to Markus Rehm.

References

1. Taylor RC, Cullen SP, Martin SJ (2008) Apoptosis: controlled demolition at the cellular level. Nat Rev Mol Cell Biol 9(3):231–241, nrm2312
2. Hellwig CT, Passante E, Rehm M (2011) The molecular machinery regulating apoptosis signal transduction and its implication in human physiology and pathophysiologies. Curr Mol Med 11(1):31–47
3. Stennicke HR, Renatus M, Meldal M, Salvesen GS (2000) Internally quenched fluorescent peptide substrates disclose the subsite preferences of human caspases 1, 3, 6, 7 and 8. Biochem J 350(Pt 2):563–568
4. Luthi AU, Martin SJ (2007) The CASBAH: a searchable database of caspase substrates. Cell Death Differ 14(4):641–650, 4402103

5. Mace PD, Shirley S, Day CL (2009) Assembling the building blocks: structure and function of inhibitor of apoptosis proteins. Cell Death Differ 17(1):46–53, cdd200945
6. Lazebnik YA, Kaufmann SH, Desnoyers S, Poirier GG, Earnshaw WC (1994) Cleavage of poly(ADP-ribose) polymerase by a proteinase with properties like ICE. Nature 371(6495):346–347. doi:10.1038/371346a0
7. Koopman G, Reutelingsperger CP, Kuijten GA, Keehnen RM, Pals ST, van Oers MH (1994) Annexin V for flow cytometric detection of phosphatidylserine expression on B cells undergoing apoptosis. Blood 84(5):1415–1420
8. Segawa K, Kurata S, Yanagihashi Y, Brummelkamp TR, Matsuda F, Nagata S (2014) Caspase-mediated cleavage of phospholipid flippase for apoptotic phosphatidylserine exposure. Science 344(6188):1164–1168. doi:10.1126/science.1252809
9. Riedl S, Rinner B, Asslaber M, Schaider H, Walzer S, Novak A, Lohner K, Zweytick D (2011) In search of a novel target—phosphatidylserine exposed by non-apoptotic tumor cells and metastases of malignancies with poor treatment efficacy. Biochim Biophys Acta 1808(11):2638–2645. doi:10.1016/j.bbamem.2011.07.026
10. Smrz D, Draberova L, Draber P (2007) Non-apoptotic phosphatidylserine externalization induced by engagement of glycosylphosphatidylinositol-anchored proteins. J Biol Chem 282(14):10487–10497. doi:10.1074/jbc.M611090200
11. Goth SR, Stephens RS (2001) Rapid, transient phosphatidylserine externalization induced in host cells by infection with Chlamydia spp. Infect Immun 69(2):1109–1119. doi:10.1128/IAI.69.2.1109-1119.2001
12. Logue SE, Elgendy M, Martin SJ (2009) Expression, purification and use of recombinant annexin V for the detection of apoptotic cells. Nat Protoc 4(9):1383–1395. doi:10.1038/nprot.2009.143
13. Zal T, Gascoigne NR (2004) Photobleaching-corrected FRET efficiency imaging of live cells. Biophys J 86(6):3923–3939. doi:10.1529/biophysj.103.022087
14. Förster T (1948) Zwischenmolekulare Energiewanderung und Fluoreszenz. Ann Phys 437 (1–2):55–75. doi:10.1002/andp.19484370105
15. Shaner NC, Steinbach PA, Tsien RY (2005) A guide to choosing fluorescent proteins. Nat Methods 2(12):905–909. doi:10.1038/nmeth819
16. Rehm M, Dussmann H, Janicke RU, Tavare JM, Kogel D, Prehn JH (2002) Single-cell fluorescence resonance energy transfer analysis demonstrates that caspase activation during apoptosis is a rapid process. Role of caspase-3. J Biol Chem 277(27):24506–24514
17. Tyas L, Brophy VA, Pope A, Rivett AJ, Tavare JM (2000) Rapid caspase-3 activation during apoptosis revealed using fluorescence-resonance energy transfer. EMBO reports 1(3):266–70. doi:10.1093/embo-reports/kvd050.
18. Hellwig CT, Kohler BF, Lehtivarjo AK, Dussmann H, Courtney MJ, Prehn JH, Rehm M (2008) Real time analysis of tumor necrosis factor-related apoptosis-inducing ligand/cycloheximide-induced caspase activities during apoptosis initiation. J Biol Chem 283(31):21676–21685. doi:10.1074/jbc.M802889200
19. O'Connor CL, Anguissola S, Huber HJ, Dussmann H, Prehn JH, Rehm M (2008) Intracellular signaling dynamics during apoptosis execution in the presence or absence of X-linked-inhibitor-of-apoptosis-protein. Biochim Biophys Acta 1783(10):1903–1913
20. Delgado ME, Olsson M, Lincoln FA, Zhivotovsky B, Rehm M (2013) Determining the contributions of caspase-2, caspase-8 and effector caspases to intracellular VDVADase activities during apoptosis initiation and execution. Biochim Biophys Acta 1833(10):2279–2292. doi:10.1016/j.bbamcr.2013.05.025
21. Albeck JG, Burke JM, Aldridge BB, Zhang M, Lauffenburger DA, Sorger PK (2008) Quantitative analysis of pathways controlling extrinsic apoptosis in single cells. Mol Cell 30(1):11–25
22. Takemoto K, Nagai T, Miyawaki A, Miura M (2003) Spatio-temporal activation of caspase revealed by indicator that is insensitive to environmental effects. J Cell Biol 160(2):235–243
23. McStay GP, Salvesen GS, Green DR (2008) Overlapping cleavage motif selectivity of caspases: implications for analysis of apoptotic pathways. Cell Death Differ 15(2):322–331
24. Hellwig CT, Ludwig-Galezowska AH, Concannon CG, Litchfield DW, Prehn JH, Rehm M (2010) Activity of protein kinase CK2 uncouples Bid cleavage from caspase-8 activation. J Cell Sci 123(Pt 9):1401–1406. doi:10.1242/jcs.061143
25. Laussmann MA, Passante E, Dussmann H, Rauen JA, Wurstle ML, Delgado ME, Devocelle M, Prehn JH, Rehm M (2011) Proteasome inhibition can induce an autophagy-dependent apical activation of caspase-8. Cell Death Differ 18(10):1584–1597, cdd201127
26. Laussmann MA, Passante E, Hellwig CT, Tomiczek B, Flanagan L, Prehn JH, Huber HJ,

Rehm M (2012) Proteasome inhibition can impair caspase-8 activation upon submaximal stimulation of apoptotic tumor necrosis factor-related apoptosis inducing ligand (TRAIL) signaling. J Biol Chem 287(18):14402–14411. doi:10.1074/jbc.M111.304378

27. Delgado ME, Dyck L, Laussmann MA, Rehm M (2014) Modulation of apoptosis sensitivity through the interplay with autophagic and proteasomal degradation pathways. Cell Death Dis 5:e1011. doi:10.1038/cddis.2013.520

28. Webb JL (1963) Effect of more than one inhibitor. In: Enzymes and metabolic inhibitors, vol 1. Academic, New York, pp 66–79

Chapter 8

Automated Ratio Imaging Using Nuclear-Targeted FRET Probe-Expressing Cells for Apoptosis Detection

Krupa Ann Mathew*, Deepa Indira*, Jeena Joseph, Prakash Rajappan Pillai, Indu Ramachandran, Shankara Narayanan Varadarajan, and Santhoshkumar Thankayyan Retnabai

Abstract

In recent years, innovative bioassays have been designed to detect intracellular caspase activation as a reliable read-out of apoptotic activity of bioactive compounds. Most anticancer drugs target cells by triggering caspase dependent protein cleavage, culminating in apoptotic cell death. Therefore, detection of caspase activation has been recognized as one of the best approaches for detecting cancer cell death as compared to assaying the ill-defined general cytotoxic activity that often manifests with off-target side effects. Among the available methods of detection, those with cells stably expressing FRET-based fluorescent probes are more suitable for studying live cells, because they yield reliable readouts due to minimal invasiveness and easy automated quantitation possibilities. We have recently reported a successful FRET-based high-throughput assay protocol for detecting caspase activation in live cells using stable cells expressing intracellular FRET probes. An important advantage of this approach with respect to other apoptosis assays is its ability to monitor caspase activation, real-time, in live cells. However, proper automated identification of individual cells (viz., segmentation) requires nuclear staining and complex image processing. Here, we discuss an advanced tool for detecting intracellular caspase activation, which surpasses the above disadvantages. The tool described here works by first generating stable cancer cell lines expressing the FRET probe inside nucleus, thereby eliminating the need for extra staining and complex processing. Such nuclear targeting enables accurate automated segmentation and quantification of caspase activation in a multipoint/multidrug manner, using automated microscopy. We then describe the step-by-step protocol for detecting caspase activation in live cells using such a nuclear-targeted FRET-based tool. In addition, we propose the adaptability of this method for high-throughput screening platforms, so as to single out potential anticancer compounds to aid further lead optimization.

Key words Apoptosis detection, Caspase activation, Cytotoxicity assay, Drug screening, Fluorescence resonance energy transfer (FRET), High-throughput screening, Ratio imaging

*These authors contributed equally to this chapter.

Perpetua M. Muganda (ed.), *Apoptosis Methods in Toxicology*, Methods in Pharmacology and Toxicology,
DOI 10.1007/978-1-4939-3588-8_8,

Abbreviations

BFP	Blue fluorescent protein
DMEM	Dulbecco's modified Eagle's medium
DMSO	Dimethyl sulfoxide
DsRed	*Discosoma* sp. red fluorescent protein
ECFP	Enhanced cyan fluorescent protein
EMCCD	Electron multiplying charge coupled device
EYFP	Enhanced yellow fluorescent protein
FACS	Fluorescence-activated cell sorting
FBS	Fetal bovine serum
FITC	Fluorescein isothiocyanate
FRET	Fluorescence resonance energy transfer
FSC	Forward-scattered light
GFP	Green fluorescent protein
IMD	Intensity modulation display
NCCS	National Centre for Cell Science
NCI	National Cancer Institute
NLS	Nuclear localization signal
PFS	Perfect focus system
QE	Quantum efficiency
ROI	Region of interest
SCAT	Sensor for activated caspases based on FRET

1 FRET Sensors for Detecting Caspase Activation

Innovative new generation cell-based antitumor drug screening technologies with high-throughput capabilities have undergone extensive experimental evaluation and optimization. Despite this, current preclinical antitumor drug screening is highly dependent on conventional cytotoxicity assays [1, 2]. The well-conceived "integrated anticancer drug screening program" adopted by NCI still utilizes sulforhodamine B (SRB) assay as the screening tool [3]. Even though this approach together with the NCI COMPARE program developed using the cytotoxicity profiling of the 60 cancer cell line panel yields valuable information about the nature of compounds [4], it still bears inherent disadvantages; these include the fact that the assay procedure involves labor intensive processing after drug-addition, and the procedure remains an end-stage screening tool. Additionally, the assay cannot discriminate cytotoxicity from cell growth inhibition or necrosis-like activity. Apoptosis assays, such as those involving caspase activation using fluorescent substrates, were initially reported as second-generation anticancer drug screening technologies. However, even though such substrates can be employed using the same 60 cancer cell line panel of the NCI model, the process is costly, involves a labor intensive staining protocol and a nonreliable probe uptake; this, thus, fails to emerge as a successful anticancer screening methodology [5].

Apoptotic cell death involves diverse hierarchical signaling events, such as translocation of Bax and Bak, Bid activation, cytochrome c release, mitochondrial membrane permeabilization, caspase activation (initiator and executioner caspases) and cleavage of diverse protein substrates [3, 6–8]. Thus far, all of these signaling events have been explored in an effort to generate user-friendly and sensitive assay platforms for anticancer drug screening or apoptosis inducing activity analysis [4, 9]. Immunofluorescence-based approaches, mitochondrial permeability sensitive dyes, and caspase-specific fluorescent substrates have been widely employed as apoptosis detection tools [10–13]. Even though the above-mentioned assays are better methods than the cytotoxicity assays, they suffer from several demerits, such as inferior sensitivity, need of complex sample processing, requirement of costly reagents, inconsistency in results, and toxicity of certain probes.

New developments in the area of fluorescent proteins offer a large number of molecules with differing spectral properties [14–16]. These spectral variants have been extensively used in the creation of sensitive probes for diverse cell signaling studies, including visualization of apoptosis in live cells [17–20]. Among these, Fluorescence Resonance Energy Transfer (FRET)-based probes are a remarkable tool for visualizing caspase activation in live cells [21–23]. Intracellular caspase activation can be detected by using either chemical probes or genetically encoded FRET-based fluorescent probes. The latter are, however, more suitable for use in live cells, because they yield reliable read-outs due to minimal invasiveness, and have easy automated quantitation possibilities. In general, the approach involves the creation of a FRET-based probe with a suitable FRET donor and acceptor fluorophore combination joined together with a caspase-specific amino acid linker. Once such a probe is introduced into a cancer cell, as long as the linker is intact within the probe, fluorescence resonance energy transfer (FRET) takes place between the donor and acceptor fluorophores. This means that upon excitation of the donor fluorophore, the emitted energy is transferred to the acceptor fluorophore, thereby generating acceptor fluorescence. When these cells are treated with an apoptosis inducing agent, intracellular caspases are activated. These activated caspases cleave the amino acid linker sequence connecting the donor and the acceptor fluorophores, leading to loss of FRET, and a consequent decrease in acceptor fluorescence (or conversely, an increase in donor fluorescence) [24–26]. The advantage of this approach over other apoptosis assays is its ability to monitor real-time caspase activation in live cells with no additional processing. A schematic representation of a FRET probe with ECFP (enhanced cyan fluorescent protein) and Venus (an improved variant of yellow fluorescent protein, YFP [27]) as the donor–acceptor combination is shown (Fig. 1a). In addition, Fig. 1b shows the ratiometric image of cells when there is no FRET loss and when FRET loss happens due to the treatment with drug.

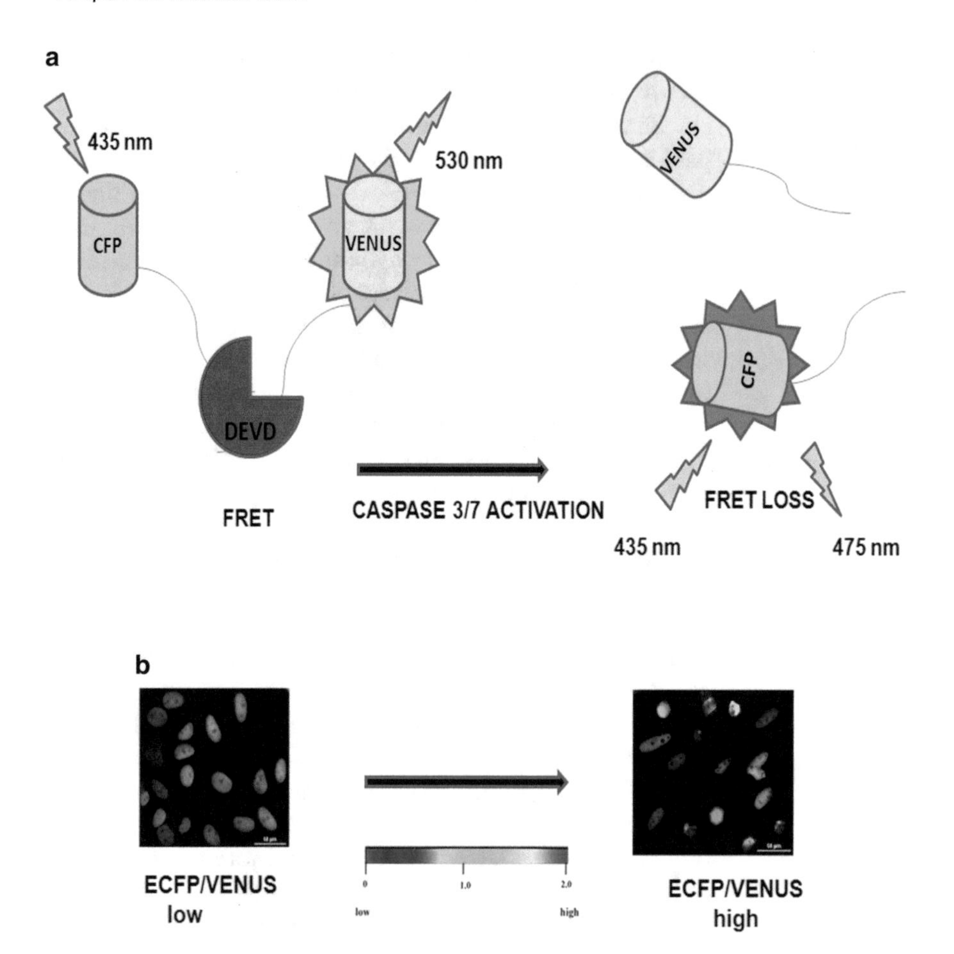

Fig. 1 (**a**) Schematic representation of the mechanism of action of FRET probe. FRET from donor fluorophore (ECFP) to acceptor fluorophore (Venus) is lost as a result of the activation of caspase 3/7 and subsequent cleavage of the caspase substrate-linker sequence, DEVD. FRET loss is indicated by an increase in ECFP emission fluorescence. (**b**) Representative ratio images of population with FRET and upon FRET loss

Several recent studies have employed FRET-based genetically encoded probes using diverse donor and acceptor fluorophore pairs for monitoring caspase activation in live cells. These include ECFP-Venus (enhanced cyan fluorescent protein–Venus) [28], BFP-GFP (blue fluorescent protein–green fluorescent protein) [17], GFP-RFP (green fluorescent protein–red fluorescent protein) [29], and YFP-DsRed (yellow fluorescent protein–*Discosoma* sp. red fluorescent protein) [30]. We have recently reported a successful image-based high-throughput assay protocol for caspase activation using stable cells expressing intracellular ECFP-DEVD-EYFP ("DEVD" for the amino acid linker sequence Asp–Glu–Val–Asp) probe for high-throughput drug screening [31], where the linker holds together the donor and acceptor fluorophores ECFP and EYFP, respectively. In order to identify individual cells in this study, we employed automated segmentation on the basis of parameters such as fluorescence intensity, circularity, and elongation. Proper segmentation in such populations also requires nuclear staining and complex segmentation procedures in cases where cell boundary is not clearly differentiable.

In order to overcome these deficits, we employed a methodology for easily and noninvasively detecting caspase activation by targeting the FRET probe inside the nucleus of a cell. We have previously reported the development of Mouse Embryonic Fibroblast (MEF) cells stably expressing caspase-specific FRET-based probe with an upstream nuclear localization signal (NLS) sequence [28, 32]. In the present chapter, we describe this methodology in detail. We first describe the protocol for generation of cells stably expressing the FRET-based nuclear-targeted caspase sensor probe. We have utilized both clonal selection and flow cytometer-based sorting to enrich cell populations with homogeneous expression of the probe to aid easy segmentation using conventional image analysis software. Further on, we have laid down the step-by-step protocol for detecting caspase activation in these stable cells, employing live cell imaging and further image analysis.

1.1 Miniaturization of Apoptosis Assay for High-Throughput Anticancer Drug Screening

Improved high-throughput approaches in natural product extraction and combinatorial chemistry have a great potential to generate large numbers of compound libraries with diverse biological activities, for lead identification in pharmaceutical industries [33–35]. Towards this end, a major and immediate task is to develop a sensitive and easy screening protocol for a specific biological activity in a high-throughput manner. Such a protocol would aid in selecting the best hits among large compound libraries for downstream drug development processes. In the case of antitumor agents, the key biological activity is the induction of apoptosis in target cancer cells [36, 37]. Therefore, miniaturization of apoptosis assays to enable high-throughput screening has been an intensive research area both for the academia and industry.

Miniaturization of apoptosis assays involves cell line development and optimization of measurement parameters in a quantifiable manner with improved automation to ensure consistent and repeatable assay results. The caspase activation assay reported here is high-throughput compatible, does not require labor intensive staining or processing, is easy to segment, and generates automated results at the end of imaging. A major advantage of this assay is the provision for repeated sampling and temporal resolution, so that kinetics of cell death induction by the compound can be generated from individual cells. Also, the use of stage-top incubators for multi-well plate allows repeated imaging of live cells over a period of time in multiple drug treated wells.

2 Materials

2.1 Cell Lines

For the current protocol, cervical cancer cell line SiHa, osteosarcoma cell line U2OS, neuroblastoma cell line U251, and non-small-cell lung carcinoma (NSCLC) cell line A549 are used. U251 and A549 were obtained from NCI, USA, U2OS from Sigma-Aldrich, USA, and SiHa from NCCS, Pune.

2.2 Plasmids

pcDNA NLS-SCAT3 (ECFP-DEVD-Venus), a FRET-based nuclear-targeted sensor probe for activated caspases, is needed for the development of the live-cell tool for apoptosis detection detailed here. It was kindly provided by Dr. Masayuki Miura of University of Tokyo, Japan [28, 32]. The NLS sequence linked to upstream of the caspase 3/7-specific FRET probe sequence (SCAT3) helps in limiting the expression of the probe to nuclear compartment only. pcDNA3 DEVD (ECFP-DEVD-EYFP), an intracellular FRET-based activated-caspase sensor probe was kindly provided by Prof. Jeremy M. Tavare and Dr. Gavin Welsh (University of Bristol, UK) [24]. The probe consists of ECFP and EYFP linked together by a caspase 3/7 substrate sequence, DEVD. The probe expresses homogenously both in nuclear and cytoplasmic compartments of cell and displays a loss of FRET between ECFP and EYFP upon caspases activation and subsequent DEVDase activity.

2.3 Reagents

Reagents required are: DMEM, phenol-red free DMEM (Dulbecco's modified Eagle's medium), Opti-MEM™, trypsin-EDTA (0.05 %), antibiotic-antimycotic cocktail (100×), G418 (geneticin) sulfate, ampicillin sodium salt (Gibco, USA); Lipofectamine® LTX reagent, Plus™ reagent (Invitrogen, USA); FBS (fetal bovine serum) (Pan-biotech GmbH, Germany); Qiagen Plasmid Midi kit (Qiagen, Netherlands); and DMSO (dimethyl sulfoxide) (Sigma, USA). Ampicillin sodium salt is stored at −20 °C as a 100 mg/ml concentrated solution, prepared in autoclaved distilled water. G418 sulfate should be prepared in autoclaved distilled water at a concentration of 100 mg/ml and stored at −20 °C, away from light. Working concentration of G418 has to be standardized for each cell line and prepared fresh before use.

For inducing apoptosis in cancer cells as part of the apoptosis detection protocol, any anticancer drug at its active concentration can be used. In the demo protocol that follows, drugs namely, tunicamycin, thapsigargin, doxorubicin, and cisplatin (Sigma-Aldrich) are used. For long term storage, prepare these drugs at a higher stock concentration in DMSO and maintain at −80 °C; prepare tunicamycin, thapsigargin, doxorubicin, and cisplatin at concentrations of 1 mM, 1 mM, 20 mg/ml, and 50 mg/ml, respectively (*see* **Note 1**).

2.4 Instrumentation and Other Requirements

The equipments required for this protocol are as follows. Nano drop UV-Visible spectrophotometer (Thermo Scientific, USA), FACS Aria II™ (BD Biosciences, USA), Glass cloning cylinders (Sigma), Nunc™ Lab-Tek™ chambered cover glass (Thermo Scientific), Stage-mountable incubation chamber (Tokai Hit®, Japan), ECLIPSE, TE2000-E inverted fluorescence microscope (Nikon® Corporation, Japan), BD™ CARV II confocal imager (BD Biosciences), Andor iXON 885, the low light EMCCD (Electron multiplying charge coupled device) camera (Andor Technology Ltd., UK), ECLIPSE T*i*-E inverted phase contrast fluorescence microscope (Nikon® Corporation), Nunc™ MicroWell 96-well optical bottom-plates with coverglass base (Thermo Scientific) and CoolSNAP HQ*2* camera (Roper Scientific GmbH, Germany).

Two imaging software are also required: IPLab™ imaging software (BD Biosciences) and NIS-elements Advanced Research (AR) imaging software, version 4.2 (Nikon® Corporation).

3 Methods

To visualize and quantify apoptosis in live cells on a real-time basis, we employ an automated ratio imaging-based approach; using a nuclear-targeted FRET-based caspase sensor probe. This approach helps in an easier, accurate, unbiased and direct determination of apoptosis signaling, which can be scaled up to a high-throughput level. To achieve this end, our group has generated cancer cell lines stably expressing FRET probe ECFP-DEVD-Venus, attached to the nuclear localization signal (NLS-SCAT3), by transfection. Stable cell clones homogenously expressing the probe were generated using antibiotic selection, fluorescence-based flow cytometer-sorting and colony isolation methods. Image acquisition and subsequent analysis has been standardized for these cells in multi-well format also. Figure 2 shows a schematic representation of this method. A detailed description of the protocol is presented in the subsequent sections.

3.1 Development of Nuclear-Targeted Caspase Sensor FRET Probe in Cancer Cell Lines

Cancer cell lines used for the development of FRET probe-expressing cells should be maintained in suitable culture conditions. SiHa, U2OS, U251, and A549 cells used in the current protocol can be grown in DMEM, supplemented with 10 % FBS and 1× antibiotic-antimycotic cocktail, within a humidified water jacketed CO_2 chamber at 37 °C as per standard cell culture protocols. Generally, in our lab, in order to obtain transfection grade plasmid, expression vectors are transformed in the DH5α strain of *Escherichia coli* by heat shock method. Plasmid DNA is prepared based on the standard methods of alkaline lysis and anion exchange separation using Qiagen Plasmid Midi Kit for plasmid

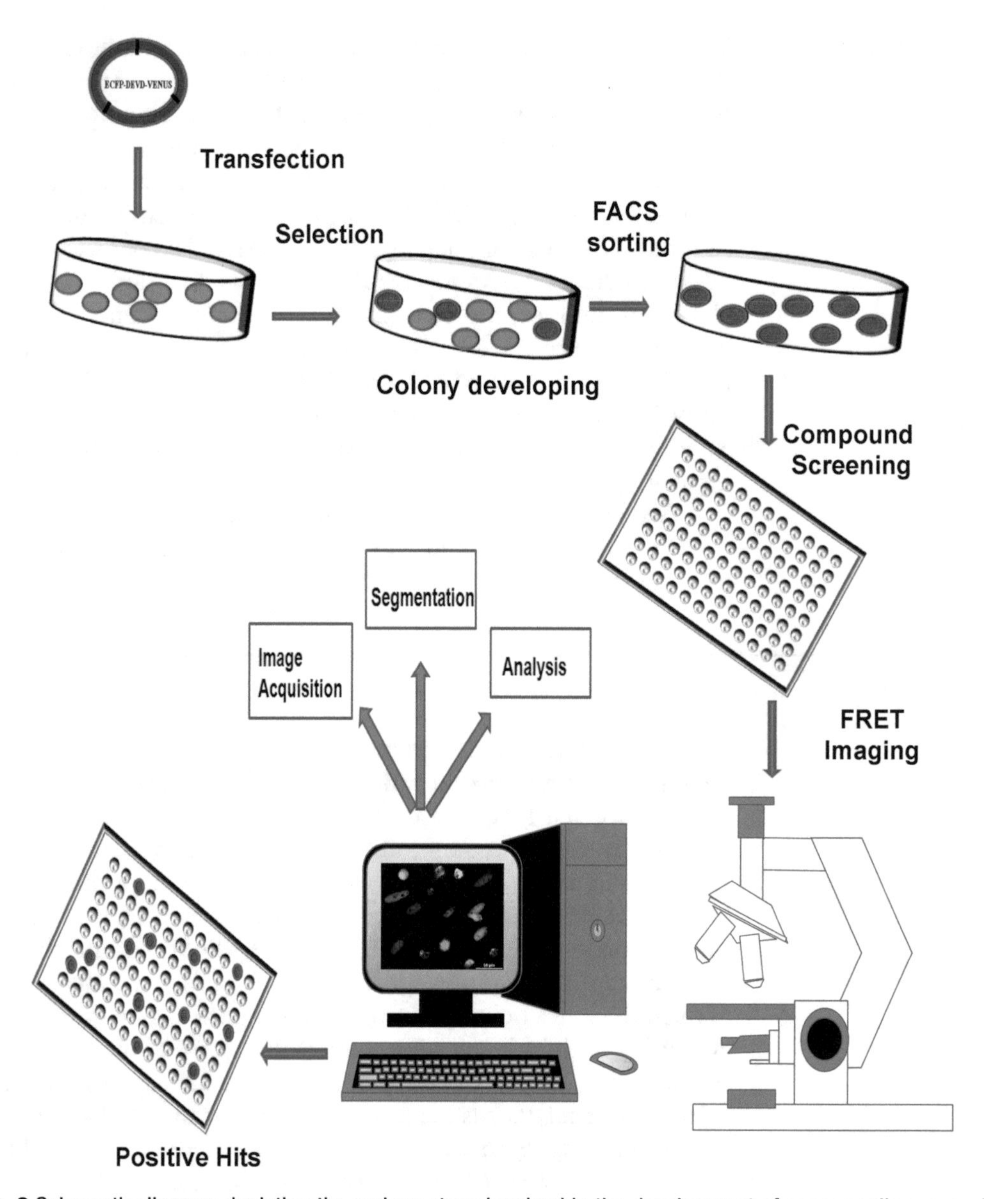

Fig. 2 Schematic diagram depicting the various steps involved in the development of cancer cells expressing nucleus-targeted FRET probe NLS-SCAT3, followed by its ratio image acquisition and analysis

isolation (*see* **Note 2**). DH5α strain is employed as it is a non-fastidious microorganism which is easy to maintain and preserve, shows good transformation efficiency and is the most frequently used *E. coli* strain for transformation. DH5α cells are made competent by the chemical method of calcium chloride treatment and transformed by heat shock method, according to the standard practice (*see* **Note 3**). Following transformation of DH5α cells with the caspase sensor FRET plasmid, pcDNA NLS-SCAT3, transformed bacterial colonies are allowed to expand in

Luria–Bertani (LB) broth containing ampicillin sodium salt at a final concentration of 100 μg/ml. The FRET plasmid, by virtue of its backbone being pcDNA, contains ampicillin resistance gene as the selectable marker for transformation.

Before proceeding to the next step, measure concentration and purity of the DNA sample using NanoDrop 1000 spectrophotometer. We recommend NanoDrop instead of conventional spectrophotometers as it requires only a very low sample volume for analysis; viz. 0.5–2 μl. Once the plasmid DNA sample with requisite concentration/amount of genetic material is available, subsequent transfection experiments can be initiated. For successful transfection experiments it is desirable for the plasmid DNA solution to have a concentration of 0.5–5 μg/μl.

3.1.1 Protocol for Transfecting Cancer Cells with NLS-SCAT3 Plasmid

In most of the transfection experiments carried out in our laboratory, lipofection is the preferred method; it is efficient in the cancer cell line panel, and is less damaging to the cells compared to other physical and chemical methods for transfection. Transfection reagents, Lipofectamine® LTX and Plus™ reagent, obtained from Invitrogen, are used. Lipofection with LTX and Plus reagents shows better efficiency than that with Lipofectamine™ 2000 (Invitrogen) alone.

1. Seed cells in 12-well plates, and allow them to grow for 24–48 h so that the cells attain 70 % confluency.
2. One hour prior to transfection, decant the culture medium, add 300 μl of serum-free Opti-MEM™ to the wells, and incubate at 37 °C.
3. In a separate sterile eppendorf tube, mix 200 μl of Opti-MEM™ and 5–7 μg of plasmid DNA, followed by addition of 3 μl of Plus reagent.
4. Incubate the mix for 5 min so that DNA–lipid complex is formed.
5. To this complex, 12 μl Lipofectamine® LTX reagent need to be added and incubated for 25 min.
6. Add the Lipofectamine–plasmid mix to the cells with mild agitation to ensure uniform spreading of the mix to the cells, and incubate at 37 °C with 5 % CO_2 in an incubator.
7. Dilute (*see* **Note 4**) the transfection mixture in the wells with fresh Opti-MEM™ containing 10 % FBS after 8 h. This will reduce the toxic effect of Lipofectamine and DNA on the cells; at the same time facilitating further transfection events, if any.

By 24 h, the transfected cells should start expressing the probes transiently, which can be visualized under fluorescence microscope using GFP filter sets. Though we could achieve good transfection efficiency by this protocol, it varies from cell to cell. We could attain a transfection efficiency of 80 %, 50 %, 90 %, and 40 % for SiHa, U2OS, U251, and A549, respectively.

3.1.2 Protocol for Selection and Generation of Stably Transfected Cells

The expression vector contain neomycin resistance gene so that cells expressing the probe can be selected using G418 antibiotic. Since G418 is an antibiotic that can kill both prokaryotic and eukaryotic cells, it is extensively used to eliminate un-transfected cells in cases where cells are transfected with plasmids possessing gene conferring resistance to it. The optimum concentration of antibiotic and the duration of antibiotic selection vary between cells and needs to be optimized for each cell line.

1. After 24 h of transfection, remove the medium containing the lipid–plasmid mix, and replace with fresh Opti-MEM™ or DMEM containing 10 % FBS supplemented with respective concentration of G418 for each cell line.
2. Replace the cells with fresh selection media every 4 days until the maximum number of un-transfected/non-expressing cells are eliminated as observed under fluorescence microscopy.
3. When the wells become confluent in between, cells can be trypsinized and re-seeded in new culture vessels. Once the cells get attached to the surface of the culture vessels, they should be resubjected to selection medium.
4. It is desirable to cryopreserve the cell population at different stages of development; with different percentage of transfected cells, specifying the percentage at each stage. This will ensure availability of transfected cell population in case of accidental loss of the cells due to contamination.

Continuous maintenance of the transfected cells under selection pressure results in an enriched population of transfected cells, stably expressing the plasmid of interest. Table 1 lists the optimized concentration and duration of antibiotic selection to generate stably expressing cells in each of the cell line utilized in this protocol.

3.1.3 Methods for Further Enrichment of Transfected Cells

After 3–4 weeks of antibiotic selection, cell population with highly heterogeneous expression of the fluorescent probe can be seen. For effective automated segmentation and imaging, it is extremely important to have an enriched cell population with homogenous

Table 1
Concentration and duration of G418 treatment used in each cell line for selection of cells transfected with NLS-SCAT3 plasmid

Sl. No.	Cell line	Concentration of G418 (μg/ml)	Duration of selection (weeks)
1	SiHa	600	3
2	U251	600	2
3	U2OS	1000	4
4	A549	1000	4–6

expression of probe. We employ either colony selection or flow cytometry-based sorting to generate stable cell population with homogenous expression of the probe.

Cell sorting by flow cytometry: All of the sterile sorting experiments conducted in our laboratory are performed using FACS Aria II™. Standard operating procedure for sterile sorting is followed for all sorting applications.

1. Select a transfected cell population with heterogeneous expression pattern of green fluorescence for FACS experiment.
2. After trypsinization, wash the cells with serum containing medium and resuspend in Opti-MEM™ containing 2 % FBS at a density of 2×10^6 cells/ml.
3. Filter the cell suspension through a 45 μm cell strainer (BD Falcon, USA) into a sterile FACS tube (BD Falcon).
4. Flush the fluidics path of the FACS machine with sterile distilled water, followed by sequential running of 70 % ethanol, sterile distilled water, and sterile medium. This help in clearing the path of any contaminating organisms or dye particles from earlier experiments.
5. The scatter plots obtained during a FACS experiment display each cell identified based on forward scatter (FSC) and side scatter signals (SSC) as individual events. The cells with low width signal of FSC and SSC identified after doublet discrimination (*see* **Note 5**) can be considered as a single population (Fig. 3a).
6. This population can be further analyzed for GFP expression in the 488 nm light path as an indication of the expression of the plasmid of interest. Two distinct populations based on the expression pattern of GFP are visible in the scatter plot (Fig. 3b).
7. Fix a tight gate for GFP high-expressing cells (gate P4) before sorting, to ensure high and homogenous expression of the fusion protein (Fig. 3b).
8. Sort the gated cells into sterile FACS tubes containing 1 ml of culture medium supplemented with 20 % FBS.
9. Centrifuge and resuspend the sorted cells in fresh medium containing 20 % FBS and seed on 6-well plates.
10. After the cells get attached to the surface of the culture vessel, replenish the wells with fresh cell culture medium containing 10 % FBS.

The sorted population shows an enrichment of cells expressing the plasmid probe compared to the unsorted population (Fig. 4).

Colony picking/isolation: As mentioned elsewhere, an alternate approach to generate cell population with highly homogenous and enriched expression of plasmid probe for imaging purpose is to use colony isolation.

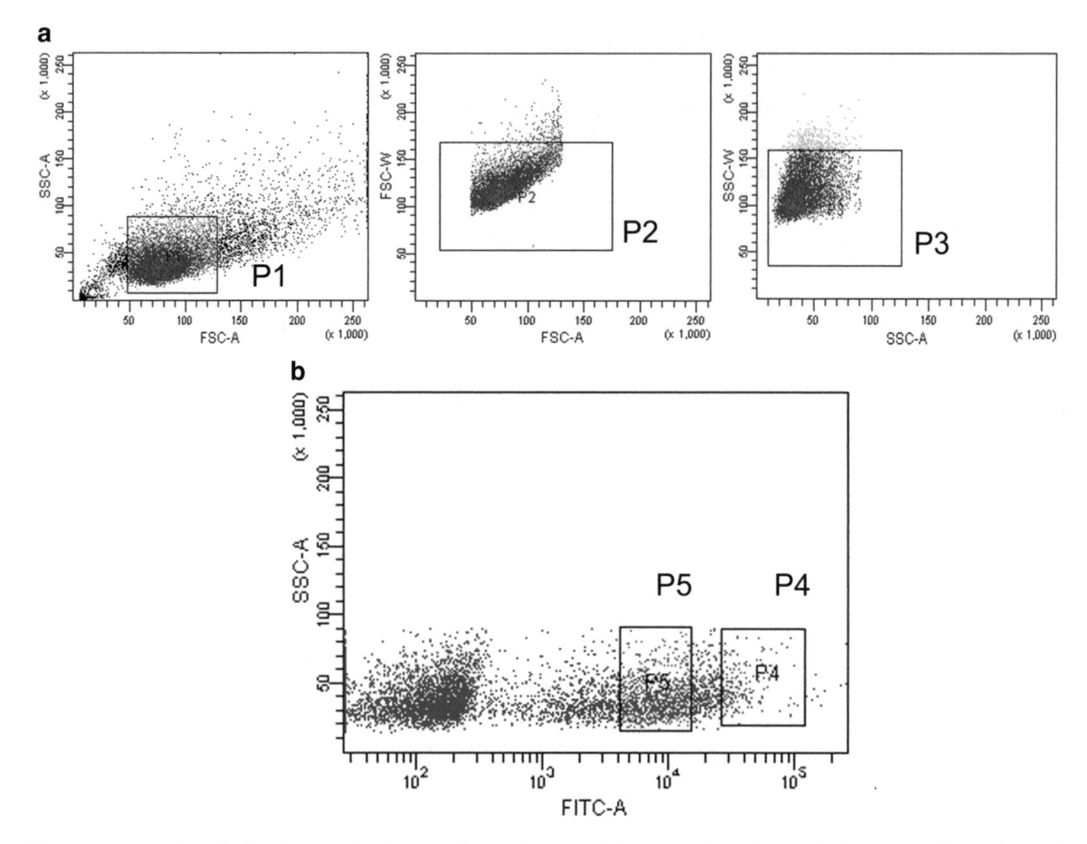

Fig. 3 Scatter plots indicating methodology of enrichment of the transfected population by FACS sorting. (**a**) Doublets were avoided and single population (gate P3) was identified as that with lower width-signal of both FSC (forward scattered light) and SSC (side scattered light) intensity. (**b**) Population P3 was subsequently analyzed for intensity of green fluorescence. Population P4, with uniform and high intensity green fluorescence, was sorted out and re-seeded. FITC filter was used for obtaining green fluorescence

1. Re-seed the transfected cells in fresh 60 mm cell culture dish after a few weeks of continuous exposure to selection condition. The cells will generate colonies in such a way that transfected and untransfected cells give rise to separate colonies.
2. Identify colonies with sufficient homogenous expression under microscope and enough spatial separation from neighboring colonies.
3. Mark the colony on the outer surface of the culture dish.
4. With the help of a sterile forceps, smear the lower edges of an autoclaved, glass-cloning cylinder with sterile high-vacuum silicone grease (Sigma). This aids in the fixing of the cloning cylinder onto the surface of the culture dish and prevents leakage of liquid from inside the cylinder.
5. Place a silicone grease-smeared cloning cylinder carefully inside the culture dish in such a way that the marked colony comes inside the cylinder.

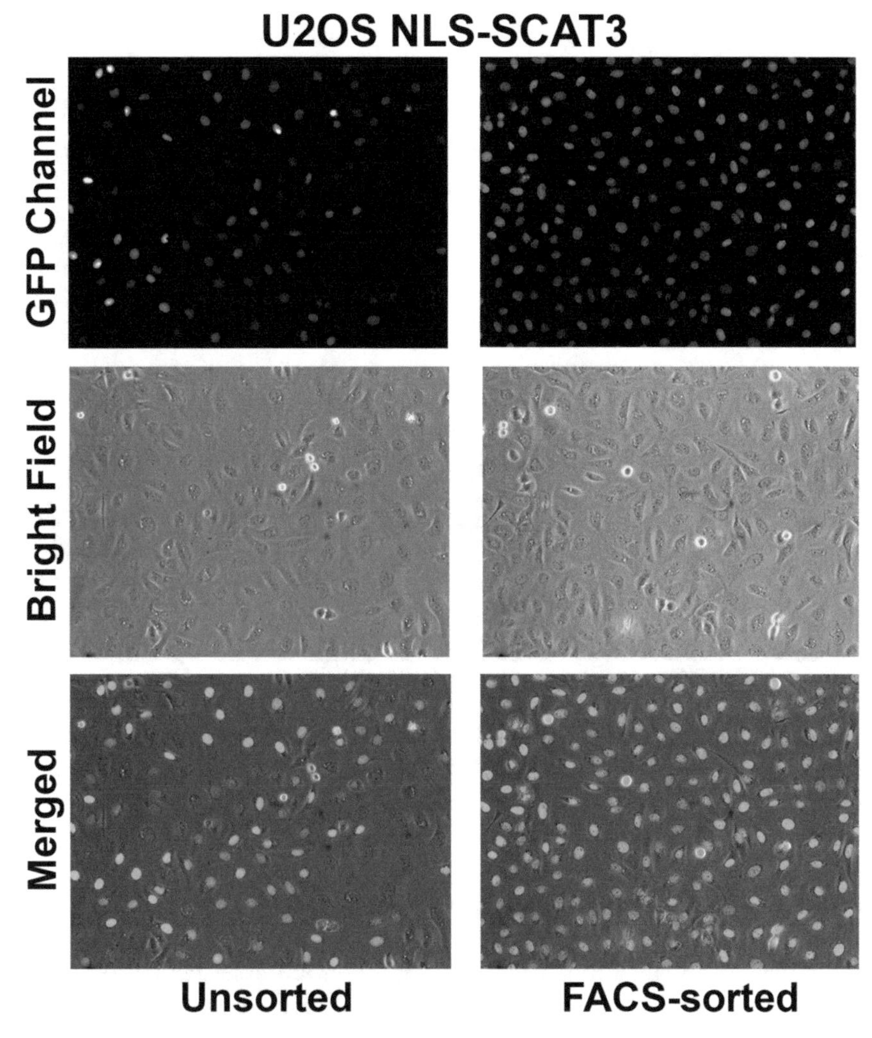

Fig. 4 U2OS cells expressing NLS-SCAT3 before and after enrichment by FACS-sorting. FACS based enrichment yielded a population with cent percentage, high, homogeneous and targeted expression of NLS-SCAT3 probe

6. Trypsinize only the cells within the cylinder. Take care not to disturb the cloning cylinder while trypsinizing the colony.
7. Seed the trypsinized colony into a new culture dish and allow the cells to expand under selection conditions.

Figure 5a is a schematic representation of the protocol described above, while Fig. 5b shows a single colony of U251 cells transfected with NLS-SCAT3 plasmid.

Cell populations showing a uniform and stable expression of the plasmid, generated in the manner described in the previous sections, are used for further FRET imaging and analysis. These transfected cell lines can also be cryopreserved at very low temperatures and used for experiments later (*see* **Note 6**).

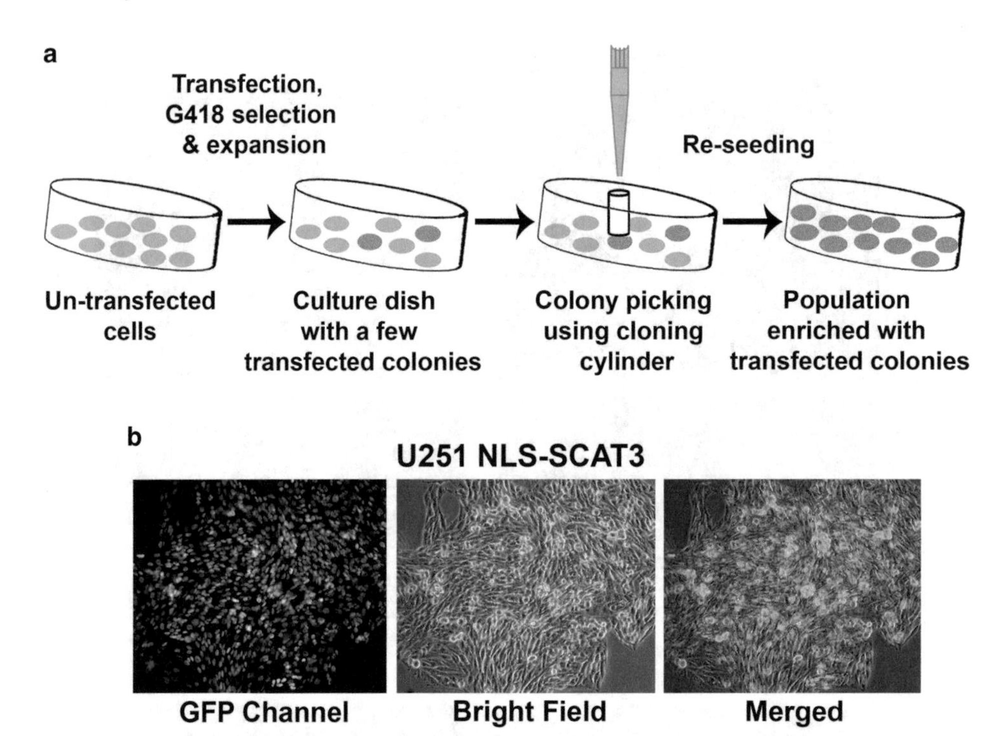

Fig. 5 (**a**) A schematic representation of the enrichment of transfected cells by colony picking technique. (**b**) Representative image of an isogenic colony with homogeneous expression pattern of NLS-SCAT3 (U251). Images of the colony from GFP, bright field, and merged channels are shown

3.2 Fluorescence Time-Lapse Live Cell Imaging for Caspase Activation

For imaging caspase activation in live cells, cells stably expressing caspase-specific FRET probe are to be grown in chambered cover glass and placed in the stage-mountable incubation chamber that maintains optimum CO_2, temperature and humidity. Imaging is carried out with a 20× Plan Apo 0.75 NA objective under the inverted fluorescence microscope, ECLIPSE TE2000-E, fitted with a perfect focus system (PFS) and automated excitation and emission filter wheels. When cells are imaged in a time-lapse manner for prolonged time periods, mild focus drift of objective can happen. PFS compensates for this focus drift and enables continuous monitoring and maintenance of preset focal length values. The liquid–solid interface at the bottom of the culture dish is utilized for determining the perfect focus. Plan Apo objective is the one in which the lens shows the highest degree of apochromatic (Apo) correction for aberrations (both spherical and chromatic aberrations) as well as flat field correction for field curvature.

For real-time imaging of caspase activation in live cells, several optimizations can be employed to minimize imaging-induced damage to cells. High speed filter wheels, fast shutter, high quantum efficiency (QE) camera, minimum excitation-light exposure, etc. are very important for this. A versatile model based on spinning-disk confocal microscopy and controlled by IPLab™ imaging software, the BD™ CARV II confocal imager, is typically employed for this purpose in our laboratory. The imaging can be done either in confocal mode with "disk in" position of spinning-disk or non-confocal mode with "disk out" position.

The low light EMCCD (Electron multiplying charge coupled device) camera, Andor iXON 885, is routinely used to acquire sufficient quality images with extremely low exposure time (*see* **Note 7**). The details of the filters to be used are 438 ± 12 nm band pass filter for ECFP excitation, 458 long pass (LP) dichroic filter and two band pass emission filters at 483 ± 15 nm for ECFP and 542 ± 27 nm for Venus FRET (Semrock Inc., USA) (*see* **Note 8**). The excitation, emission and dichroic filter wheels were independently controlled through IPLab™ software to collect FRET ratio images. So as to minimize photo bleaching of the fluorescence signal due to repeated and continuous imaging, excitation intensity was reduced to 1 % by intensity iris control.

1. Seed cancer cells stably expressing caspase-specific FRET probe, NLS-SCAT3 (here, SiHa NLS-SCAT3), in 8-well chambered cover glass and grow under standard cell culture conditions for 24 h. It is desirable for the cell population to be at 60–70 % confluency.
2. Prepare working concentrations of thapsigargin (2 μM) and cisplatin (50 μg/ml) in phenol red-free DMEM containing 5 % FBS.
3. Remove the spent medium in the wells and replace with drug-containing DMEM. Add fresh phenol red-free culture medium supplemented with 5 % FBS in control wells also.
4. Place the chambered cover glass in the stage-mountable incubation chamber and focus the cell with the 20× Plan Apo 0.75 NA objective of ECLIPSE, TE2000-E inverted fluorescence microscope.
5. Identify the optimum exposure time required for capturing the emission signal from both donor and acceptor fluorophores so that the resultant image is noise free and clear (*see* **Note 9**). Try to define lowest exposure time possible without hindering the image quality so that photobleaching and injury to the cell are minimized. Filters for all channels are accessed using the "linking device" function.
6. In the window for image acquisition ("multi dimensional acquire" function) there are options for confirming fluorescent channels to be acquired as well as for specifying the total time duration and interval between each image acquisition.
7. After setting all the parameters, run the experiment for the indicated time period (for 24 h), at 15 min interval.
8. Once image capturing is over, the information will be obtained as a sequence of images from both CFP and YFP channels at each time point. The images can either be analyzed in IPLab™

software or be exported to NIS-elements Advanced Research (AR) 4.2 imaging software for ratio analysis.

9. Obtain the ratio view of each image field from the emission intensity information of donor and acceptor channels using the software. Define ECFP signal as the numerator and EYFP signal as the denominator in the "ratio properties" window. Also specify the range of the ratio scale to span from a minimum value of 0 to a maximum value of 2. The image obtained after defining these parameters is called the IMD [38] (Intensity Modulation Display) (*see* **Note 10**).
10. The background noise in these ratio images can be reduced by defining the background offset value using either a constant or a ROI (region of interest) in the image and subsequently subtracting background using this offset value. A region in the image which is devoid of any cells and hence any fluorescence signal throughout the whole sequence can be identified to define background ROI.

Ratio image obtained in this way demonstrates that, in the drug-treated cells, caspase activation lead to the cleavage of DEVD sequence in the NLS-SCAT3 probe, resulting in the loss of FRET between ECFP and Venus with consequent increase in ECFP intensity and decrease in EYFP intensity. The caspase activation occurring in cells treated with thapsigargin for 24 h is clearly observed as a ratio change in the cells compared to control cells (Fig. 6). In an IMD cells displayed in colors towards the upper limit of the color

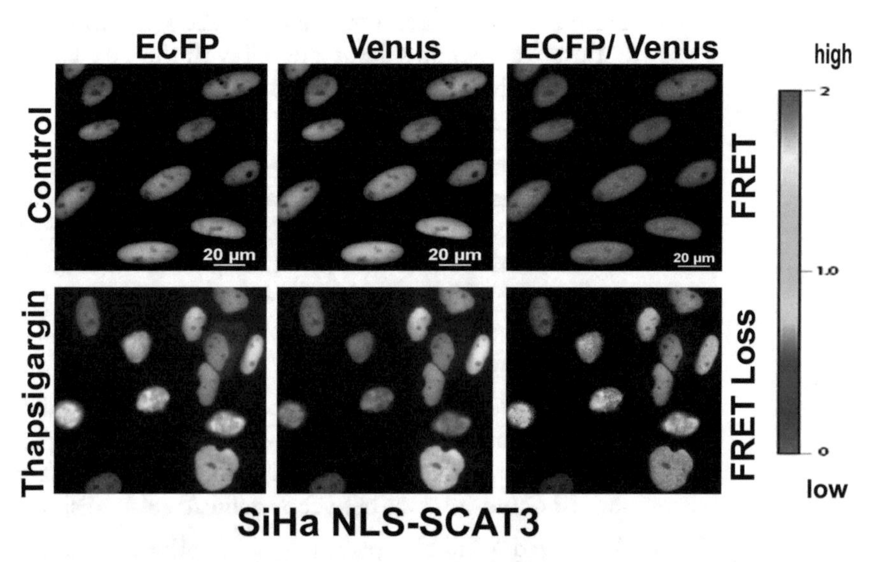

Fig. 6 Ratio imaging of SiHa NLS-SCAT3 cells indicates that probe cleavage effectively represents the apoptotic changes occurring in cells. Ratio images were generated by imaging cells before and after treatment with thapsigargin (2 μM) for 24 h. Compared to the cells in the untreated population, cells in the drug treated population show FRET loss and a corresponding increase in ECFP/Venus ratio. This is depicted by a color change in accordance with the color scale generated

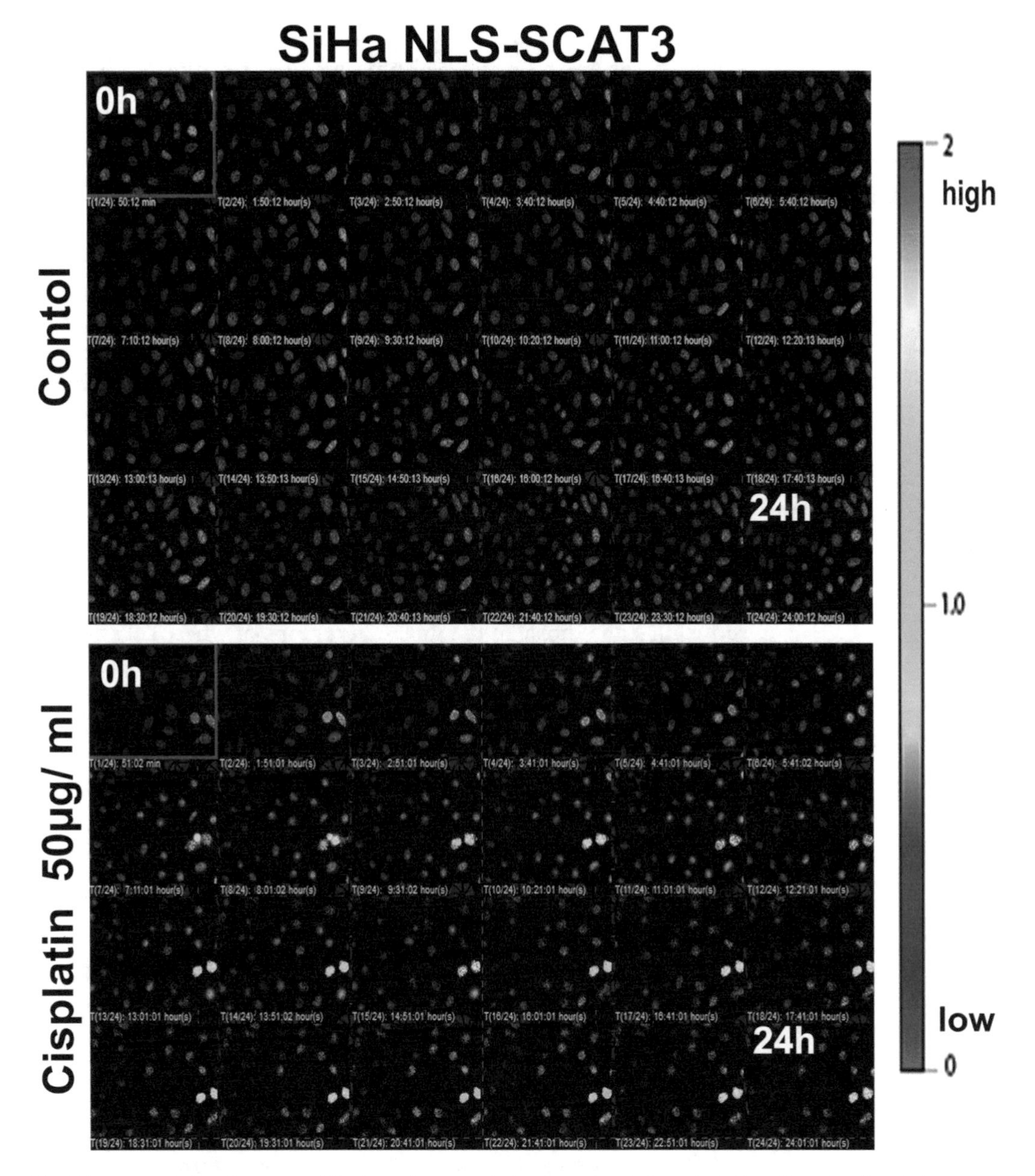

Fig. 7 Representative tile view of the time-lapse ratio images at an interval of 1 h with and without drug treatment (50 μg/ml cisplatin, 24 h)

scale (i.e., towards red) exhibit a higher FRET ratio and hence is assumed to possess activated caspases. Hence, the yellow-colored nuclei observed in the thapsigargin treated well are of those cells showing caspase activation; in contrast to the blue-colored cells observed in control well, without any caspase activation (Fig. 6). Ratio change in NLS-SCAT3-expressing cells due to FRET loss can be monitored over time in a time-lapse manner as evident in Fig. 7. NLS-SCAT3, as a FRET-based probe to monitor caspase activation, thus proves to be very efficient for visualizing caspase activation in live cells in a real-time manner. Continuous imaging does not affect the image quality or cell viability.

3.3 Post-acquisition Segmentation and Analysis

Post-acquisition analysis of the images obtained from cells stably expressing the FRET-based caspase sensor probe, NLS-SCAT3, is routinely done in our lab using NIS-elements software 4.0. The NIS-elements software offers better segmentation and post-acquisition analysis facilities. The nuclear-targeted (NLS) expression of the probe facilitates easy identification, segmentation and monitoring of individual cells. The very first step of post acquisition image analysis is generation of the ratio view of individual images. So as to get a clear and discernable ratiometric view, the ratio properties have to be defined. Specify CFP signal as the numerator, YFP signal as the denominator and the ratio scale such that the ratio values range from a minimum value of 0 to a maximum value of 2. The ratio image will be displayed in pseudo colors as an IMD.

The next step in image analysis is background correction for reducing signal noise in the image in regions of where cells are not present. An area in the ratio image, where there is no signal from both CFP and YFP channels throughout the whole image sequence (i.e., where there are no cells) is to be identified to draw a background ROI. After defining the background ROI, the background signal is subtracted by using the signal of this ROI as the offset value. The fourth row of images in Fig. 8a is a representation of how the ratio view of the acquired image will look like after background subtraction (*see* **Note 11**). Individual images in the panel shows the gradual change over time of the FRET ratio in SiHa NLS-SCAT3 cells after treatment with 2 μM thapsigargin.

Automated identification of individual cells in a population is achieved by arriving at proper segmentation settings. There are various approaches by which cells can be segmented in NIS-elements software. The first approach is to directly use the "segmentation" function in the "binary" drop-down menu. Here cells are identified as objects to be segmented based on parameters such as intensity, circularity and elongation of the fluorescence signal. Values for each of these parameters can be edited till the software identifies maximum number of cells in the image and draws the

Fig. 8 (**a**) Representative images at different time points (0, 8, 16, and 24 h) of cells subjected to drug treatment (2 μM thapsigargin). The *fifth row* shows automated segmentation of respective images generated based on intensity and dimensions of NLS-SCAT3 expression. The segmentation data were used to accurately identify the ROIs in the image, as indicated in the *sixth row* of images in the panel. (**b**) The ROIs generated were used to collect ratio (ECFP/Venus) data in a time-lapse manner over a period of 24 h. The data was used to create graphs indicating the change in ratio for each ROI ID (ROI identifier), over time (0, 8, 16 and 24 h)

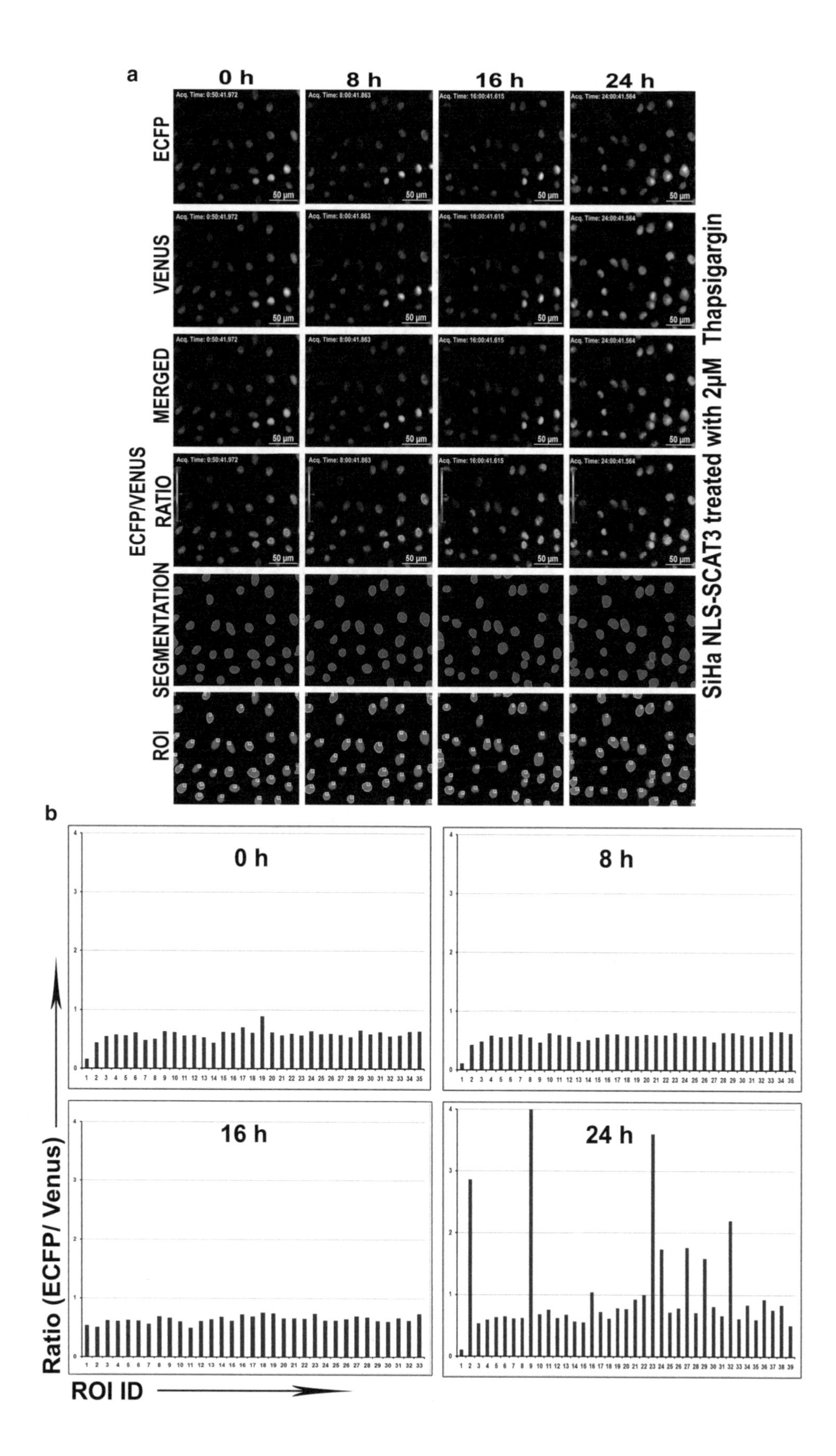
a
0 h
8 h
16 h
24 h
ECFP
VENUS
MERGED
ECFP/VENUS RATIO
SEGMENTATION
ROI
50 µm
SiHa NLS-SCAT3 treated with 2µM Thapsigargin
b
0 h
8 h
16 h
24 h
Ratio (ECFP/ Venus)
ROI ID

segments in coherence with the shape of the nuclear signal. The segmentation defined as such can be applied to the whole image sequence. The fifth row of images in Fig. 8a represents the images after segmentation.

Alternately, cells can be segmented by defining the threshold value. This function for defining threshold can be accessed from the "binary" drop-down menu of the NIS-elements tool bar. Cells are identified using "Per channel" option for demarcating the cell nucleus based on intensity, size, and circularity of the fluorescence signal and optimized using "smooth," "clean," and "separate" functions. The segments can be modified by optimizing pixel, dilation width, level, and erosion factor values, using functions of the same names available in "binary" menu. The threshold function, when defined using the "intensity option," will segment the background of the image. The "invert" function available in the "binary" menu will literally invert this segmentation so that objects in the image can be identified. The "autothreshold" function in "binary" menu can also be exploited to identity cells and define segments based on the brightness of the objects in field. Here also values of the functions such as pixel, dilation width, level, and erosion factor can be manipulated to optimize the segmentation (*see* **Note 12**).

After optimizing segmentation, segmented binary should be copied/converted to ROIs. "Copy binary to ROI" function available in the "ROI" drop-down menu can be used for this purpose. Each and every segment in the image will be converted into individual ROIs. This will help in extracting the quantitative information of the segments over time. The sixth row of images in Fig. 8a indicates the conversion of segments into ROIs. The quantitative information of ROIs such as signal intensity of ECFP and Venus channels, and ratio values can be obtained from the "time measurement" function in "analysis control" menu. The mean signal intensity of both the fluorescent channels as well as the ECFP/Venus ratio values for each segments/ROIs, at each time period, is found in the "data" option in this function. This information can be exported to MS Excel software so as to generate a graphical representation of the information. The quantitative data of the ECFP/Venus ratio gives an indication of the caspase activation status of each and every ROI and in turn each and every cell.

Figure 8b shows the graphs generated from the quantitative information of the ROIs defined on the segmented images of Fig. 8a. The graphs indicate that the number of cells showing augmented FRET ratio increases with time due to drug treatment (2 μM thapsigargin for 24 h). By correlating the ratio images (Fig. 8a) with the graphs (Fig. 8b), it is evident that post-acquisition segmentation and defining of ROI (based on the NLS tag) helps in automated identification of cells with caspase activation.

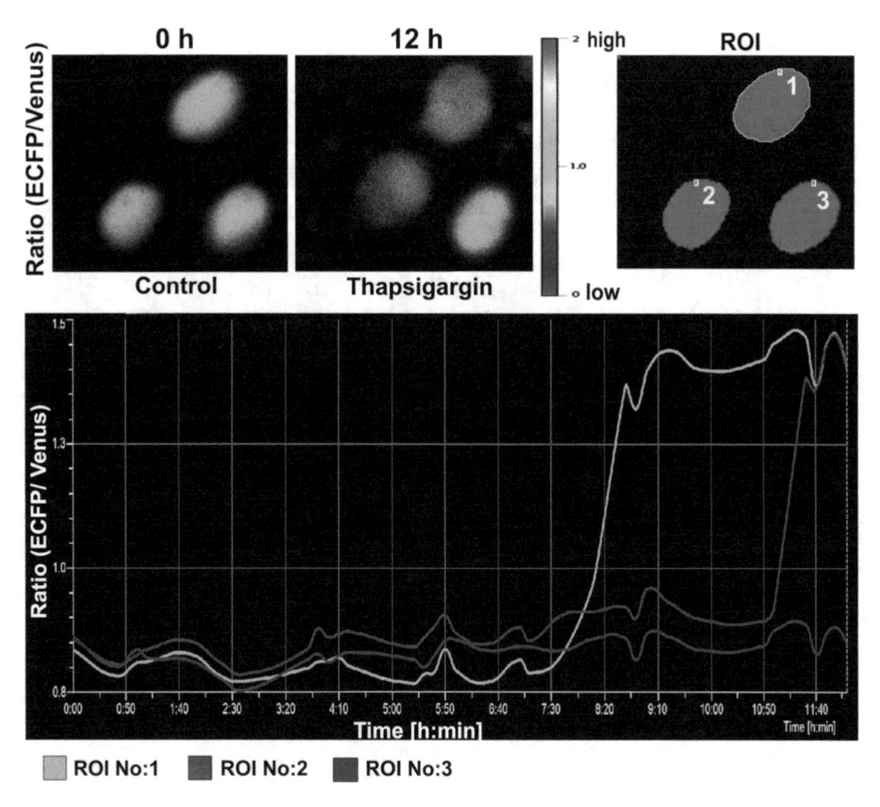

Fig. 9 Monitoring of individual ROIs/cells in a time lapse manner illustrates that NLS-SCAT3-based segmentation is accurate and closely represents ratio change over time. SiHa NLS-SCAT3 was treated with the indicated drug and three individual cells were closely monitored after segmentation and copying of the binary to ROI (ROI IDs: 1, 2, and 3). Ratio image indicates that ROI 1 and 2 exhibit a higher ratio (*yellow color*), while ROI 3 shows no ratio change (*green color*). A similar trend is obtained from the ratio data acquired over time, as indicated in the graph

NLS-directed automated segmentation and defining of ROIs can be utilized in making continuous monitoring of ratio changes easier. Tracking of ROIs over time is possible and this aids in obtaining an accurate quantitation of the alteration of ratio values. Figure 9 demonstrates that precise automated segmentation based on NLS also assists in data acquisition of individual cells over time in a time-lapse manner.

The nuclear-targeted expression of SCAT3 in NLS-SCAT3-expressing cells is so precise and strong that each cell can be accurately defined into individual ROIs directly, without the need for prior segmentation of the image. ROIs can be directly defined on images using the "autodetect all ROI" function in the "simple ROI editor" menu. The quantitative information obtained from ROIs defined directly on NLS-expressing cells is consistent with that obtained after segmentation. NLS-SCAT3 probe has proved to be of greater utility than DEVD probe as

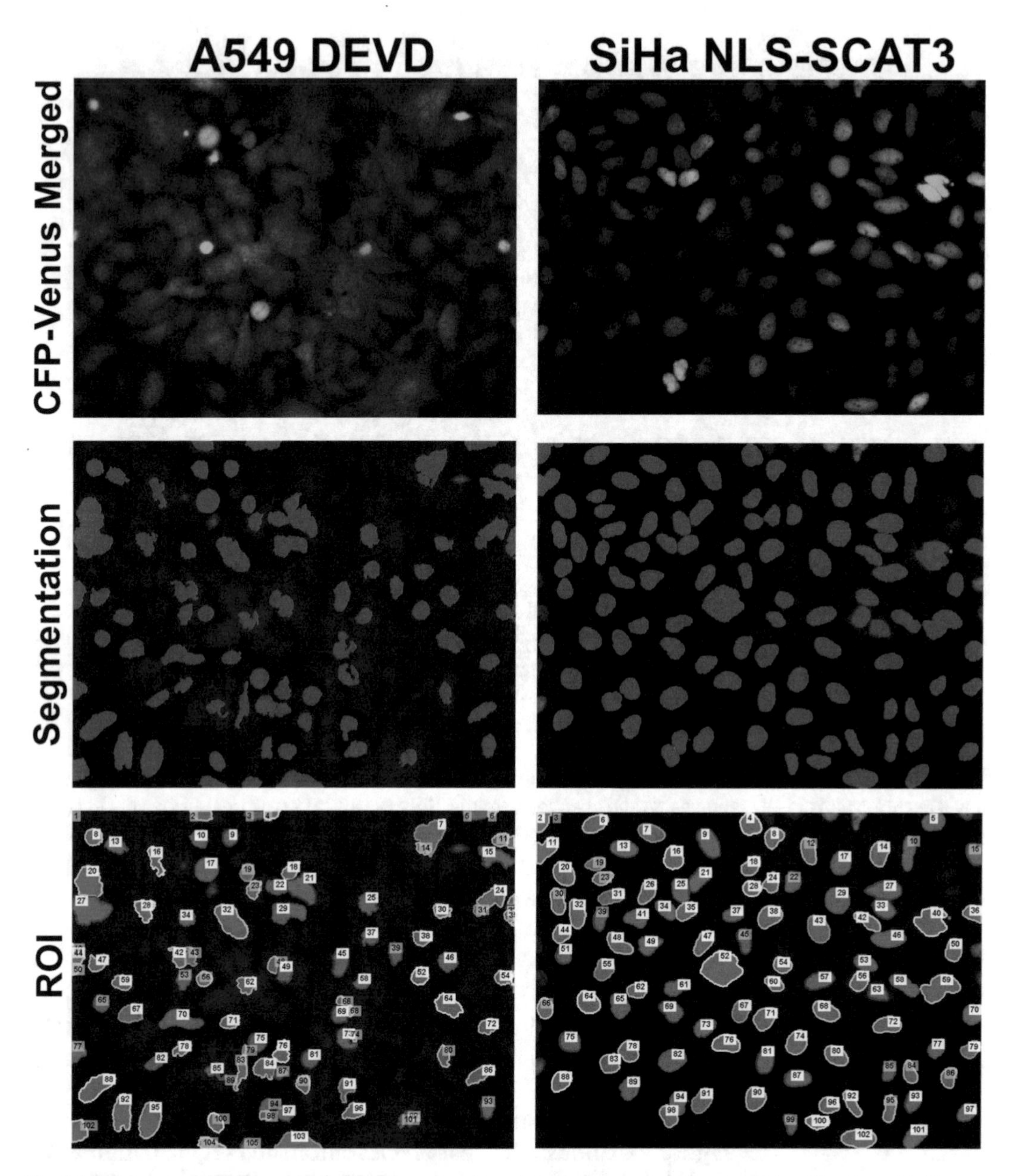

Fig. 10 Efficiency of DEVD and NLS-SCAT3 probes in automated segmentation and identification of ROIs was compared using A549 DEVD and SiHa NLS-SCAT3 cells. As seen in the figure, NLS-SCAT3 probe is more effective in the identification of individual cells via automated segmentation than DEVD probe

the former allows easier, accurate segmentation and identifies almost all the cells in the field. DEVD, being a probe homogenously expressed in the cell, is less efficient in differentiating individual cells in an image. This is evident from the comparative analysis of the segmented images from SiHa NLS-SCAT3 and A549 DEVD (Fig. 10).

3.4 Automated Microscopy and Image Acquisition in Multi-well Plates

Ratio imaging-compatible cellular tool for caspase activation, consisting of nucleus targeted FRET-based sensor probe, can be adapted for high-throughput imaging in multi-well plates in an automated manner. For this, cells expressing the sensor should be seeded in Nunc™ MicroWell 96-well optical bottom-plates and imaged using fully motorized inverted phase contrast fluorescence microscope, ECLIPSE T*i*-E, with focus drift compensation (PFS) controlled using NIS-element AR 4.2 software, with the 20× Plan Apo objective (NA 0.75). Glass-bottom plates give better signal-to-noise ratio during fluorescence imaging as the bottom plate is very thin (170 μM), is optically clear and is devoid of auto fluorescence property (*see* **Note 13**). Using NIS-element AR 4.2 software, multiple distinct imaging spots can be selected within each well and also in multiple-wells in the multidimensional acquisition setting. Here, PFS is extremely helpful in maintaining focus when images are being captured from different *XY* positions.

For imaging of ECFP-DEVD-Venus, the microscope should be configured with an excitation filter of 438 ± 12 nm band pass for ECFP, 458 long pass (LP) dichroic filter and dual emission filter set with band pass at 483 ± 15 nm and 542 ± 27 nm for ECFP and Venus FRET respectively. The dichroic and excitation filters are to be placed in the motorized fluorescent turret within the fluorescent filter cube (Nikon® Corporation) of the microscope. The dual emission filters may be placed in front of the camera in a motorized emission filter-wheel (8-position filter wheel from Nikon® Corporation). Dual emission was collected by changing the emission filter wheel position, while keeping the dichroic filter in-position using NIS-elements AR software in a ratio mode acquisition-setting. Images can be collected using cooled CCD camera, CoolSNAP HQ2.

1. Multi-welled culture vessels such as 96-well glass-bottom plate are to be seeded with FRET probe-expressing cells (here, SiHa NLS-SCAT3) and maintained in CO_2 incubator till the cell population reaches 60–70 % confluency.
2. Prepare the drugs to be added at their respective working concentrations in phenol red-free DMEM containing 5 % FBS (5 μM for tunicamycin, 2 μM for thapsigargin, 10 μg/ml for doxorubicin and 50 μg/ml for cisplatin).
3. Add the drug-containing DMEM into cells in triplicate wells. Also maintain control wells in fresh phenol red-free culture medium supplemented with 5 % FBS.
4. Position the 96-well plate in the stage-mountable incubation chamber and adjust the microscope (ECLIPSE T*i*-E) so that the cells are accurately focused using the 20× objective lens, under white light.

5. In the NIS-elements AR software, open the ND acquisition window and specify the channels in which the images are to be acquired.
6. Define the exposure time for capturing both donor and acceptor emission fluorescence signal in a range of 200–400 ms so as to have similar level of emission intensities. The output image should be noise free and clear with optimum signal intensity (*see* **Note 9**).
7. Identify at least three imaging fields in each of the well and confirm each in the "*XY* position" listing in the ND acquisition window.
8. Before running, specify the total time duration of the whole imaging session (24 h) as well as the interval between subsequent imaging (1 h).
9. The image sequence will be saved in .nd2 file format. Each and every image obtained from both the color channels (CFP and YFP), all imaging fields (*XY* position) as well as different time points will be included in the image sequence.
10. Carry out the ratio analysis of the images after getting ratio view, subtracting background, defining binary (segmentation and ROIs) and applying the binary to the whole sequence. The quantitative date of ROI/object can be used to generate a graphical representation of the percentage of cells with caspase activation after drug treatment.

Representative ratio images and graphs shown in Fig. 11 substantiates that, the FRET-based caspase sensor probe-expressing live-cell tool developed for monitoring caspase activation in live cells, effectively indicates the ratio change due to caspase activation and is highly compatible for automated, multi-point, time-lapse imaging as well as automated segmentation and ratio analysis.

4 Future Prospects

The present chapter describes in detail a live cell intracellular caspase activation assay using FRET-based probe; that has potential implications in anticancer drug screening. For any anticancer drug screening technique to be successful as a preliminary screening technology, a critical requirement is the availability of ready-to-use sensor cell lines—capable to detecting the desired endpoint—that have sensitivity and reproducibility of results apart from high-throughput adaptability. As described in previous sections, with increasing numbers of genetically encoded fluorescent proteins and probes along with efficient transfection protocols and cell sorting capabilities available, it is possible to generate fluorescent probe-expressing stable cellular tools for detecting caspase activation in all the 60 cell lines of the NCI-60 platform.

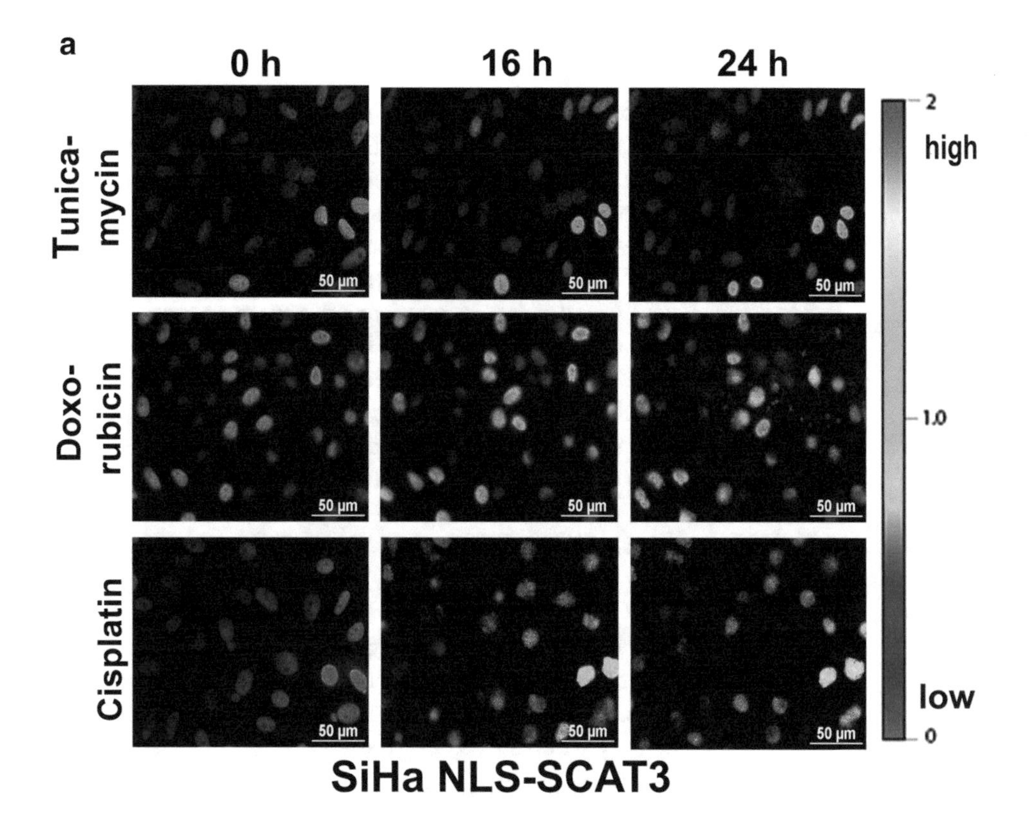

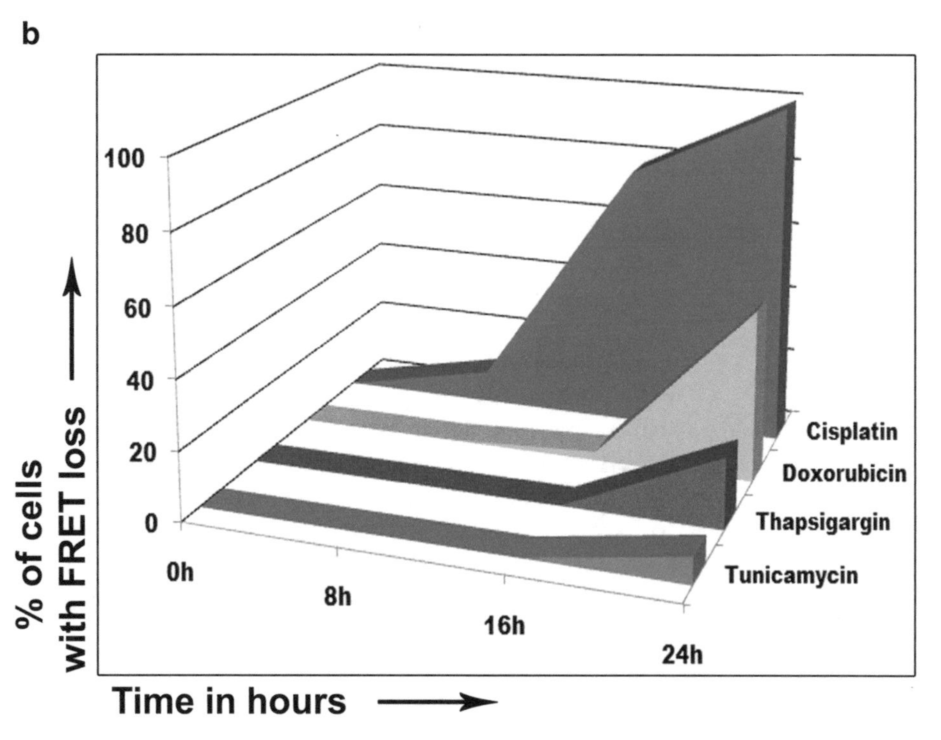

Fig. 11 SiHa NLS-SCAT3 cells were seeded on to 96-well plates, subjected to treatment with multiple drugs and imaged for FRET loss in a time-lapse dependent manner. (**a**) Representative ratio images of SiHa NLS-SCAT3 cells subjected to tunicamycin (5 μM), doxorubicin (10 μg/ml), and cisplatin (50 μg/ml) at 0, 16, and 24 h are shown. (**b**) Representative graphs of selected drugs, showing percentage of cells with FRET loss obtained using NLS-SCAT3 probe-directed segmentation, are also given

We describe the protocol for detection of caspase activation using FRET-based probes in cancer cell lines stably expressing them inside the nuclear compartment. Overall, this approach possesses several advantages over the current sulforhodamine B (SRB) assay such as being able to distinguish apoptotic cell death continuously and reliably, and being noninvasive and less labor-intensive. Currently, our lab is focusing on developing cells with stable expression of caspase-specific FRET-based sensor probe in an extended panel of cancer cell lines. Once this is successfully done, such a panel could be a reliable tool for high-throughput anticancer drug screening, with superior capabilities for the systematic characterization of caspase activation kinetics.

5 Notes

1. We routinely prepare thapsigargin stock solution in DMSO and store at −20 °C. We recommend storing thapsigargin stock solution in brown colored glass vials as the drug seems to lose its activity when stored in plastic vials.
2. We prefer Midi kit over Mini or Mega kits (Qiagen) as it yields pure plasmid DNA (a maximum of 100 μg) enough for further adequate number of repetitive transfection experiments (a minimum of 100 ng for a single well in a 96-well plate/1–4 × 10^4 cells) and has convenient-to-handle sample and reagent volumes as well.
3. During transformation experiments, in order to ensure finding transformed bacterial colonies irrespective of the transformation efficiency, the bacterial cells will be seeded on antibiotic-containing LB agar plates at two different concentrations. After transformation by heat shock method and growth for 1 h in 1 ml of antibiotic free LB broth, about 100 μl of broth of transformed colonies will be directly spread-plated on an ampicillin positive LB agar plate. Rest of the transformed colonies will be pelleted down, resuspended in 100 μl of broth and spread-plated on the second ampicillin positive LB agar plate. The first plating event generates isolated colonies of transformed bacterial cells when the transformation efficiency is fairly good, while the second plating event increases the chance of obtaining isolated colonies of transformed bacterial cells even when the transformation efficiency is very poor.
4. Too much of the transfection reagent and plasmid DNA are toxic to the mammalian cells used for transfection. Though the time period up to which cells should be exposed to the lipid–DNA transfection complexes for maximum transfection efficiency varies with cell lines (24–72 h), toxic effect of the transfection complexes can be reduced by diluting the reaction mixture with additional medium after 12 h.

5. Forward scatter (FSC) and side scatter (SSC) signal information as well as doublet discrimination technique helps in locating a single population in the FACS profile. While forward scatter signal gives an indication of the size of the cell, side scatter signal gives an indication of the granularity of the cell. Cells from a single colony may have comparable cell size and cellular granularity. Live, single cells are represented by an identifiable separate population with medium signal values in a scatter plot generated by FSC signal against SSC signal. This population is gated tight and further used for doublet discrimination (first scatter plot in Fig. 3a). Presence of cell aggregates, doublets, or triplets can interfere with analysis of fluorescence signal from each event. Doublet/aggregates are identified as events with higher width value in the scatter plot graphs generated by taking width value against area values of both FSC and SSC signals. These events are omitted first in FSC-width (FSC-W) versus area (FSC-A) graph and then in SSC-width (SSC-W) versus are (SSC-A) graph (second and third scatter plots respectively in Fig. 3a); such that the events finally gated represent a single population of live, single cells (third scatter plot in Fig. 3a).
6. When a cell line possessing a stable, homogeneous expression of an exogenous protein is revived after a long period of cryopreservation, there may be a slight decline in the level of expression as well as the percentage of cells expressing the plasmid. Hence, to enrich the cell population for homogeneous, stable expression of the protein, it is advisable to subject the cells to selection medium for a few days before it is being used for further experiments.
7. An EMCCD camera will enable very fast imaging with minimum exposure time and multiplexing of signals from different fluorescent channels. If confocal mode is employed for imaging, with "disk in" position of spinning-disk, the fluorescence signal will dramatically decrease. In order to compensate for this, better, sensitive and high QE camera needs to be employed; preferably EMCCD camera than normal interlined CCD camera. Position of the spinning disk is controlled by the "disk slider" option in IPLab™ software.
8. A band-pass filter allow light of a particular frequency/wavelength range, while block (attenuates) the rest of the frequencies. Dichroic filters, also called beam splitters selectively allows the passage of light of a narrow range of wavelength and reflects other wavelength spectra. A dichroic filter is identified by its cutoff value for reflection. A long pass dichroic filter sharply reflects back all the wavelengths below the cutoff value, at the same time allowing all the wavelengths above this value to pass through. The dichroic filter used in FRET experiment (452 LP)

should reflect excitation wavelength signal (here, 438 nm), at the same time transmit emission wavelength signal from donor and acceptor fluorophores (here, 483 and 582 nm).

9. While taking ratio images for FRET analysis using NLS-SCAT3 probe, exposure time was fixed to have almost same level of emission intensities for both ECFP and Venus channels that ranges from 200 to 400 ms. While working with NIS-elements, the emission intensities are adjusted in such a way that cells in the untreated samples are displayed in a color (purple or blue) that corresponds to the lower level of ratio values in the color scale (discussed in **Note 10**).
10. The accurate display of the ratio images affects the subsequent analysis of ratio information and representation of caspase activation. Once the ratio properties like numerator, denominator, range, and background are defined, the software will automatically generate a ratio image based on a color scale. Called the IMD (Intensity Modulation Display), it is generated from both the fluorescence intensity values and ratio level values. In this mode of display, ratio images are generated by presenting the ratio information in a color scale ranging from purple to red. Different colors in the color scale indicate corresponding ratio values in the ratio scale ranging from 0 to 2.
11. Sometimes it may be difficult to find a region in the image field that is free of any fluorescence signal throughout the image sequence. This may be due to cells getting more confluent over time or due to movement of the cells or due to high magnification during single cell imaging. An alternate way of background correction can be followed in such situations. There is an option for background correction in the drop-down menu of the "Image" function of the main tool bar of NIS-elements software. A constant value can be specified, which can be used for background correction all throughout the image sequence.
12. All the approaches described for bringing about accurate segmentation are equally effective and give comparable output. At times, two equally fluorescing nuclei staying close to each other may however be identified as a single segment. Keeping the "separate objects" tab active during segmentation will improve segmentation and help in identifying two closely placed objects as separate.
13. 96-well optical-bottom imaging plates are available from various manufactures (Nunc, BD Falcon, Greiner Bio-One, etc.). Our lab has found comparable image quality upon using plates from these different manufactures. Optical-bottom part of such plates can be made of either glass or polymer with optical properties similar to glass. Though glass-bottom plates provide clearer images even for oil objectives, they are less suitable for

growing cells that require highly adherent growth-surfaces. Polymer-bottom optical plates provide a better adherent growth-surface for cells.

Acknowledgements

This study was supported by research grants from Department of Biotechnology, Government of India. KAM was supported by research fellowship from ICMR (Indian Council of Medical Research, Govt. of India), DI and JJ were supported by research fellowship from CSIR (Council of Scientific and industrial Research, Govt. of India), and SNV was supported by research fellowship from UGC (University Grants Commission, Govt. of India). We thank Professor M. Radhakrishna Pillai, Director, RGCB, for his constant support and encouragement. We are grateful to all members of cancer research program laboratory and staff of the institutional FACS facility for helping throughout the work. We thank Prof. Jeremy M. Tavare and Dr. Gavin Welsh (University of Bristol, UK) for kindly sharing the plasmid, pcDNA3 ECFP-DEVD-EYFP; and Dr. Masayuki Miura (University of Tokyo, Japan) for the plasmid, pcDNA NLS-SCAT3.

References

1. Suggitt M, Bibby MC (2005) 50 years of preclinical anticancer drug screening: empirical to target-driven approaches. Clin Cancer Res 11(3):971–981
2. Alley MC, Scudiero DA, Monks A, Hursey ML, Czerwinski MJ, Fine DL, Abbott BJ, Mayo JG, Shoemaker RH, Boyd MR (1988) Feasibility of drug screening with panels of human tumor cell lines using a microculture tetrazolium assay. Cancer Res 48(3):589–601
3. Lovell JF, Billen LP, Bindner S, Shamas-Din A, Fradin C, Leber B, Andrews DW (2008) Membrane binding by tBid initiates an ordered series of events culminating in membrane permeabilization by Bax. Cell 135(6):1074–1084
4. Bates SE, Fojo AT, Weinstein JN, Myers TG, Alvarez M, Pauli KD, Chabner BA (1995) Molecular targets in the National Cancer Institute drug screen. J Cancer Res Clin Oncol 121(9–10):495–500
5. Tian H, Ip L, Luo H, Chang DC, Luo KQ (2007) A high throughput drug screen based on fluorescence resonance energy transfer (FRET) for anticancer activity of compounds from herbal medicine. Br J Pharmacol 150(3):321–334
6. Goldstein JC, Waterhouse NJ, Juin P, Evan GI, Green DR (2000) The coordinated release of cytochrome c during apoptosis is rapid, complete and kinetically invariant. Nat Cell Biol 2(3):156–162
7. Seervi M, Joseph J, Sobhan PK, Bhavya BC, Santhoshkumar TR (2011) Essential requirement of cytochrome c release for caspase activation by procaspase-activating compound defined by cellular models. Cell Death Dis 2:e207. doi:10.1038/cddis.2011.90
8. Wei MC, Zong W-X, Cheng EH-Y, Lindsten T, Panoutsakopoulou V, Ross AJ, Roth KA, MacGregor GR, Thompson CB, Korsmeyer SJ (2001) Proapoptotic BAX and BAK: a requisite gateway to mitochondrial dysfunction and death. Science 292(5517):727–730
9. Sharma SV, Haber DA, Settleman J (2010) Cell line-based platforms to evaluate the therapeutic efficacy of candidate anticancer agents. Nat Rev Cancer 10(4):241–253
10. Vichai V, Kirtikara K (2006) Sulforhodamine B colorimetric assay for cytotoxicity screening. Nat Protoc 1(3):1112–1116
11. Pop C, Salvesen GS, Scott FL (2008) Caspase assays: identifying caspase activity and sub-

strates *in vitro* and *in vivo*. In: Khosravi-Far R, Zakeri Z, Lockshin RA, Piacentini M (eds) Programmed cell death, the biology and therapeutic implications of cell death—Part B, vol 446, Methods in enzymology. Elsevier, Amsterdam, pp 351–367

12. Packard BZ, Komoriya A (2008) Intracellular protease activation in apoptosis and cell-mediated cytotoxicity characterized by cell-permeable fluorogenic protease substrates. Cell Res 18(2):238–247
13. Petrovsky A, Schellenberger E, Josephson L, Weissleder R, Bogdanov A Jr (2003) Near-infrared fluorescent imaging of tumor apoptosis. Cancer Res 63(8):1936–1942
14. Shimomura A, Johnson FH, Saiga Y (1962) Extraction, purification and properties of aequorin, a bioluminescent protein from the luminous hydromedusan, *Aequorea*. J Cell Comp Physiol 59(3):223–239
15. Hoffman RM (2015) Application of GFP imaging in cancer. Lab Invest 95(4):432–452
16. Haseloff J (1999) GFP variants for multispectral imaging of living cells. In: Sullivan KF, Kay SA (eds) Green fluorescent proteins, vol 58, Methods in cell biology. Elsevier, Amsterdam, pp 139–151
17. He L, Wu X, Meylan F, Olson DP, Simone J, Hewgill D, Siegel R, Lipsky PE (2004) Monitoring caspase activity in living cells using fluorescent proteins and flow cytometry. Am J Pathol 164(6):1901–1913
18. Huitema C, Eltis LD (2010) A fluorescent protein-based biological screen of proteinase activity. J Biomol Screen 15(2):224–229
19. Tait SW, Bouchier-Hayes L, Oberst A, Connell S, Green DR (2009) Live to dead cell imaging. In: Erhardt P, Toth A (eds) Apoptosis: methods and protocols, vol 559, 2nd edn, Springer protocols: Methods in molecular biology. Humana, New York, pp 33–48
20. Puigvert JC, de Bont H, van de Water B, Danen EH (2010) High-throughput live cell imaging of apoptosis. In: Bonifacino JS, Dasso M, Harford JB, Lippincott-Schwartz J, Yamada KM (eds) Current protocols in cell biology, vol 47. Wiley Online Library, New York, pp 18.10.1–18.10.13
21. Figueroa RA, Ramberg V, Gatsinzi T, Samuelsson M, Zhang M, Iverfeldt K, Hallberg E (2011) Anchored FRET sensors detect local caspase activation prior to neuronal degeneration. Mol Neurodegener 6(1):35. doi:10.1186/1750-1326-6-35
22. Angres B, Steuer H, Weber P, Wagner M, Schneckenburger H (2009) A membrane-bound FRET-based caspase sensor for detection of apoptosis using fluorescence lifetime and total internal reflection microscopy. Cytometry A 75A(5):420–427
23. Elphick LM, Meinander A, Mikhailov A, Richard M, Toms NJ, Eriksson JE, Kass GE (2006) Live cell detection of caspase-3 activation by a Discosoma-red-fluorescent-protein-based fluorescence resonance energy transfer construct. Anal Biochem 349(1):148–155
24. Tyas L, Brophy VA, Pope A, Rivett AJ, Tavare JM (2000) Rapid caspase-3 activation during apoptosis revealed using fluorescence-resonance energy transfer. EMBO Rep 1(3): 266–270
25. Jones J, Heim R, Hare E, Stack J, Pollok BA (2000) Development and application of a GFP-FRET intracellular caspase assay for drug screening. J Biomol Screen 5(5):307–317
26. Rehm M, Düßmann H, Jänicke RU, Tavaré JM, Kögel D, Prehn JH (2002) Single-cell fluorescence resonance energy transfer analysis demonstrates that caspase activation during apoptosis is a rapid process. J Biol Chem 277(27):24506–24514
27. Nagai T, Ibata K, Park ES, Kubota M, Mikoshiba K, Miyawaki A (2002) A variant of yellow fluorescent protein with fast and efficient maturation for cell-biological applications. Nat Biotechnol 20(1):87–90
28. Takemoto K, Nagai T, Miyawaki A, Miura M (2003) Spatio-temporal activation of caspase revealed by indicator that is insensitive to environmental effects. J Cell Biol 160(2):235–243
29. Shcherbo D, Souslova EA, Goedhart J, Chepurnykh TV, Gaintzeva A, Shemiakina II, Gadella TW, Lukyanov S, Chudakov DM (2009) Practical and reliable FRET/FLIM pair of fluorescent proteins. BMC Biotechnol 9:24. doi:10.1186/1472-6750-9-24
30. Kawai H, Suzuki T, Kobayashi T, Sakurai H, Ohata H, Honda K, Momose K, Namekata I, Tanaka H, Shigenobu K, Nakamura R, Hayakawa T, Kawanishi T (2005) Simultaneous real-time detection of initiator- and effector-caspase activation by double fluorescence resonance energy transfer analysis. J Pharmacol Sci 97(3):361–368
31. Joseph J, Seervi M, Sobhan PK, Retnabai ST (2011) High throughput ratio imaging to profile caspase activity: potential application in multiparameter high content apoptosis analysis and drug-screening. PLoS One 6(5):e20114. doi:10.1371/journal.pone.0020114
32. Seervi M, Sobhan PK, Mathew KA, Joseph J, Pillai PR, Santhoshkumar TR (2014) A high-throughput image-based screen for the identification of Bax/Bak-independent caspase

activators against drug-resistant cancer cells. Apoptosis 19(1):269–284

33. Bhattacharyya S (2001) Combinatorial approaches in anticancer drug discovery: recent advances in design and synthesis. Curr Med Chem 8(12):1383–1404
34. Lam KS (1997) Application of combinatorial library methods in cancer research and drug discovery. Anticancer Drug Des 12(3): 145–167
35. Shoemaker RH, Scudiero DA, Melillo G, Currens MJ, Monks AP, Rabow AA, Covell DG, Sausville EA (2002) Application of high-throughput, molecular-targeted screening to anticancer drug discovery. Curr Top Med Chem 2(3):229–246
36. Reed JC (1999) Dysregulation of apoptosis in cancer. J Clin Oncol 17(9):2941–2953
37. Sellers WR, Fisher DE (1999) Apoptosis and cancer drug targeting. J Clin Invest 104(12): 1655–1661
38. Tsien RY, Harootunian AT (1990) Practical design criteria for a dynamic ratio imaging system. Cell Calcium 11:93–100

Chapter 9

Antibody-Based Proteomic Analysis of Apoptosis Signaling

Matthew P. Stokes, Hongbo Gu, and Jeffrey C. Silva

Abstract

Reagents that assess activation of apoptosis and associated signaling pathways are critical for greater understanding of the molecular basis of programmed cell death. The advent of proteomic technologies to probe these events allows monitoring of hundreds to thousands of proteins, as well as sites of posttranslational modification involved in apoptosis at one time. This view of apoptosis at a network level is a powerful tool in studying known apoptotic pathways, as well as elucidating novel signaling events that affect or are affected by apoptotic signaling. The following is a detailed method for successful proteomic profiling of apoptosis using antibody-based enrichment methods along with a liquid chromatography–tandem mass spectrometry analytical platform.

Key words Apoptosis, Proteomics, Posttranslational modification, Antibody, LC-MS/MS, Phosphorylation, Caspase

Abbreviations

IAP	Immunoaffinity purification
LC	Liquid chromatography
MS/MS	Tandem mass spectrometry
MeCN	Acetonitrile
MS	Mass spectrometry
PTM	Posttranslational modification
TFA	Trifluoroacetic acid

1 Introduction

The apoptotic program and its associated signaling networks are a critical component of cellular and organismal growth, development, and disease [1–5]. Methods and reagents that probe these pathways are therefore important for the study of a wide variety of

Electronic supplementary material: The online version of this chapter (doi:10.1007/978-1-4939-3588-8_9) contains supplementary material, which is available to authorized users.

Perpetua M. Muganda (ed.), *Apoptosis Methods in Toxicology*, Methods in Pharmacology and Toxicology,
DOI 10.1007/978-1-4939-3588-8_9, © Springer Science+Business Media New York 2016

biological systems. Cell Signaling Technology has developed a number of reagents for the study of apoptosis signaling. These include antibodies that target apoptotic proteins, as well as site-specific antibodies that target sites of posttranslational modification on apoptotic proteins (see pathway diagram, Fig. 1). These reagents can be used for assays such as western blotting, immunoprecipitation, flow cytometry, and immunohistochemistry. Antibodies have also been developed that specifically recognize the cleaved, activated form of caspases, the cysteine proteases responsible for the effects of initiation of the apoptotic program [6–8]. These antibodies provide a direct view of the activation of apoptosis in cells and tissues.

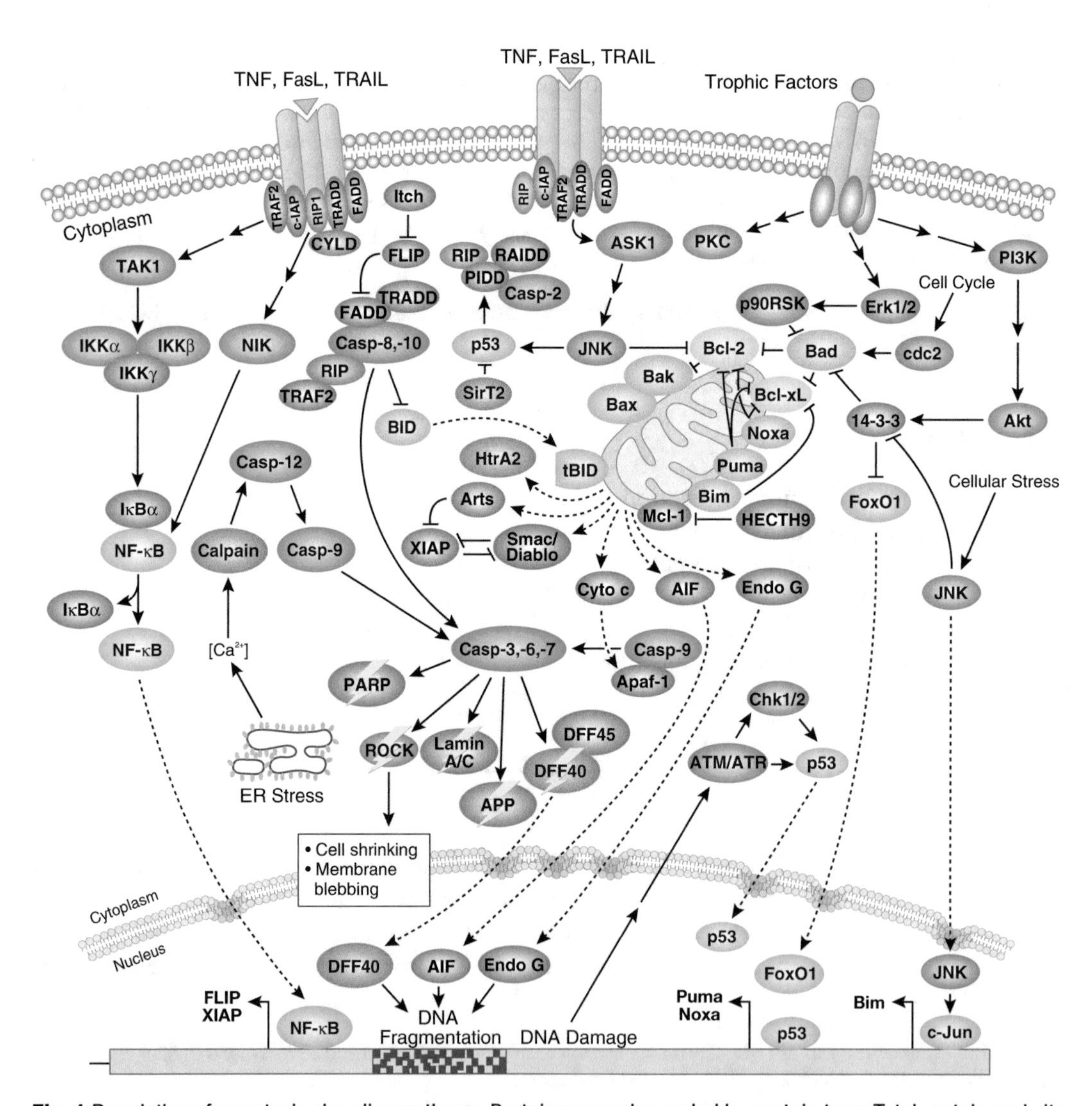

Fig. 1 Regulation of apoptosis signaling pathway. Proteins are color-coded by protein type. Total protein and site specific antibodies are available for most proteins in the pathway. For full list of associated antibodies and color-coding key, see http://www.cellsignal.com/common/content/content.jsp?id=pathways-apoptosis-regulation

Beyond single antibody methods for analysis of apoptotic signaling, Cell Signaling Technology has developed novel, liquid chromatography–tandem mass spectrometry (LC-MS/MS) based methods that combine the specificity and enrichment of immunoaffinity capture of peptides with the speed and dynamic range of the tandem mass spectrometer. These methods, collectively termed PTMScan, have been used to profile a wide variety of cell lines and tissue types in development and disease [9–16]. The PTMScan Method is outlined in Fig. 2. In PTMScan, cells, tissues, or other biological materials are prepared under denaturing conditions, digested to peptides with enzymes such as trypsin or LysC, purified over reverse-phase columns, and subjected to immunoaffinity purification (IAP) using antibody bound to Protein A or Protein G beads. Bound peptides are then eluted off the beads, desalted over C18, and run in LC-MS. The relative abundance of the same peptide ion across multiple samples can be determined using a label-free quantitative method that reports the integrated peak area in the MS1 channel (corresponding to the intact tryptic peptide), or labeling methods such as SILAC, reductive amination, or isobaric tags [9, 15, 17–24]. Thus, PTMScan provides quantitative analysis of hundreds to thousands of posttranslationally modified or motif-containing peptides in a single experiment.

One of the key factors contributing to the success of the PTMScan method was the development of motif antibodies. In contrast to a standard antibody development program in which one peptide or protein antigen is used to elicit an immune response, motif antibodies are raised by injecting degenerate peptide

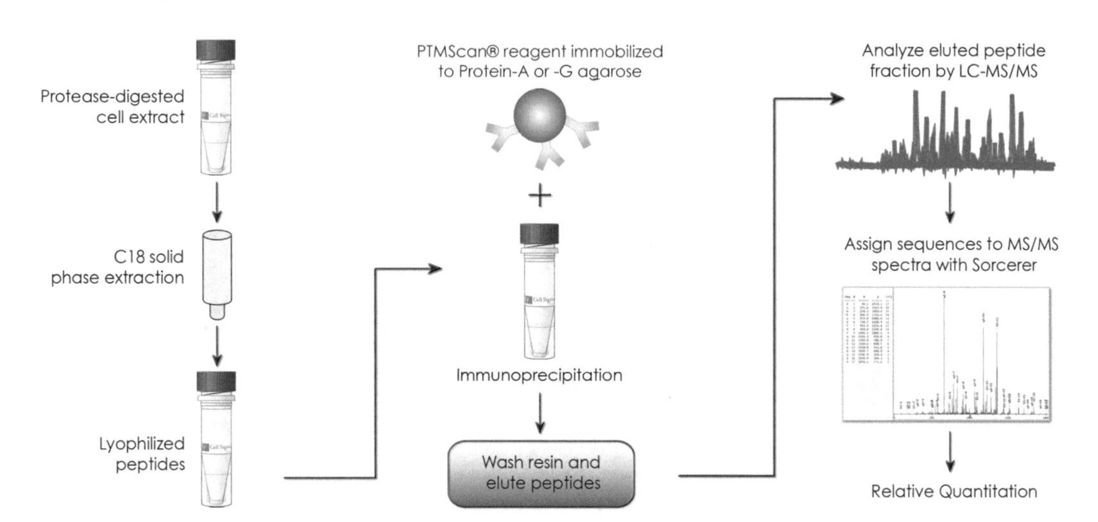

Fig. 2 The PTMScan method. Cells or tissues are harvested, digested to peptides, purified over reverse phase columns, and immunoprecipitated with the appropriate antibody. IP'd peptides are run in LC-MS/MS, identified by database search, and relative quantification is performed

libraries, in which one or a few key residues are fixed and present in all peptides in the library, while the other amino acids are varied [25]. This strategy has allowed development of antibodies that recognize phosphotyrosine, a series of antibodies that recognize consensus kinase substrate motifs, and other posttranslational modifications such as acetylation, ubiquitination, methylation, and lysine succinylation (www.cellsignal.com) [9, 10, 13, 14].

A motif antibody that specifically detects sites of caspase cleavage at an aspartic acid residue [2, 7, 8] has also been developed. The antibody preferentially immunoprecipitates peptides with a DE(T/S/A)D sequence at the C-terminus (Cell Signaling Technology, #8698/12810). This antibody allows simultaneous detection and quantitation of peptides from caspases themselves, as well as putative caspase substrates, providing a global view of the apoptotic program [26]. Figure 3 shows a representative western blot using the cleaved caspase substrate motif antibody in untreated

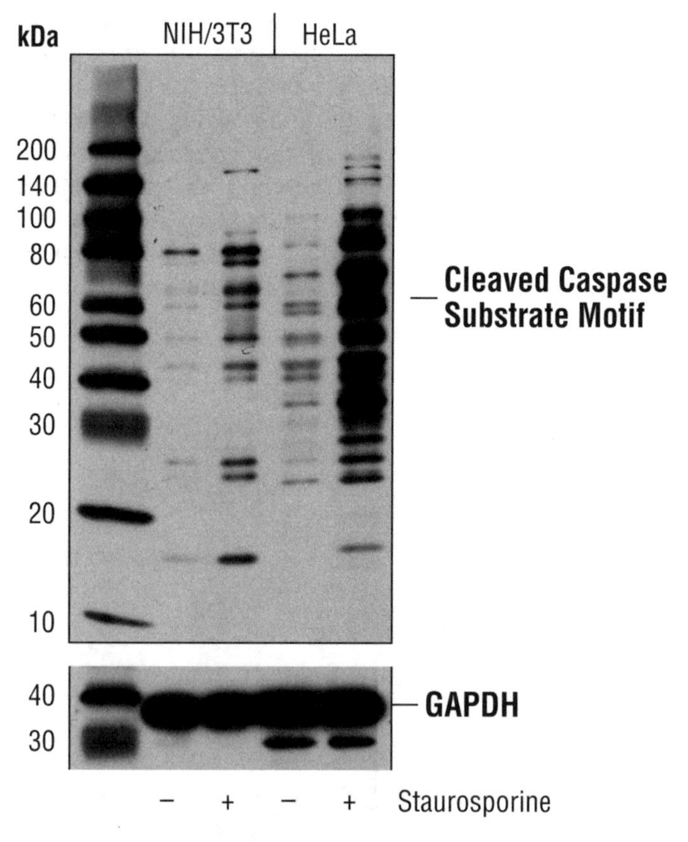

Fig. 3 Western blot analysis using the cleaved caspase substrate motif antibody #8698. NIH-3T3 or HeLa cells were treated with DMSO or Staurosporine, harvested, and run in western blot using the cleaved caspase substrate motif antibody (Cell Signaling Technology, #8698) or GAPDH antibody (Cell Signaling Technology, #5174)

and staurosporine treated cells. Supplemental Table S1 shows a typical PTMScan result using the cleaved caspase substrate motif antibody to immunoprecipitate peptides from Jurkat cells treated with Etoposide.

In addition to PTMScan using motif antibodies (collectively termed PTMScan Discovery), an LC-MS method utilizing cocktails of site-specific antibodies, termed PTMScan Direct, has also been developed [15, 16]. Whereas PTMScan Discovery will allow identification and quantification of peptides that share a common sequence motif (such as the caspase cleavage substrate motif DE(T/S/A)D), PTMScan Direct reagents have been designed to immunoprecipitate peptides from proteins that reside in the same signaling pathway or from the same protein type, such as kinases. Six different PTMScan Direct reagents have been developed, among them a reagent that probes apoptotic and autophagolytic pathways (PTMScan Direct: Apoptosis/Autophagy). This method allows simultaneous identification and quantification of roughly 300 sites on 100 proteins involved in apoptotic and autophagolytic signaling.

The following method describes in detail the protocol for performing immunoaffinity purification and liquid chromatography/tandem mass spectrometry (IAP LC-MS/MS) in five sections: (a) Cell Lysis and Protein Digestion, (b) Sep-Pak® C18 Purification of Lysate Peptides, (c) Immunoaffinity Purification (IAP) of Peptides, (d) Concentration and Purification of Peptides for LC-MS Analysis, and (e) LC-MS/MS Analysis of Peptides.

2 Materials

1. HEPES (Sigma, H-4034)
2. Sodium pyrophosphate (Sigma, S-6422)
3. β-glycerophosphate (Sigma, G-9891)
4. Urea, Sequanal Grade (Thermo Scientific, 29700)
5. Sodium orthovanadate (Sigma, S-6508)
6. Coomassie Plus Protein Assay Reagent (Thermo Scientific, 1856210)
7. Dithiothreitol (American Bioanalytical, AB-00490)
8. Iodoacetamide (Sigma, I-6125)
9. Trypsin-TPCK (Worthington, LS-003744)
10. Trifluoroacetic acid, Sequanal Grade (Thermo Scientific, 28903)
11. Acetonitrile (Thermo Scientific, 51101)
12. Sep-Pak® Classic C18 columns, 0.7 mL (Waters, WAT051910)

13. MOPS (Sigma, M3183)
14. Burdick and Jackson water (Honeywell, AH365-4)
15. Ammonium bicarbonate (Sigma, A6141)
16. Trypsin (Promega, V5113)

NOTE: Prepare solutions for cell lysis, Sep-Pak purification, and IAP enrichment with Milli-Q or equivalent grade water. Prepare solutions for subsequent steps with HPLC grade water (e.g., Burdick and Jackson water).

3 Methods

3.1 Cell Lysis and Protein Digestion

3.1.1 Stock Solutions

1. 200 mM HEPES/NaOH, pH 8.0: Dissolve 23.8 g HEPES in approximately 450 mL water, adjust pH with 5 M NaOH to 8.0, and bring to a final volume of 500 mL. Filter through a 0.22 μM filter (as used for cell culture), store at −20 °C, use for up to 6 months.
2. Sodium pyrophosphate: Make 50× stock (125 mM, MW = 446): 1.1 g/20 mL. Store at 4 °C, use for up to 1 year.
3. β-glycerophosphate: Make 1000× stock (1 M, MW = 216): 2.2 g/10 mL. Divide into 100 μL aliquots and store at −80 °C, use for up to 1 year.
4. Sodium orthovanadate (Na_3VO_4): Make 100× stock (100 mM, MW = 184): 1.84 g/100 mL. Sodium orthovanadate must be depolymerized (activated) according to the following protocol:
   (a) For a 100 mL solution, fill up with water to approximately 90 mL. Adjust the pH to 10.0 using 1 M NaOH with stirring. At this pH, the solution will be yellow.
   (b) Boil the solution until it turns colorless and cool to room temperature (put on ice for cooling).
   (c) Readjust the pH to 10.0 and repeat step 2 until the solution remains colorless and the pH stabilizes at 10.0 (usually it takes two rounds). Adjust the final volume with water.
   (d) Store the activated sodium orthovanadate in 1 mL aliquots at −80 °C, use for up to 1 year. Thaw one aliquot for each experiment; do not refreeze thawed vial.
5. Urea lysis buffer: 20 mM HEPES pH 8.0, 9 M urea, 1 mM sodium orthovanadate, 2.5 mM sodium pyrophosphate, 1 mM β-glycerophosphate.

   NOTE: The urea lysis buffer should be prepared fresh prior to each experiment.

NOTE: The urea lysis buffer should be used at room temperature. Placing the urea lysis buffer on ice will cause the urea to precipitate out of solution.

6. Dithiothreitol (DTT): Make 1.25 M stock (MW = 154): 19.25 g/100 mL. Divide into 200 μL aliquots, store at −80 °C for up to 1 year.
7. Iodoacetamide solution: Dissolve 95 mg of iodoacetamide (formula weight = 184.96 mg/mmol) in water to a final volume of 5 mL. After weighing the powder, store in the dark and add water only immediately before use. The iodoacetamide solution should be prepared fresh prior to each experiment.
8. Trypsin-TPCK: Store dry powder for up to 1 year at 4 °C. Parafilm cap of trypsin container (Worthington) to avoid collecting moisture, which can lead to degradation of the reagent. Prepare 1 mg/mL stock in 1 mM HCl. Divide into 1 mL aliquots, store at −80 °C for up to 1 year.

3.1.2 Preparation of Cell Lysate

1. Grow approximately 2×10^8 cells for each experimental condition (enough cells to produce approximately 20 mg of soluble protein).

   NOTE: Cells should be washed with 1× PBS before lysis to remove any media containing protein contaminants. Elevated levels of media-related proteins will interfere with the total protein determination.
2. Suspension cells: Harvest cells by centrifugation at 130 rcf (*g*), for 5 min at room temperature. Carefully remove supernatant, wash cells with 20 mL of cold PBS, centrifuge and remove PBS wash and add 10 mL urea lysis buffer (room temperature) to the cell pellet. Pipet the slurry up and down a few times (do not cool lysate on ice as this may cause precipitation of the urea).
3. Adherent cells: Wash the plates with 10 mL each cold 1× PBS, harvest each plate by scraping sequentially with the same 10 mL of urea lysis buffer (10 mL buffer to plate 1, scrape, add this same 10 mL buffer to plate 2, etc.)

   NOTE: If desired, the PTMScan protocol may be interrupted at this stage. The harvested cells can be frozen and stored at −80 °C for several weeks.
4. Using a microtip, sonicate at 15 W output with 3 bursts of 15 s each. Cool on ice for 1 min between each burst. Clear the lysate by centrifugation at 20,000 rcf (*g*) for 15 min at room temperature and transfer the protein extract (**supernatant**) into a new tube.

   NOTE: Sonication fragments DNA and reduces sample viscosity. Ensure that the sonicator tip is submerged in the lysate. If the sonicator tip is not submerged properly, it may induce foaming and degradation of your sample (refer to the manufacturer's instruction manual for the sonication apparatus).

NOTE: The protein extract is the supernatant from this spin, separated from cellular debris in the pellet.

NOTE: Centrifugation should be performed in an appropriate container rated for at least 20,000 rcf (*g*), such as Oak Ridge tubes.

5. Normalize each sample by total protein. Add 2–5 μL of each supernatant to 1 mL Coomassie Plus Reagent, mix and incubate 5 min, read in a spectrophotometer at 595 nm. Calculate protein concentration for each sample based on a BSA standard curve (0.625, 1.25, 2.5, 5.0, 10.0 μg BSA standards), and use the same amount of protein for each sample. Equalize volumes with urea lysis buffer, store any reserve lysate at −80 °C.

   NOTE: It is critical to start with the same amount of total protein for each sample to ensure accuracy of quantitation and the ability to compare results across multiple samples.

3.1.3 Reduction, Alkylation, and Digestion of Proteins

6. Add 1/278 volume of 1.25 M DTT to the cleared cell supernatant (e.g., 36 μL of 1.25 M DTT for 10 mL of protein extract), mix well and place the tube into a 55 °C incubator for 30 min.
7. Cool the solution on ice briefly until it has reached room temperature.
8. Add 1/10 volume of iodoacetamide solution to the cleared cell supernatant, mix well, and incubate for 15 min at room temperature in the dark.

   NOTE: A small portion of each sample can be reserved at this point undigested for follow-up western blotting or other testing. Store reserve material at −80 °C.
9. Add 4× volume 20 mM HEPES pH 8.0 to the reduced/alkylated lysate, mix, and add 1:100 dilution of 1 mg/mL Trypsin-TPCK stock solution. Mix overnight at room temperature.

   NOTE: Check digestion using 25 μL of each sample into 1 mL Coomassie Plus Reagent. Protein concentration should be low/zero at this point.

3.2 Sep-Pak® C18 Purification of Lysate Peptides

3.2.1 Solutions and Reagents

NOTE: Prepare solutions with Milli-Q or equivalent grade water. Organic solvents (trifluoroacetic acid, acetonitrile) should be of the highest grade. All percentage specifications for solutions are vol/vol.

1. 20 % trifluoroacetic acid (TFA): Add 10 mL TFA to 40 mL water.
2. 0.1 % TFA: Add 5 mL 20 % TFA to 995 mL water.
3. 0.1 % TFA, 40 % acetonitrile: Add 400 mL acetonitrile and 5 mL 20 % TFA to 500 mL water.

4. 0.1 % TFA, 5 % acetonitrile: Add 50 mL acetonitrile and 5 mL 20 % TFA to 945 mL water.

   NOTE: Purification of peptides is performed at room temperature on 0.7 mL Sep-Pak columns from Waters Corporation, WAT051910.

   NOTE: Sep-Pak® C18 purification utilizes reversed-phase (hydrophobic) solid-phase extraction. Peptides and lipids bind to the chromatographic material. Large molecules such as DNA, RNA, and most proteins, as well as hydrophilic molecules such as many small metabolites are separated from peptides using this technique. Peptides are eluted from the column with 40 % acetonitrile (MeCN) and separated from lipids and proteins, which elute at approximately 60 % MeCN and above.

   NOTE: About 20 mg of protease-digested peptides can be purified from one Sep-Pak column. Purify peptides immediately after proteolytic digestion.

   NOTE: Before loading the peptides from the protein digest on the column, the digest must be acidified with TFA for efficient peptide binding. The acidification step helps remove fatty acids from the digested peptide mixture.

3.2.2 SEP-PAK® Protocol

1. Add 1/20 volume of 20 % TFA to the digest for a final concentration of 1 % TFA. Check the pH by spotting a small amount of peptide sample on a pH strip (the pH should be under 3). After acidification, allow precipitate to form by letting stand for 15 min.
2. Centrifuge the acidified peptide solution for 15 min at 1780 rcf (*g*) at room temperature to remove the precipitate. Transfer peptide-containing supernatant into a new 50 mL conical tube without dislodging the precipitated material.

   NOTE: Application of all solutions should be performed by gravity flow.
3. Connect a 10 cc reservoir (remove 10 cc plunger) to the SHORT END of the Sep-Pak column.
4. Pre-wet the column with 5 mL 100 % MeCN.

   NOTE: Each time solution is applied to the column air bubbles form in the junction where the 10 cc reservoir meets the narrow inlet of the column. These must be removed with a gel-loader tip placed on a P-200 micropipettor, otherwise the solution will not flow through the column efficiently. Always check for appropriate flow.
5. Wash sequentially with 1, 3, and 6 mL of 0.1 % TFA.
6. Load acidified and cleared peptide solution.

   NOTE: In rare cases, if the flow rates decrease dramatically upon (or after) loading of sample, the purification procedure can be accelerated by **gently** applying pressure to the column using the 10 cc plunger after cleaning it with organic solvent. Again make sure to remove air bubbles from the narrow inlet of the column before doing so. Do not apply vacuum (as advised against by the manufacturer).

7. Wash sequentially with 1, 5, and 6 mL of 0.1 % TFA.
8. Wash with 2 mL of 5 % MeCN, 0.1 % TFA.
9. Place columns above new 50 mL tubes to collect eluate. Elute peptides with a sequential wash of 3 × 3 mL 0.1 % TFA, 40 % acetonitrile.
10. Freeze the eluate on dry ice (or −80 °C freezer) for 4 h to overnight and lyophilize frozen peptide solution for a minimum of 2 days to assure TFA has been removed from the peptide sample.

    NOTE: The lyophilization should be performed in a standard lyophilization apparatus. DO NOT USE a Speed-Vac apparatus at this stage of the protocol.

    NOTE: The lysate digest may have a much higher volume than the 10 cc reservoir will hold (up to 50–60 mL from adherent cells) and therefore the peptides must be applied in several fractions. If available a 60 cc syringe may be used in place of a 10 cc syringe to allow all sample to be loaded into the syringe at once.

    NOTE: Lyophilized, the digested peptides are stable at −80 °C for several months (seal the closed tube with Parafilm for storage).

3.3 Immunoaffinity Purification (IAP) of Peptides

NOTE: Prepare solutions with Milli-Q or equivalent grade water. Trifluoroacetic acid should be of the highest grade. All percentage specifications for solutions are vol/vol.

3.3.1 Solutions and Reagents

1. 1× MOPS IAP Buffer: Add 10.5 g MOPS and 1.42 g Na_2HPO_4 (anhydrous) to 800 mL water. pH to 7.2 with NaOH. Add 2.92 g NaCl, mix until dissolved, add water to 1 L, filter through 0.22 μm filter and store at 4 °C.

3.3.2 IAP Protocol

1. Centrifuge the tube containing lyophilized peptide in order to collect all material to be dissolved. Add 1.4 mL IAP buffer. Resuspend pellets mechanically by pipetting repeatedly with a P-1000 micropipettor taking care not to introduce excessive bubbles into the solution. Transfer solution to a 1.7 mL Eppendorf tube.

    NOTE: After dissolving the peptide, check the pH of the peptide solution by spotting a small volume on pH indicator paper (The pH should be close to neutral, or no lower than 6.0). In the rare case that the pH is more acidic (due to insufficient removal of TFA from the peptide under suboptimal conditions of lyophilization), titrate the peptide solution with

1 M Tris base solution that has not been adjusted for pH. 5–10 μL is usually sufficient to neutralize the solution.

2. Clear solution by centrifugation for 5 min at 10,000 rcf (*g*) at 4 °C in a microcentrifuge. The insoluble pellet may appear considerable. This will not pose a problem since most of the peptide will be soluble. Cool on ice.
3. Wash motif antibody-bead slurry sequentially, five times with 1 mL of 1× PBS and resuspend as a 50 % slurry in PBS to remove the glycerol contaminating buffer.
4. Transfer the peptide solution into the microfuge tube containing motif antibody beads. Pipet sample directly on top of the beads at the bottom of the tube to ensure immediate mixing. Avoid creating bubbles upon pipetting.
5. Incubate for 2 h on a rotator at 4 °C. Before incubation, seal the microfuge tube with Parafilm in order to avoid leakage.
6. Centrifuge at 2000 rcf (*g*) for 30 s and transfer the supernatant with a P-1000 micropipettor to a labeled Eppendorf tube to save for future use. Store at −80 °C Flow-through material can be used for subsequent IAPs.

   NOTE: In order to recover the beads quantitatively, do not spin the beads at lower *g*-forces than what is specified in this procedure. Avoid substantially higher *g*-forces as well, since this may cause the bead matrix to collapse. All centrifugation steps should be performed at the recommended speeds throughout the protocol.

   NOTE: If the cells were directly harvested from culture medium without PBS washing, some Phenol Red pH indicator will remain (it co-elutes during the Sep-Pak® C18 purification of peptides) and color the peptide solution yellow. This coloration has no effect on the immunoaffinity purification step.

   NOTE: All subsequent wash steps are done at 0–4 °C.

   NOTE: In all wash steps, the supernatant should be removed reasonably well. Avoid removing the last few microliters, except in the last step, since this may cause inadvertent carryover of the beads.
7. Add 1 mL of IAP buffer to the beads, mix by inverting tube five times, centrifuge for 30 s, and remove supernatant with a P-1000 micropipettor.
8. Repeat step 7 once for a total of TWO IAP buffer washes.

   NOTE: All steps from this point forward should be performed with solutions prepared with Burdick and Jackson or other HPLC grade water.
9. Add 1 mL chilled Burdick and Jackson water to the beads, mix by inverting tube five times, centrifuge for 30 s, and remove supernatant with a P-1000 micropipettor.

10. Repeat step 9 two times for a total of THREE water washes.
    NOTE: After the last wash step, remove supernatant with a P-1000 micropipettor as before, then centrifuge for 5 s to remove fluid from the tube walls, and carefully remove all remaining supernatant with a gel loading tip attached to a P-200 micropipettor.
11. Add 55 μL of 0.15 % TFA to the beads, tap the bottom of the tube several times (do not vortex), and let stand at room temperature for 10 min, mixing gently every 2–3 min.
    NOTE: In this step, the peptides of interest will be in the **eluent**.
12. Centrifuge for 30 s at 2000 rcf (*g*) in a microcentrifuge and transfer supernatant to a new 1.7 mL Eppendorf tube.
13. Add 50 μL of 0.15 % TFA to the beads, and repeat the elution/centrifugation steps. Combine both eluents in the same 1.7 mL tube. Briefly centrifuge the eluent to pellet any remaining beads and carefully transfer eluent to a new 1.7 mL tube taking care not to transfer any beads.

3.4 Concentration and Purification of Peptides for LC-MS Analysis

NOTE: We recommend concentrating peptides using the following protocol by Rappsilber et al. [27].

NOTE: We recognize there are many other routine methods for concentrating peptides using commercial products such as ZipTip® (see link provided below) and StageTips (see link provided below) that have been optimized for peptide desalting/concentration. Regardless of the particular method, we recommend that the method of choice be optimized for recovery and be amenable for peptide loading capacities of at least 10 μg.

StageTips: Proxeon catalog #SP201

ZipTip®: Millipore catalog #ZTC18S096

3.4.1 Solutions and Reagents

NOTE: Prepare solutions with HPLC grade water. Organic solvents (trifluoroacetic acid, acetonitrile) should be of the highest grade. Stage tips can also be homemade using a blunt point 18-gauge needle to bore 2-4 discs of Empore C18 (3M #2215-C18) and pack into micropipette tips.

1. Wetting solution (0.1 % trifluoroacetic acid, 50 % acetonitrile): Add 0.1 mL trifluoroacetic acid to 49.9 mL water, then add 50 mL acetonitrile.
2. Wash solution (0.1 % trifluoroacetic acid): Add 0.1 mL trifluoroacetic acid to 99.9 mL water.
3. Elution solution (0.1 % trifluoroacetic acid, 40 % acetonitrile): Add 0.1 mL trifluoroacetic acid to 59.9 mL water, then add 40 mL acetonitrile.
4. 1 M ammonium bicarbonate: Dissolve 7.9 g ammonium bicarbonate in 80 mL water. Adjust volume to 100 mL, and freeze at –80 °C in single use aliquots.

5. Digestion buffer (50 mM ammonium bicarbonate, 5 % acetonitrile): Add 10 μL 1 M Ammonium Bicarbonate pH 10.0 and 10 μL Acetonitrile to 180 μL water. Prepare fresh for each experiment.

   NOTE: Organic solvents are volatile. Tubes containing small volumes of these solutions should be prepared immediately before use and should be kept capped as much as possible, because the organic components evaporate quickly.

3.4.2 StageTip Protocol

1. Equilibrate the StageTip by passing 50 μL of Wetting Solution through (once) followed by 50 μL of Solvent D two times.
2. Load sample by passing IP eluent through the StageTip. Load IAP eluent in two steps using 50 μL in each step.
3. Wash the StageTip by passing 55 μL of Solvent D through two times.
4. Elute peptides off the StageTip by passing 10 μL of Solvent E through two times, pooling the resulting eluent.
5. Dry down the StageTip eluent in a vacuum concentrator (Speed-Vac) and redissolve the peptides in an appropriate solvent for LC-MS analysis such as 5 % acetonitrile, 0.125 % formic acid.

   NOTE: The antibodies for IAP are not covalently bound to the beads, and thus a small amount will leach off beads into the eluate during elution. Acetonitrile concentrations greater than 40 % will lead to more antibody being released during elution and may interfere with results. Do NOT use elution buffers with greater than 40 % acetonitrile.

   NOTE: In solution digestion of peptides can be used to digest any antibody remaining in the sample after StageTip purification.
6. Dilute a stock solution of sequencing grade trypsin (Promega) with digestion buffer from 0.4 μg/μL to a final concentration of 25 ng/μL.
7. Resuspend the dried, LysC digested peptides generated from the StageTip concentration protocol above with 10 μL of trypsin solution (25 ng/μL, 250 ng total). Vortex 5 s to redissolve the peptides and centrifuge the sample to collect peptides/trypsin solution at the bottom of the microfuge tube as the final step.
8. Incubate the solution at 37 °C for 2 h.
9. After trypsin digestion, add 1 μL of 5 % TFA and 20 μL 0.1 % TFA to the digest solution. Vortex to mix and microfuge to collect peptide solution at the bottom of the microfuge tube.
10. Transfer the acidified peptide solution to a newly conditioned StageTip (perform step 1 of the StageTip protocol), rinse the

0.5 mL Eppendorf tube with 20 μL of 0.1 % TFA once and apply the rinse solution to the StageTip.

11. Perform the StageTip desalting of the peptide digest and elute the peptides into an HPLC insert. Dry purified peptides under vacuum prior to LC-MS analysis (as described above).

3.5 LC-MS/MS Analysis of Peptides

1. Resuspend vacuum-dried, immunoaffinity-purified peptides in 0.125 % formic acid. 15 mg of starting material will generate sufficient peptide for two to three injections on the instrument.
2. Separate on a reverse phase column (75 μm inner diameter × 10 cm) packed into a PicoTip emitter (~8 μm diameter tip) with Magic C18 AQ (100 Å × 5 μm). Elute PTM peptides using a 90 min gradient of acetonitrile (5–40 %) in 0.125 % formic acid delivered at 280 nL/min.

   NOTE: Run samples on a high mass accuracy instrument (ppm accuracy in the MS1 channel) to ensure high quality peptide identifications and accurate quantification. We use the LTQ-Orbitrap VELOS or ELITE systems (Thermo Scientific).
3. LC-MS/MS Instrument parameter settings: MS Run Time 120 min, MS1 Scan Range (300.0–1500.00), Top 20 MS/MS, Min Signal 500, Isolation Width 2.0, Normalized Coll. Energy 35.0, Activation-Q 0.250, Activation Time 20.0, Lock Mass 371.101237, Charge State Rejection Enabled, Charge State 1+ Rejected, Dynamic Exclusion Enabled, Repeat Count 1, Repeat Duration 35.0, Exclusion List Size 500, Exclusion Duration 40.0, Exclusion Mass Width Relative to Mass, Exclusion Mass Width 10 ppm.
4. Database search the files generated in the run versus the correct species database with a reverse decoy database included to estimate false discovery rates [28]. Search settings: mass accuracy of parent ions: 50 ppm, mass accuracy of product ions: 1 Da, up to four missed cleavages, up to four variable modifications, max charge = 5, variable modifications allowed on methionine (oxidation, +15.9949), serine, threonine, and tyrosine (phosphorylation, +79.9663). Semi-tryptic peptides allowed (K or R residue on one side of peptide only). Results can be further narrowed by MError (usually ±3 ppm) and the presence of the intended motif (phosphorylation, caspase cleavage, validated Apoptosis/Autophagy peptide, etc.).
5. Quantification can be performed via a number of methods such as SILAC, reductive amination, isobaric tags, or label-free. For label free quantification, we use Progenesis (Nonlinear Dynamics) to retrieve the integrated peak area for each peptide in the MS1 channel. This software only requires one MS/MS event for a particular peptide across all samples, eliminating "holes" (no data) in the study due to LC-MS duty cycle limitations.

References

1. Favaloro B, Allocati N, Graziano V, Di Ilio C, De Laurenzi V (2012) Role of apoptosis in disease. Aging 4:330–349
2. Kaufmann T, Strasser A, Jost PJ (2012) Fas death receptor signalling: roles of Bid and XIAP. Cell Death Differ 19:42–50
3. Fuchs Y, Steller H (2011) Programmed cell death in animal development and disease. Cell 147:742–758
4. Degterev A, Yuan J (2008) Expansion and evolution of cell death programmes. Nat Rev Mol Cell Biol 9:378–390
5. Jacobson MD, Evan GI (1994) Apoptosis. Breaking the ICE. Curr Biol 4:337–340
6. Cohen GM (1997) Caspases: the executioners of apoptosis. Biochem J 326(Pt 1):1–16
7. Fischer U, Janicke RU, Schulze-Osthoff K (2003) Many cuts to ruin: a comprehensive update of caspase substrates. Cell Death Differ 10:76–100
8. Alenzi FQ, Lotfy M, Wyse R (2010) Swords of cell death: caspase activation and regulation. Asian Pac J Cancer Prev 11:271–280
9. Guo A, Gu H, Zhou J, Mulhern D, Wang Y, Lee KA, Yang V, Aguiar M, Kornhauser J, Jia X, Ren J, Beausoleil SA, Silva JC, Vemulapalli V, Bedford MT, Comb MJ (2014) Immunoaffinity enrichment and mass spectrometry analysis of protein methylation. Mol Cell Proteomics 13:372–387
10. Lee KA, Hammerle LP, Andrews PS, Stokes MP, Mustelin T, Silva JC, Black RA, Doedens JR (2011) Ubiquitin ligase substrate identification through quantitative proteomics at both the protein and peptide levels. J Biol Chem 286(48):41530–41538
11. Moritz A, Li Y, Guo A, Villen J, Wang Y, MacNeill J, Kornhauser J, Sprott K, Zhou J, Possemato A, Ren JM, Hornbeck P, Cantley LC, Gygi SP, Rush J, Comb MJ (2010) Akt-RSK-S6 kinase signaling networks activated by oncogenic receptor tyrosine kinases. Sci Signal 3:ra64
12. Rikova K, Guo A, Zeng Q, Possemato A, Yu J, Haack H, Nardone J, Lee K, Reeves C, Li Y, Hu Y, Tan Z, Stokes M, Sullivan L, Mitchell J, Wetzel R, Macneill J, Ren JM, Yuan J, Bakalarski CE, Villen J, Kornhauser JM, Smith B, Li D, Zhou X, Gygi SP, Gu TL, Polakiewicz RD, Rush J, Comb MJ (2007) Global survey of phosphotyrosine signaling identifies oncogenic kinases in lung cancer. Cell 131:1190–1203
13. Rush J, Moritz A, Lee KA, Guo A, Goss VL, Spek EJ, Zhang H, Zha XM, Polakiewicz RD, Comb MJ (2005) Immunoaffinity profiling of tyrosine phosphorylation in cancer cells. Nat Biotechnol 23:94–101
14. Stokes MP, Comb MJ (2008) A wide-ranging cellular response to UV damage of DNA. Cell Cycle 7:2097–2099
15. Stokes MP, Farnsworth CL, Moritz A, Silva JC, Jia X, Lee KA, Guo A, Polakiewicz RD, Comb MJ (2012) PTMScan direct: identification and quantification of peptides from critical signaling proteins by immunoaffinity enrichment coupled with LC-MS/MS. Mol Cell Proteomics 11:187–201
16. Stokes MP, Silva JC, Jia X, Lee KA, Polakiewicz RD, Comb MJ (2012) Quantitative profiling of DNA damage and apoptotic pathways in UV damaged cells using PTMScan direct. Int J Mol Sci 14:286–307
17. Dayon L, Hainard A, Licker V, Turck N, Kuhn K, Hochstrasser DF, Burkhard PR, Sanchez JC (2008) Relative quantification of proteins in human cerebrospinal fluids by MS/MS using 6-plex isobaric tags. Anal Chem 80: 2921–2931
18. Hsu JL, Huang SY, Chow NH, Chen SH (2003) Stable-isotope dimethyl labeling for quantitative proteomics. Anal Chem 75:6843–6852
19. Ibarrola N, Kalume DE, Gronborg M, Iwahori A, Pandey A (2003) A proteomic approach for quantitation of phosphorylation using stable isotope labeling in cell culture. Anal Chem 75:6043–6049
20. Ong SE, Blagoev B, Kratchmarova I, Kristensen DB, Steen H, Pandey A, Mann M (2002) Stable isotope labeling by amino acids in cell culture, SILAC, as a simple and accurate approach to expression proteomics. Mol Cell Proteomics 1:376–386
21. Paardekooper Overman J, Yi JS, Bonetti M, Soulsby M, Preisinger C, Stokes MP, Hui L, Silva JC, Overvoorde J, Giansanti P, Heck AJ, Kontaridis MI, den Hertog J, Bennett AM (2014) PZR coordinates Shp2 Noonan and LEOPARD syndrome signaling in zebrafish and mice. Mol Cell Biol 34:2874–2889
22. Unwin RD, Pierce A, Watson RB, Sternberg DW, Whetton AD (2005) Quantitative proteomic analysis using isobaric protein tags enables rapid comparison of changes in transcript and protein levels in transformed cells. Mol Cell Proteomics 4:924–935
23. Viner RI, Zhang T, Second T, Zabrouskov V (2009) Quantification of post-translationally modified peptides of bovine alpha-crystallin

using tandem mass tags and electron transfer dissociation. J Proteomics 72:874–885

24. Wiese S, Reidegeld KA, Meyer HE, Warscheid B (2007) Protein labeling by iTRAQ: a new tool for quantitative mass spectrometry in proteome research. Proteomics 7:340–350
25. Zhang H, Zha X, Tan Y, Hornbeck PV, Mastrangelo AJ, Alessi DR, Polakiewicz RD, Comb MJ (2002) Phosphoprotein analysis using antibodies broadly reactive against phosphorylated motifs. J Biol Chem 277:39379–39387
26. Pham VC, Pitti R, Anania VG, Bakalarski CE, Bustos D, Jhunjhunwala S, Phung QT, Yu K, Forrest WF, Kirkpatrick DS, Ashkenazi A, Lill JR (2012) Complementary proteomic tools for the dissection of apoptotic proteolysis events. J Proteome Res 11:2947–2954
27. Rappsilber J, Ishihama Y, Mann M (2003) Stop and go extraction tips for matrix-assisted laser desorption/ionization, nanoelectrospray, and LC/MS sample pretreatment in proteomics. Anal Chem 75:663–670
28. Lundgren DH, Martinez H, Wright ME, Han DK (2009) Protein identification using Sorcerer 2 and SEQUEST. Curr Protoc Bioinformatics Chapter 13, Unit 13 13

Chapter 10

Novel Electrochemical Biosensor for Apoptosis Evaluation

Peng Miao and Jian Yin

Abstract

Apoptosis evaluation is one of the most important tasks of toxicology. By using a peptide as the recognition element, and assembling apoptotic cells on a solid surface, we have established a novel electrochemical method for the detection of apoptosis levels. Such a peptide-based electrochemical biosensor is simple, cost-effective, convenient, and sensitive. Since the results obtained are well in line with other standard methods, this method holds a great potential towards the analysis of apoptosis and its applications. In this chapter, we introduce a general overview of this technical approach for detecting apoptotic cells. We discuss its advantages over the ordinary methods. We also provide practical guidelines for designing studies, and summarize the step-by-step protocols used in our lab for sample preparation, electrode modification, and accurate electrochemical quantification of apoptotic cells.

Key words Analytical chemistry, Apoptosis, Electrochemical analysis, Phosphatidylserine, Peptide, Differential pulse voltammetry, Electrochemical impedance spectroscopy, Cyclic voltammetry

1 Introduction

Apoptosis is a process of programmed cell death, distinguished from necrosis. Apoptosis occurs in multicellular organisms, and is involved in development, cellular maintenance, and the sculpturing of organs and tissues [1, 2]. Recently, the topic of apoptotic death has attracted more and more attentions of toxicologists, since various published reports have revealed that chemicals or drugs might cause cells to die by apoptosis, resulting in occurrence of organ injury or illness [3–5]. Toxicologists are thus currently faced with the task of quantifying apoptotic levels, as well as mechanisms by which chemicals or drugs interact with apoptotic signaling factors, and apply these results to safety evaluation or to assess relevance of environmental exposures [6].

Since cells undergoing apoptosis are accompanied by notable morphological changes and biochemical events, various means of detecting apoptotic cells have been explored and made available over the time. For example, based on the observation of morphologic features of apoptosis, including chromatin condensation, cell

Perpetua M. Muganda (ed.), *Apoptosis Methods in Toxicology*, Methods in Pharmacology and Toxicology,
DOI 10.1007/978-1-4939-3588-8_10, © Springer Science+Business Media New York 2016

shrinkage, and pyknosis, light microscopy and electron microscopy are the gold standards for definitive identification of apoptotic cells [7]. However, these procedures are time-consuming and require professional skills and complex sample preparation techniques. Moreover, quantification of apoptotic cells by such methods is difficult, and this hinders their applications to a great extent. Several alternative approaches have thus been developed based on the biochemical events, such as DNA fragmentation [8], activation of caspases [9, 10], and release of cytochrome c [11, 12]. Among these, cleavage of genomic DNA into multiple fragments of 180–200 bp is the most typical feature. However, this procedure involves gel electrophoresis and multiple steps, and is normally used for qualitative analysis. Although the detection of caspase activities is more quantitative, the analysis is usually performed in a cell lysate system, where the complex contents may interfere with the accurate detection of apoptotic cells. The exposure of phosphatidylserine (PS) on the outer surface of the cell membrane occurs in the early stage of apoptosis, and is thus utilized as another universal biomarker for apoptosis characterization [13]. For most approaches based on this biomarker, annexin V is used to recognize externalized PS in the presence of Ca^{2+} [14]. Annexin V-based flow cytometric assays are now being widely utilized in the detection of apoptosis [15]. Unfortunately, this method has certain disadvantages, such as high cost and inconvenient operation. There is an urgent need, therefore, to develop advanced methods for apoptosis evaluation.

Optimized peptides for PS targeting based on the original sequence of PS-binding site in PS decarboxylase have recently been exploited to replace annexin V [16]. The advantages are as follows: First, the peptides are less expensive than the fluorescent reagents used for annexin V-based methods. Second, the specific binding of peptides is more convenient because there is no requirement for Ca^{2+}. Third, the lower molecular weight of peptides can increase the binding efficacy with PS. Infact, in some solid phase reaction based methods, peptides have higher stability, and the succinct structures can accelerate the self-assembly on the solid surface. This is in contrast to annexin V, which does not retain well its native structure and biological functions without certain treatments.

It is well-known that electrochemical techniques exhibit attractive merits in terms of inherent simplicity, low cost, fast-response, high sensitivity, and convenience of analysis [17]. Common electrochemical techniques include cyclic voltammetry (CV), differential pulse voltammetry (DPV), anodic stripping voltammetry (ASV), electrochemical impedance spectroscopy (EIS), and electrochemiluminescence (ECL). Due to the development of the functionalization of electrode surface in recent years, electrochemical techniques have been widely used to detect various analysts at extremely low concentrations [18, 19]. Many endeavors have also been carried out on the fabrication of novel electrochemical

Table 1
Recently developed electrochemical methods for apoptosis evaluation

Techniques	Targets	Strategies	Reference
DPV	Etoposide induced apoptotic cells	Microfluidic chip based cytosensor	[22]
CV	Caspase-3	Tetrapeptide motif of DEVD	[23]
ASV	Caspase-3	DEVD recognition and nanomaterials labels	[24]
EIS	Cytochrome c and caspase-9	Real-time in situ immunosensor	[25]
ECL	Drug-treated cells	Antibody-PS	[26]
ASV	Camptothecin induced apoptotic cells	Annexin V-PS	[13]
EIS	Liposome induced apoptotic cells	Annexin V-PS	[27]
DPV	H_2O_2 induced apoptotic cells	Peptide-PS	[28]

DPV differential pulse voltammetry, *CV* cyclic voltammetry, *ASV* anodic stripping voltammetry, *EIS* electrochemical impedance spectroscopy, *ECL* electrochemiluminescence

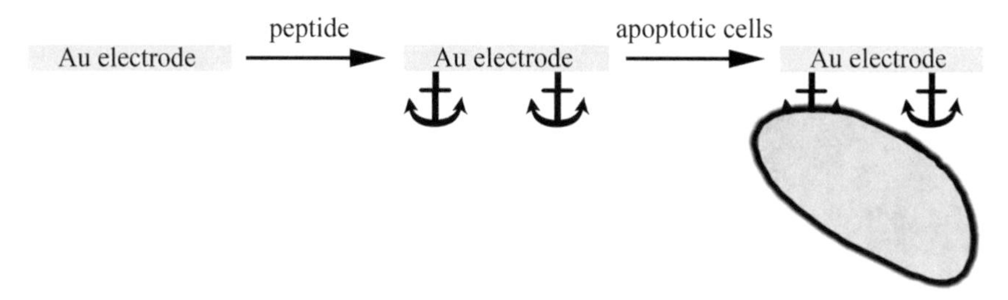

Fig. 1 Schematic illustration of the PS specific peptide-based electrochemical method for the detection apoptosis

methods with better performance and easier operation towards the detection of apoptotic cells [20, 21]. Table 1 lists several electrochemical biosensors for the detection of apoptotic cells.

In this work, we have designed a PS-specific peptide to assemble apoptotic cells on a solid surface, and achieve further electrochemical evaluation of apoptosis [28]. The peptide bridge possesses excellent electric conductivity [29]. As shown in Fig. 1, the gold electrode is first functionalized with a positively charged peptide, which selectively binds to the externalized PS on apoptotic cells, localizing the apoptotic cells onto the electrode surface. A significant decline of the electrochemical responses appears, due to the resulting increase of steric hindrance and shield of the positive charges of peptide attracting $Fe(CN)_6^{3-/4-}$; this feature can be used to evaluate the amount of attached apoptotic cells. Since the high molecular weighted annexin V may hinder electric transfer rate, the peptide used here is more suitable for PS recognition in electrochemical systems. By combining the advantages of electrochemical techniques and peptide-based recognition process, the biosensor offers attractive merits in inherent simplicity, convenient operation, low cost, and high sensitivity. Therefore, it possesses potential applications towards apoptosis evaluation, therapeutic effect assessment, and deeper cellular biological studies.

2 Materials

2.1 Reagents

PS-specific peptide (FNFRLKAGAKIRFGRGC) and control peptide (AFGNRGRAAKNFHARGC) are synthesized and purified by China peptides Co., Ltd. (Shanghai, China). Fetal bovine serum can be purchased from Hangzhou Sijiqing Biological Engineering Material Co., Ltd. (Hangzhou, China). DMEM medium can be purchased from Gibco (Gaithersburg, USA). Tris(2-carboxyethyl) phosphinehydro-chloride (TCEP), mercaptohexanol (MCH), ethylenediaminetetraacetic acid (EDTA), trypsin, penicillin, and streptomycin can be purchased from Sigma (USA). H_2O_2 used to induce apoptosis can be obtained from Sinopharm Chemical Reagent Co., Ltd. (Shanghai, China). Sulfuric acid, nitric acid, ethanol, sodium hydroxide, disodium hydrogen phosphate, sodium dihydrogenphosphate, potassium ferrocyanide, potassium ferricyanide, potassium nitrate, HEPES, and Tris–HCl can be provided by Nanjing Chemical Reagent Co., Ltd. (Nanjing, China). P5000 silicon carbide paper, suede and alumina can be obtained from Tianjin Aidahengsheng Technology Co., Ltd. (Tianjin, China). All the other chemicals are of analytical grade.

2.2 Cells

MCF-7 cells (ATCC) can be obtained from Institute of Biochemistry and Cell Biology, Chinese Academy of Sciences (Shanghai, China).

2.3 Apparatus

All electrochemical measurements are carried out on an electrochemical analyzer (CHI660C, CH Instrument, China). A three-electrode system is constructed for the experiments, which constitutes a reference electrode (saturated calomel electrode), an auxiliary electrode (platinum wire electrode), and a working electrode (gold electrode modified with the peptide). All electrodes used here can be easily obtained from CH Instrument (China).

2.4 Buffer Solutions

Water used to prepare solutions is purified with a Milli-Q purification system (Barnstead, USA) and the resistance of 18 MΩ cm is reached. The buffer solutions employed in this work are as follows. Peptide immobilization buffer: 20 mM HEPES containing 10 mM TCEP (pH 6.2). Electrode washing buffer: 10 mM Tris–HCl solutions (pH 7.4). Buffer solutions for EIS, CV, and DPV are 5 mM $Fe(CN)_6^{3-/4-}$ with 1 M KNO_3.

2.5 Materials for Caspase-3 Assay

A colorimetric activity assay kit (Beyotime, Haimen, Jiangsu, China) is used to measure the caspase-3 activity of cell lysate, and the colorimetric signals are read by a microplate reader (Thermo Multiskan Ascent 354, USA; Thermo Labsystems, Helsinki, Finland).

2.6 Materials for Fluorescence Microscopy Observation and Flow Cytometry

Hoechst 33342 can be purchased from Sigma (USA). Propidium iodide (PI) is available from Shanghai Sangon Biological Technology & Services Co., Ltd. (Shanghai, China). Fluorescence microscopy (Axio observer A1, Zeiss, Germany) is applied to observe the cultured and treated MCF-7 cells directly. Cyan flow cytometer (BD FACSCalibur) and ModFit software are employed to analyze the DNA content of the treated cells.

3 Methods

3.1 Cell Culture and H_2O_2 Treatment

MCF-7 cells are cultured in DMEM medium containing 10 % (v/v) fetal bovine serum, 2 mM L-glutamine, 100 U/mL penicillin, and 100 μg/mL streptomycin on a six-well plate at 37 °C in 5 % CO_2 atmosphere.

1. Cells are washed before use with phosphate buffered saline (PBS, 1.5 mM KH_2PO_4, 8 mM K_2HPO_4, 135 mM NaCl, 2.7 mM KCl, pH 7.4) after reaching 80 % of confluence.
2. Cells are then incubated at 37 °C in DMEM medium containing 25, 50, 80, and 100 μM H_2O_2 to induce apoptosis.
3. At 30 min post-exposure, the cells are washed twice with PBS, and then detached by 0.25 % trypsin (1 mM EDTA).
4. Cells are subsequently resuspended in PBS at a concentration of 10^6 cells/mL, and kept on ice for analysis of H_2O_2 induced apoptosis.

3.2 Electrode Modification

A gold electrode with the diameter of 2 mm is used as the working electrode.

1. The electrode is firstly cleaned in piranha solution (98 % H_2SO_4:30 % H_2O_2 = 3:1) for 5 min in order to eliminate any adsorbed materials on the electrode (*Caution: Piranha solution dangerously attacks organic matter!*). It is then rinsed with double-distilled water.
2. The electrode is carefully polished with P5000 silicon carbide paper.
3. The electrode is further polished with 1, 0.3, and 0.05 μm alumina slurries.
4. It is washed with double-distilled water, and dipped into ethanol for a 5 min sonication. The ethanol is later changed to

double-distilled water, and another 5 min sonication is carried out to clean the electrode.

5. The electrode is treated with 50 % HNO_3 for 30 min. A further electrochemical cleaning procedure is performed in 0.5 M H_2SO_4 with scanning between 0 and 1.6 V for 20 cycles.
6. The electrode is rinsed with double-distilled water and dried with nitrogen.
7. The pretreated electrode is incubated in 100 μM peptide solution (20 mM HEPES, 10 mM TCEP, pH 6.2) for 16 h. Due to the interaction between the thiol group of the C-terminus cystine residue and the gold surface [30], the peptide could be immobilized on the gold electrode surface.
8. The electrode is rinsed and then incubated in 1 mM MCH for 30 min to achieve a well aligned peptide monolayer, which is beneficial for further cell attachment [31].

3.3 Electrochemical Measurement Using Electrochemical Analyzer

The H_2O_2 treated cells are used as samples to interact with the modified gold electrode for 1 h. After that, the electrode is washed with electrochemical washing buffer, followed by EIS, CV, and DPV measurements, respectively.

EIS Characterization. EIS is obtained upon application of the biasing potential 0.204 V, amplitude 5 mV, and frequency range from 1 to 100,000 Hz. Impedance spectra usually consist of a linear portion at lower frequencies corresponding to diffusion and a semicircle portion at higher frequencies which reflects the electron transfer limited process [32]. No obvious semicircle domain of impedance spectrum is observed on bare electrode, and a tiny one appears on the peptide modified electrode due to the balance of the steric hindrance and positive charges to attract $Fe(CN)_6^{3-/4-}$. After the incubation of nonapoptotic cells (cells without H_2O_2 treatment), the resulted larger semicircle domain indicates that nonapoptotic cells may still cause certain nonspecific adsorption on the peptide modified electrode. In the case of apoptotic cells, specific interaction between the peptide and externalized PS on the cells contributes to a much larger impedance spectrum. The R_{ct} are depicted in Fig. 2 and the results confirm well the proposed sensing mechanism.

CV Measurement. The electrochemical properties of the modified electrode are then studied by CV, another powerful electrochemical tool. The scan range is set from −0.3 to 0.6 V and the scan rate is 100 mV/s. Bare gold electrode exhibit a pair of well-defined redox peak in $Fe(CN)_6^{3-/4-}$ solutions. After the modification of peptide and subsequent MCH, the current peak drops. Another tiny decrease of current peak is observed after the incubation of nonapoptotic cells, demonstrating that nonspecific adsorption exists but is indistinctive. In the case of apoptotic cells, the CV

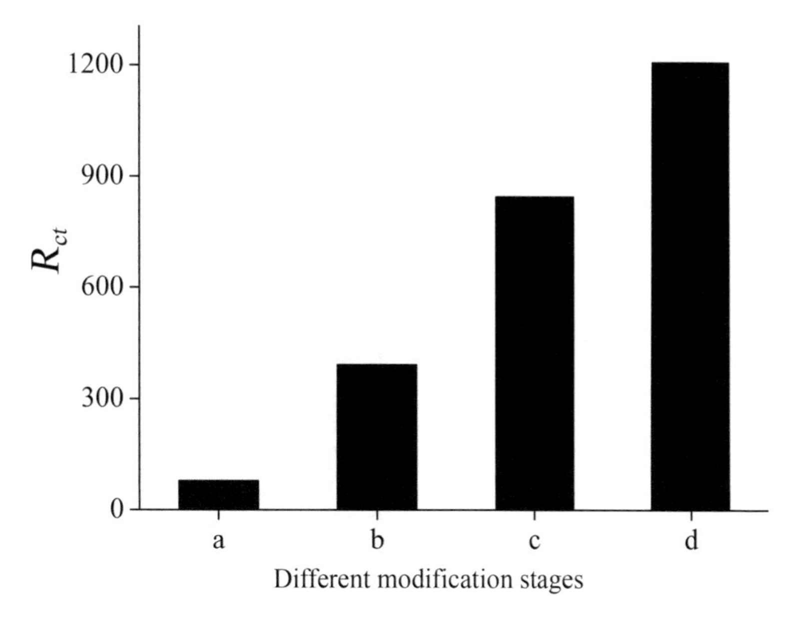

Fig. 2 R_{ct} values of the gold electrode at different modification stages: (*a*) bare electrode, (*b*) peptide modified electrode, after further incubation with (*c*) non-apoptotic and (*d*) apoptotic cells

current peak drops to a large extent, which confirms effective interfacial charge transfer resistance from captured apoptotic cells on the electrode.

DPV Determination. To validate the electrochemical approach for the detection of apoptosis, we have then conducted DPV, a more sensitive electrochemical technique, to obtain electrochemical signals. The scan range is from 0.6 to −0.3 V and the pulse amplitude is 50 mV. The voltammetric wave of the peptide modified electrode may have a sharp peak at around 0.22 V, since the positively charged electrode surface can effectively attract electrochemical species. However, after interacting with MCF-7 cells treated with H_2O_2 of various concentrations, the peak drops. More H_2O_2 will induce more apoptotic cells, which can cause larger interfacial charge transfer resistance and lower DPV peak (Fig. 3). The signal intensity is linearly related to the concentration of H_2O_2, and this electrochemical approach can successfully distinguish H_2O_2 (as low as 3.0 μM) induced apoptosis from nonapoptotic cells (S/N = 3).

Control experiments are also carried out by introducing the control peptide to replace the PS-specific peptide. The experiment results show nearly unchanged peak currents, demonstrating that control peptide cannot capture apoptotic cells to block the electron transfer. Only the designed PS-specific peptide has high affinity towards the recognition of exposed membrane PS.

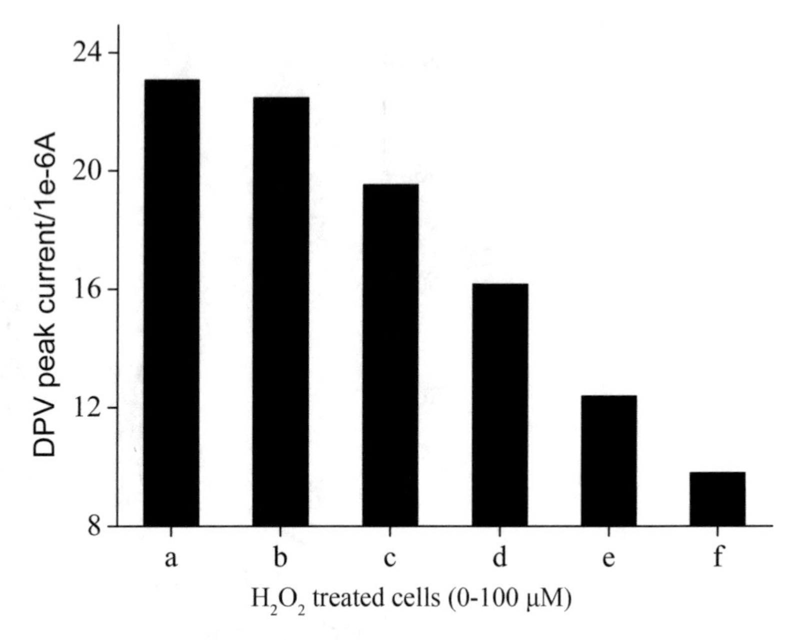

Fig. 3 DPV peak currents of (*a*) peptide modified electrode, (*b*–*f*) peptide modified electrodes after incubation with H_2O_2 (0, 25, 50, 80, 100 μM) treated cells

3.4 Other Methods

Caspase-3 Activity Measurements. Since caspase-3 activity can be used to measure the extent of apoptosis [33], a caspase-3 activity assay can be used to validate the electrochemical sensor assay. A colorimetric activity assay utilizing a kit is described below.

1. After treatment with various concentrations of H_2O_2 for 30 min, the suspended cells (1×10^6 per well) are collected by centrifugation at $1200 \times g$ for 5 min at 4 °C, homogenized in 1 mL of lysis buffer (6 M urea, 2 M thiourea, 65 mM DTT, 4 % CHAPS, and 40 mM Tris base dissolved in PBS) and incubated on ice for 15 min.
2. The lysates are centrifuged at $16{,}000 \times g$ at 4 °C for 15 min, and the protein content is determined using the Bradford method.
3. 40 μL of the cell lysates (0.70–1.72 mg/L protein) is transferred to a microplate and added to the 60 μL of PBS containing 0.2 mM Ac-DEVD-pNA. Furthermore, 40 μL lysis buffer is added to 60 μL of PBS containing 0.2 mM Ac-DEVD-pNA, which is used as a blank control.
4. After incubation at 37 °C for 2 h, the samples are measured at 405 nm on a microplate reader.

Experimental results of this caspase-3 assay suggest that the enzyme activity increases with more H_2O_2 added in a dose-dependent manner which correlates well with the proposed electrochemical approach.

Fluorescence Microscopy Observation. To further check the reliability of the proposed electrochemical method, the cells are observed directly by fluorescence microscopy. Briefly, 10^6 suspended cells are washed with PBS, and cultured in 1 mL PBS containing 10 ng Hochest 33342 and 10 ng PI for 20 min at 4 °C in the dark. The apoptotic cells (stained by Hochest 33342, since they have condensed nuclei) and necrotic cells (stained strongly by PI over Hochest 33342) are respectively viewed under a fluorescence microscopy, but untreated cells cannot be stained by Hoechst 33342 and PI. After a treatment with 50 μM H_2O_2, over half of cells exhibit a condensed chromatin and are stained by Hoechst 33342, while they are seldom stained by PI; this is a distinct characteristic of apoptosis.

Flow Cytometry Assay. To detect the percentage of apoptotic cells after H_2O_2 treatment, and validate the results of the electrochemical sensor apoptosis assay, a flow cytometric method using propidium iodide can be utilized [34]. This assay is performed as follows:

1. 10^6 suspended cells are washed with PBS and fixed in 1 mL of 70 % ethanol for 12 h at −20 °C while slowly vortexing.
2. The fixed cells are washed with PBS, and treated with 1 mL PBS containing 10 ng PI and 50 ng Rnase A for 20 min at 37 °C in the dark.
3. DNA content of these cells is analyzed by Cyan flow cytometer (BD FACSCalibur) and ModFit software.

The percentage of distributions of apoptotic cells (sub G0) are calculated for comparison with the electrochemical method. The results are also much consistent with that of the proposed electrochemical method.

4 Notes

4.1 Cell Protection

Cells undergoing apoptosis would gradually lose their membrane integrity; they can thus be easily disrupted during the washing and centrifugation processes. In this respect, 4 % bovine serum albumin or 1 % serum is added to the washing PBS buffer for the protection of cell structures, and the cells are kept at 4 °C during sample preparation.

4.2 Electrode Cleaning

The procedure of electrode cleaning is critical for the success of experiments, and uniform treated bare electrodes guarantee the sensing stability. The reagents utilized, such as H_2O_2 and H_2SO_4, should be fresh so they can have sufficient ability to remove any absorbed impurities on the electrode. Moreover, adequate polishing duration and strength on silicon carbide paper and alumina slurries are required. The final CV curves in the H_2SO_4 can be used to check whether the electrode has been cleaned.

4.3 EIS Data Processing

EIS data obtained from CHI660C electrochemical workstation is analyzed by Zsimpwin software. An equivalent circuit consisting of resistors and capacitors is constructed, simulating the actual electrochemical process on the electrode. R_{ct} values are then calculated and used to reflect interfacial charge transfer resistance.

4.4 Data Interpretation

1. Figure 1 outlines the principle of the mechanism of this electrochemical approach for apoptosis evaluation. First, the peptide is immobilized onto the electrode surface via thiol–gold interaction. In the presence of apoptotic cells with externalized PS on the cell membrane, the peptide on the electrode captures the cells onto the electrode surface; this alters the electrochemical response, which reflects the extent of apoptosis. R_{ct} value represents the impedance.
2. As shown in Fig. 2, after modified with peptide, R_{ct} increases due to the balance of the steric hindrance and positive charges of peptide to attract $Fe(CN)_6^{3-/4-}$. However, after certain nonspecific adsorption of nonapoptotic cells on the electrode, the attraction is blocked with the steric hindrance from the cells. R_{ct} grows continuously. In the case of apoptotic cells, which are specially captured by the peptide, significant increase of R_{ct} is observed.
3. Figure 3 depicts the DPV curves of peptide modified electrode before and after the incubation with H_2O_2 treated cells. Since larger amount of H_2O_2 creates more apoptotic cells, resulting in the increase of steric hindrance and shield of the positive charges of peptide attracting $Fe(CN)_6^{3-/4-}$, electrochemical signals may be blocked much more significantly.
4. As shown in Fig. 4, the results of electrochemical measurement is consistent with that of caspase-3 assay. Caspase-3 activity and decreased DPV peak current both increase with more H_2O_2 in a dose-dependent manner.

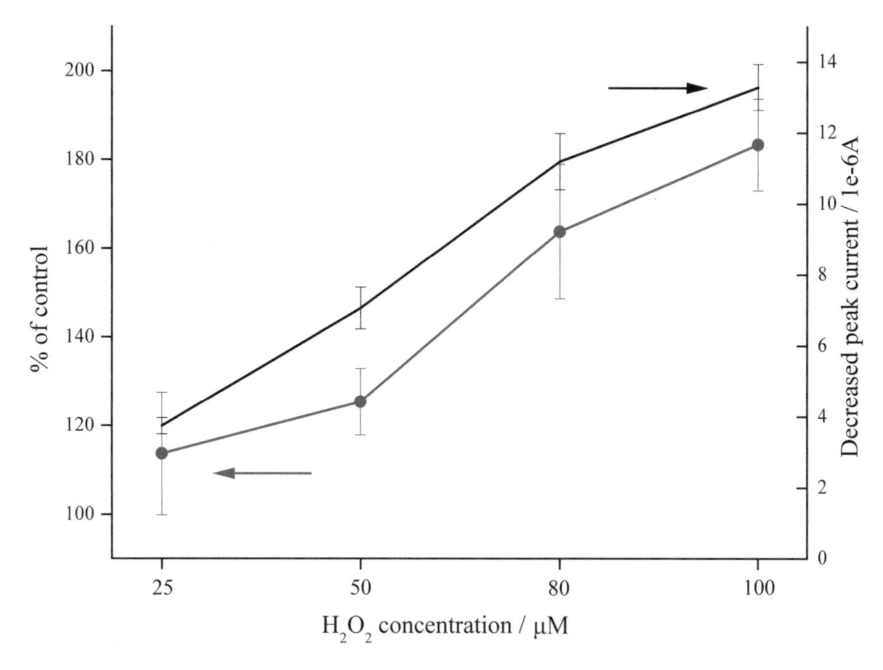

Fig. 4 Caspase-3 assay and DPV performance of cell lysates which are previously treated with H_2O_2 (25–100 μM). The *left vertical coordinate* stands for the caspase-3 activity of the cells versus that without H_2O_2 treatment. The *right vertical coordinate* stands for the decreased DPV peak current of the cells

Acknowledgments

This work was supported by the National Natural Science Foundation of China (Grant No. 21307154 and 31400847) and the National Key Instrument Developing Project of China (Grant No. ZDYZ2013-1).

Glossary

CV Cyclic voltammetry
DPV Differential pulse voltammetry
ECL Electrochemiluminescent
EDTA Ethylenediaminetetraacetic acid
EIS Electrochemical impedance spectroscopy
MCH Mercaptohexanol
PBS Phosphate buffered saline
PI Propidium iodide
TCEP Tris(2-carboxyethyl)phosphinehydro-chloride

References

1. Taylor RC, Cullen SP, Martin SJ (2008) Apoptosis: controlled demolition at the cellular level. Nat Rev Mol Cell Biol 9:231–241
2. Boulianne B, Rojas OL, Haddad D, Zaheen A, Kapelnikov A, Nguyen T, Li CL, Hakem R, Gommerman JL, Martin A (2013) AID and caspase 8 shape the germinal center response through apoptosis. J Immunol 191: 5840–5847
3. Areti A, Yerra VG, Naidu V, Kumar A (2014) Oxidative stress and nerve damage: role in chemotherapy induced peripheral neuropathy. Redox Biol 2:289–295
4. Choong G, Liu Y, Templeton DM (2014) Interplay of calcium and cadmium in mediating cadmium toxicity. Chem Biol Interact 211:54–65
5. Vaux DL (2002) Apoptosis and toxicology—what relevance? Toxicology 181–182:3–7
6. El-Naga RN (2014) Pre-treatment with cardamonin protects against cisplatin-induced nephrotoxicity in rats: impact on NOX-1, inflammation and apoptosis. Toxicol Appl Pharm 274:87–95
7. Seco-Rovira V, Beltran-Frutos E, Ferrer C, Saez FJ, Madrid JF, Pastor LM (2014) The death of sertoli cells and the capacity to phagocytize elongated spermatids during testicular regression due to short photoperiod in Syrian hamster (Mesocricetus auratus). Biol Reprod 90:107
8. Vanden Berghe T, Grootjans S, Goossens V, Dondelinger Y, Krysko DV, Takahashi N, Vandenabeele P (2013) Determination of apoptotic and necrotic cell death in vitro and in vivo. Methods 61:117–129
9. Feng H, Yin SH, Tang AZ, Cai HW, Chen P, Tan SH, Xie LH (2010) Caspase-3 activation in the guinea pig cochlea exposed to salicylate. Neurosci Lett 479:34–39
10. Liu J, Yao YZ, Ding HF, Chen RA (2014) Oxymatrine triggers apoptosis by regulating Bcl-2 family proteins and activating caspase-3/caspase-9 pathway in human leukemia HL-60 cells. Tumor Biol 35:5409–5415
11. Mendez J, Cruz MM, Delgado Y, Figueroa CM, Orellano EA, Morales M, Monteagudo A, Griebenow K (2014) Delivery of chemically glycosylated cytochrome c immobilized in mesoporous silica nanoparticles induces apoptosis in HeLa cancer cells. Mol Pharm 11:102–111
12. Gorla M, Sepuri NBV (2014) Perturbation of apoptosis upon binding of tRNA to the heme domain of cytochrome c. Apoptosis 19: 259–268
13. Pan Y, Shan W, Fang H, Guo M, Nie Z, Huang Y, Yao S (2014) Sensitive and visible detection of apoptotic cells on Annexin-V modified substrate using aminophenylboronic acid modified gold nanoparticles (APBA-GNPs) labeling. Biosens Bioelectron 52:62–68
14. Li XH, Link JM, Stekhova S, Yagle KJ, Smito C, Krohn KA, Tait JF (2008) Site-specific labeling of annexin V with F-18 for apoptosis imaging. Bioconjug Chem 19:1684–1688
15. Dong HP, Holth A, Ruud MG, Emilsen E, Risberg B, Davidson B (2011) Measurement of apoptosis in cytological specimens by flow cytometry: comparison of Annexin V, caspase cleavage and dUTP incorporation assays. Cytopathology 22:365–372
16. Burtea C, Laurent S, Lancelot E, Ballet S, Murariu O, Rousseaux O, Port M, Vander Elst L, Corot C, Muller RN (2009) Peptidic targeting of phosphatidylserine for the MRI detection of apoptosis in atherosclerotic plaques. Mol Pharm 6:1903–1919
17. Qiu Y, Yi S, Kaifer AE (2011) Encapsulation of tetrathiafulvalene inside a dimeric molecular capsule. Org Lett 13:1770–1773
18. Miao P, Ning L, Li X (2013) Gold nanoparticles and cleavage-based dual signal amplification for ultrasensitive detection of silver ions. Anal Chem 85:7966–7970
19. Miodek A, Castillo G, Hianik T, Korri-Youssoufi H (2013) Electrochemical aptasensor of human cellular prion based on multiwalled carbon nanotubes modified with dendrimers: a platform for connecting redox markers and aptamers. Anal Chem 85:7704–7712
20. Yue QL, Xiong SQ, Cai DQ, Wu ZY, Zhang X (2014) Facile and quantitative electrochemical detection of yeast cell apoptosis. Sci Rep 4:4373
21. Liu T, Zhu W, Yang X, Chen L, Yang RW, Hua ZC, Li GX (2009) Detection of apoptosis based on the interaction between Annexin V and phosphatidylserine. Anal Chem 81:2410–2413
22. Cao JT, Zhu YD, Rana RK, Zhu JJ (2014) Microfluidic chip integrated with flexible PDMS-based electrochemical cytosensor for dynamic analysis of drug-induced apoptosis on HeLa cells. Biosens Bioelectron 51:97–102
23. Xiao H, Liu L, Meng FB, Huang JY, Li GX (2008) Electrochemical approach to detect apoptosis. Anal Chem 80:5272–5275
24. Zhang JJ, Zheng TT, Cheng FF, Zhu JJ (2011) Electrochemical sensing for caspase 3 activity and inhibition using quantum dot functional-

ized carbon nanotube labels. Chem Commun 47:1178–1180

25. Wen QQ, Zhang X, Cai J, Yang PH (2014) A novel strategy for real-time and in situ detection of cytochrome c and caspase-9 in Hela cells during apoptosis. Analyst 139:2499–2506
26. Wu YF, Zhou H, Wei W, Hua X, Wang LX, Zhou ZX, Liu SQ (2012) Signal amplification cytosensor for evaluation of drug-induced cancer cell apoptosis. Anal Chem 84:1894–1899
27. Tong CY, Shi BX, Xiao XJ, Liao HD, Zheng YQ, Shen GL, Tang DY, Liu XM (2009) An annexin V-based biosensor for quantitatively detecting early apoptotic cells. Biosens Bioelectron 24:1777–1782
28. Miao P, Yin J, Ning LM, Li XX (2014) Peptide-based electrochemical approach for apoptosis evaluation. Biosens Bioelectron 62:97–101
29. Miao P, Ning LM, Li XX, Li PF, Li GX (2012) Electrochemical strategy for sensing protein phosphorylation. Bioconjug Chem 23: 141–145
30. Miao P, Wang BD, Han K, Tang YG (2014) Electrochemical impedance spectroscopy study of proteolysis using unmodified gold nanoparticles. Electrochem Commun 47:21–24
31. Miao P, Liu L, Li Y, Li GX (2009) A novel electrochemical method to detect mercury (II) ions. Electrochem Commun 11:1904–1907
32. Miao P, Liu L, Nie YJ, Li GX (2009) An electrochemical sensing strategy for ultrasensitive detection of glutathione by using two gold electrodes and two complementary oligonucleotides. Biosens Bioelectron 24:3347–3351
33. Zhou SW, Zheng TT, Chen YF, Zhang JJ, Li LT, Lu F, Zhti JJ (2014) Toward therapeutic effects evaluation of chronic myeloid leukemia drug: electrochemical platform for caspase-3 activity sensing. Biosens Bioelectron 61: 648–654
34. Riccardi C, Nicoletti I (2006) Analysis of apoptosis by propidium iodide staining and flow cytometry. Nat Protoc 1:1458–1461

Chapter 11

Microplate-Based Whole Zebrafish Caspase 3/7 Assay for Screening Small Molecule Compounds

Wen Lin Seng, Dawei Zhang, and Patricia McGrath

Abstract

In this research, using a commercially available human specific caspase 3/7 chemiluminescent test kit (Caspase 3/7 Glo, Promega, Madison, WI), developed for cell based assays, we describe a microplate-based whole zebrafish assay format to identify potential small molecule caspase inhibitors and activators. Based on the high degree of evolutionary conservation among species, we show that human specific 3/7 substrate cross reacts with zebrafish. Using untreated zebrafish, optimum assay conditions (including substrate concentration, number of zebrafish per microwell, and incubation time to generate a linear reaction) are determined. Robustness and reproducibility of the assay are established using a characterized caspase 3/7 inhibitor (z-VAD-fmk) and an activator (staurosporine). Next, the whole zebrafish microplate assay format is validated using three additional characterized caspase 3/7 inhibitors, two additional caspase 3/7 activators, and one control compound that has no effect on zebrafish apoptosis. Compared to other whole animal assay formats, chemiluminescence provides high sensitivity and low background. Next, results are compared with published results in mammalian cell based assays and animal models and show that the overall predictive success rate is 100 %. Compound effects on apoptosis are further confirmed visually by whole mount staining with acridine orange (AO), a live dye. Results support the high degree of conservation of key pathways in zebrafish and humans. The microplate-based whole zebrafish caspase 3/7 assay format represents a rapid, reproducible, predictive animal model for identifying potential inhibitors and activators. Use of zebrafish as an alternative animal model to identify potential apoptosis modulators can accelerate the drug discovery process and reduce costs.

Key words Acridine orange, Assay, Apoptosis, Caspase, Chemiluminescence, ELISA, Methods, Microplate, Whole zebrafish

Abbreviations

ac-DEVD-cho	Acetyl-aspartic acid-glutamic acid-valine-aspartic acid-aldehyde
ac-DNLD-cho	Acetyl-aspartic acid-aspartic acid-leucine-aspartic acid-aldehyde
Ala (A)	Alanine
AO	Acridine orange
Asp (D)	Aspartic acid
Asp-Glu-Val-Asp	Aspartic acid–glutamic acid–valine–aspartic acid
AVMA	American Veterinary Medical Association

Perpetua M. Muganda (ed.), *Apoptosis Methods in Toxicology*, Methods in Pharmacology and Toxicology,
DOI 10.1007/978-1-4939-3588-8_11, © Springer Science+Business Media New York 2016

CV	Coefficient of variation
DMSO	Dimethyl sulfoxide
dpf	Days post fertilization
ECVAM	European Centre for the Validation of Alternative Methods
Glu (E)	Glutaric acid
h	hours
hpf	Hours post-fertilization
HUVEC	Human umbilical vein endothelial cells
M	Mean
MESAB	Ethyl 3-aminobenzoate methanesulfonate
mg	milligram
min	Minutes
q-VD-oph	Quinoline-valine-aspartic acid-oxo-pentanoic acid hydrate
RLU	Relative luminescence units
ROI	Region of Interest
S	Seconds
SD	Standard deviation
S/N	Signal/noise
Val (V)	Valine
μl	*Microliter*
μM	*Micromole*
z-VAD-fmk	*N*-Benzyloxycarbonyl-valine-alanine-aspartic acid (*O*-Me)-fluoromethylketone

1 Introduction

Apoptosis is a genetically programmed biochemical cascade that culminates in cell death. During development, apoptosis is involved in morphologic patterning, organogenesis, limb formation, and nervous system development. Uncontrolled apoptosis is associated with several diseases, including cancers, neurodegeneration, autoimmunity, and cardiac failure [1]. Although gene products that regulate apoptosis appear to be excellent potential drug targets, few successful therapies have been developed. Furthermore, there are no rapid in vivo assays for identifying potential candidate molecules.

Caspases, a family of cysteine proteases, play a major role in apoptosis. The caspase family is separated into two groups: (1) initiator caspases that cleave and activate zymogen forms, and (2) effector caspases that cleave other cell proteins and trigger the apoptotic process; caspase 3/7 is a major effector [2]. Catalytic cysteine and histidine diad, required for catalysis by cysteine proteases, is universally conserved in caspase 3/7 among species [3]. Phylogenetic analysis shows that zebrafish caspase 3 has 60 % similarity in amino acid (aa) sequence to mammalian caspase 3 [4]. Mammalian caspase 3 and 7 recognize the tetra-peptide motif Asp-x-x-Asp and share similar substrate specificity [5]. Although the presence of C-terminal Asp is required for substrate recognition, variation in the additional three aa sequences can be tolerated [6]. The proluminescent Caspase-Glo 3/7 substrate, specific for human cells, contains the tetrapeptide aspartic acid–glutamic acid–valine–aspartic acid (Asp-Glu-Val-Asp) (DEVD)-aminoluciferin substrate and a proprietary thermostable luciferase (Ultra-Glo™ Recombinant Luciferase). Cell lysis occurs following addition of reconstituted substrate. Lysed cells release caspase-3/7, which cleaves the substrate and releases free aminoluciferin, which is then oxidized by the proprietary thermostable luciferase and produces a stable "glow-type" luminescent signal. Luminescence is proportional to level of caspase enzyme activity.

A unique feature of this assay format is that developmentally regulated apoptosis is present in zebrafish through 2 days post-fertilization (dpf), and this parameter is used to assess caspase 3/7 inhibition in early stage zebrafish. Caspase 3/7 activation is also assessed at this stage. An important advantage of the zebrafish animal model for compound screening is that the morphological and molecular characteristics of tissues and organs are either identical or similar to other vertebrates, including humans [7]. Conventional cell and single-tissue lysate-based apoptosis assays do not address the complexity of the in vivo physiological environment [8]. Because zebrafish develop rapidly, drug effects can be observed in days compared to weeks or months required for rodent studies [9]. A compelling advantage of zebrafish for this research is that their small size permits analysis of whole animals in 96-well microplates [10]; using cell based assay methods and instrumentation, reagents can be added directly to wells containing zebrafish. In addition, human specific Caspase-Glo 3/7 substrate cross reacts with zebrafish, permitting development of a microplate-based whole animal assay. We and others have used these advantages to successfully develop whole zebrafish ELISAs for angiogenesis [11], CYP3A4 [12], human cancer cell xenotransplant [13], and zebrafish microplate assays for identifying neuroprotectants [14] and radioprotectants [15, 16]. Use of zebrafish as an alternative animal model to identify potential caspase 3/7 modulators can accelerate the drug discovery process, reduce costs, and provide more predictive results than conventional cell based assays.

2 Materials

1. *Zebrafish.* AB zebrafish are generated using conventional breeding methods [9].
2. *Caspase 3/7 Substrate.* Caspase-Glo 3/7 test kit (Promega, Madison, WI) contains a lyophilized luminogenic caspase-3/7 substrate comprised of a tetrapeptide DEVD and a buffer optimized for: caspase activity, luciferase activity and cell lysis.

3. *Microplates.* 6-well microplates (VWR Scientific, Bridgeport, NJ) are used for compound treatment. Black, clear bottomed 96-well microplates (Corning Life Sciences, Corning, NY) are used for the caspase-3/7 assay.
4. *Compounds.* Acridine orange (*AO*), ac-DNLD-cho, ac-DEVD-cho, q-VD-oph, staurosporine, gambogic acid, borrelidin, and buthioine sulfoximine are available from Sigma-Aldrich, St-Louis, MO. Z-VAD-fmk can be purchased from Bachem Bioscience, King of Prussia, PA.
5. *Fluorescence microscopy.* Fluorescence microscopy is performed using a Zeiss M2Bio (Carl Zeiss Microimaging Inc., Thornwood, NY) equipped with a green FITC filter (excitation: 488 nm, emission: 515 nm), a chilled CCD camera (Axiocam MRM, Carl Zeiss Microimaging Inc.), a 10× eye piece and a 10× objective achromat.
6. *Image analysis software.* Axiovision software Rel 4.0 (Carl Zeiss Microimaging Inc.) and Adobe Photoshop 7.0 (Adobe, San Jose, CA) are used to analyze captured images.

3 Methods

3.1 Preparation of Zebrafish for Assay Development

1. *Zebrafish generation*: Phylonix AB zebrafish embryos are generated by natural pairwise mating using conventional methods [9]. Four to five pairs are set up for each mating, and, on average 50–100 embryos are generated. Embryos are maintained in fish water (5 g of Instant Ocean Salt with 3 g of $CaSO_4$ in 25 l of distilled water) at 28 °C, cleaned and sorted for viability at 6 and 24 h post-fertilization (hpf). Because zebrafish embryos receive nourishment from an attached yolk, no additional maintenance is required. Breeding methods follow Office of Laboratory Animal Welfare (OLAW), National Institutes of Health (NIH) guidelines and recommendations by the American Veterinary Medical Association's (AVMA) Panel on Euthanasia.
2. *Zebrafish stage*: In zebrafish, caspase 3 activation has been observed from 6 to 32 hpf [17–19]. In addition, developmentally regulated apoptosis is present in both the nasal placodes and hatching gland through 2 dpf [20], and decreases thereafter. Therefore, 2 dpf represents a convenient stage to assess effects of potential apoptosis inhibitors and activators.
3. *Zebrafish dechorionation*: To facilitate compound delivery and assay processing, chorion membranes, present up to 2 dpf, are removed by incubating zebrafish in fish water containing pronase (Sigma-Aldrich, St. Louis, MO) (0.5 mg/ml final concentration) for 2–3 min. To remove all traces of pronase, zebrafish are washed several times in fish water prior to assay development.

3.2 Development of Microplate-Based Whole Zebrafish Caspase 3/7 Assay

Prior to developing the whole zebrafish caspase 3/7 assay, the following are prepared: (1) zebrafish (one to four per well) are incubated with substrate in buffer (supplied by manufacturer), (2) substrate is incubated in buffer without zebrafish, and (3) zebrafish are incubated in buffer without substrate. Compared to controls (groups 2 and 3), significantly increased ($P<0.05$) chemiluminescence signal is detected in group 1 (Table 1). This study demonstrates that substrate specific for human caspase 3/7 cross reacts with zebrafish.

1. *Optimum caspase 3/7 substrate volume*: Since Caspase-Glo 3/7 kit was developed for use in human cell based assays, appropriate substrate volume is determined for the whole zebrafish assay. Two dpf zebrafish are incubated with 25, 50, or 100 μl substrate from 15 to 120 min, and RLU is measured in 15 min intervals, ten wells per condition. Using 50 (red line) and 100 μl (blue line) of caspase 3/7 substrate, RLU signal increases sharply from 15 to 45 min and then plateaus, indicating rapid reactivity followed by saturation (Fig. 1). In contrast, using 25 μl substrate (green line), signal for RLU increases slowly and is linear from 45 to 90 min, the longest range exhibiting a linear reaction. Therefore, 25 μl is the optimum substrate volume for assay development.
2. *Optimum number of zebrafish per well*: After substrate volume is determined, optimum number of zebrafish per well is also established. One to four zebrafish are manually deposited into individual wells of black, clear bottomed 96-well microplates containing 100 μl fish water per well, and 25 μl substrate volume, and enzyme reactions are performed from 15 to 195 min; RLU is measured in 15 min intervals. Wells containing buffer alone are used to assess buffer background signal. Wells containing substrate in buffer without zebrafish are used to assess substrate background signal. Wells containing zebrafish incubated with buffer without substrate are used to assess zebrafish background signal; ten wells per condition are used. Because the first and last columns are located at the edge of each microplate, uneven evaporation can occur during incubation, contributing to high assay variation (edge effect). Therefore, only 80 wells per plate are used. To ensure thorough mixing of substrate and fish water, microplates are gently shaken on a low speed orbital shaker (Fisher Scientific, Pittsburgh, PA) for 30 s at room temperature. After thorough mixing, RLU are measured using a Synergy multi-mode microplate reader (BioTek, Winooski, VT). Results in Table 1 show that signal of zebrafish incubated with buffer alone without substrate range from 262 to 422 for one to four zebrafish, and no significant difference is observed for varying number of zebrafish per well. This result shows that background is caused

Table 1
Mean Relative Luminescence Units (RLU) for endogenous caspase 3/7 activity quantitated in untreated 2 dpf zebrafish ($n = 10$)

Zebrafish/well		Incubation time (min) 15	30	45	60	75	90	105	120	135	150	165	180	195
RLU (with substrate)	1	4,060	8,257	10,186	11,435	12,256	12,645	13,078	14,135	14,616	15,856	16,120	17,092	15,773
	2	6,521	12,087	14,950	16,975	18,599	19,526	20,071	22,008	22,721	24,764	26,681	28,300	27,185
	3	8,776	15,316	19,019	21,682	24,153	25,839	27,549	31,008	32,346	34,280	36,114	38,957	37,929
	4	10,347	17,580	21,613	24,336	27,376	29,773	31,896	36,147	36,293	38,848	38,528	42,647	41,426
RLU (with buffer alone)	1	271	309	327	352	362	376	385	394	399	414	405	422	422
	2	267	299	303	342	357	348	364	360	373	386	386	384	375
	3	262	288	314	332	347	360	357	367	363	371	368	370	373
	4	270	295	310	338	342	350	358	368	368	375	381	360	379
S/N ratio	1	15	27	31	32	34	34	34	36	37	38	40	41	37
	2	24	40	49	50	52	56	55	61	61	64	69	74	72
	3	33	53	61	65	70	72	77	84	89	92	98	105	102
	4	38	60	70	72	80	85	89	98	99	104	101	118	109

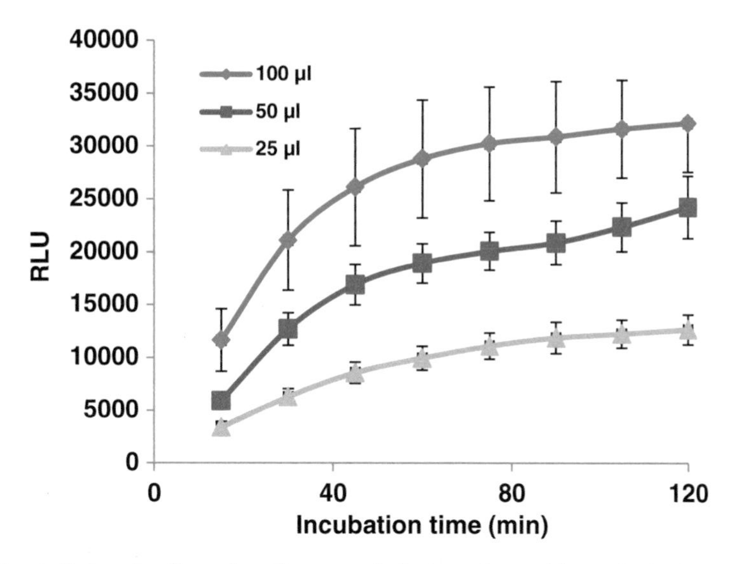

Fig. 1 Determination of optimum substrate volume. Three dechorionated untreated 2 dpf zebrafish are placed in black, clear bottomed 96-well microplates containing 100 μl fish water and incubated with 25 (*green line*), 50 (*red line*), and 100 μl (*blue line*) of substrate. RLU are measured in 15 min intervals from 15 to 120 min. *X* axis represents substrate incubation time. *Y* axis represents RLU. Each point represents mean ± SD ($n = 10$)

by buffer not by zebrafish. Wells containing substrate without zebrafish do not exhibit signal above buffer background (data not shown), indicating that substrate alone does not generate signal. Signal from zebrafish incubated with substrate ranges from 4,060 to 42,647, which represents level of activated endogenous caspase 3/7 in 2 dpf zebrafish; RLU signal increases both with incubation time (15–195 min) and number of zebrafish per well (one to four) (Table 1). Signal/noise (S/N) ratio ranges from 15 to 118; S/N ≥ 3 is one of the important criteria for an acceptable assay [21].

Kinetic curves for each condition are generated by plotting RLU signal against incubation time. Using one or two zebrafish/well, kinetic curves plateau at 45 and 75 min, respectively (some data are omitted to improve graph readability). However, using three or four zebrafish/well, a linear range is observed between 45 and 180 min. The difference in RLU signal for three or four zebrafish/well is insignificant (SDs are overlapping) (Fig. 2). These results define baseline level of endogenous caspase 3/7 activity and demonstrate that three zebrafish per well is optimum for assay development.

3.3 Assay Robustness and Reproducibility

Following conventional assay development, to determine assay robustness and reproducibility, a high signal generating condition (H) and a low signal generating condition (L) are used to determine if the assay can unequivocally distinguish the two conditions

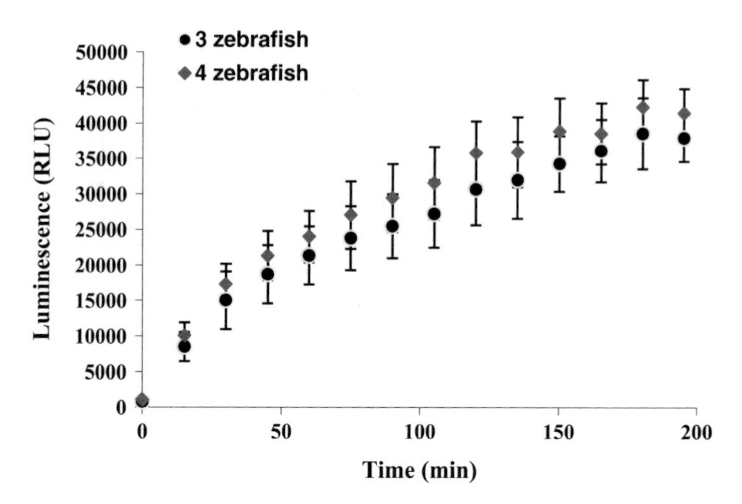

Fig. 2 Kinetic curves for interaction of caspase 3/7 enzyme with substrate in untreated 2 dpf zebrafish (3–4 zebrafish/well). *X* axis represents substrate incubation time. *Y* axis represents RLU signal. Each point represents mean ± SD ($n = 10$)

[22]. Mean (M) and standard deviation (SD) of H and L signals are used to calculate Z' factor following formula (1), where the absolute value for the M difference between H and L are used.

$$Z' = 1 - \left(3\left(\mathrm{SD_H} + \mathrm{SD_L}\right) / \left(M_\mathrm{H} - M_\mathrm{L}\right)\right) \qquad (1)$$

To assess assay reproducibility, the coefficient of variation (CV = SD/Mean × 100 %) is calculated.

3.3.1 Compound Treatment for Generating High and Low Signal Generating Conditions

Since compound effects are assessed on developmentally regulated apoptosis present up to 2 dpf, and zebrafish are treated for 24 h, 1 dpf zebrafish are deposited in 6-well microplates containing fish water, 30 zebrafish per well. Staurosporine, a caspase activator, and Z-VAD-fmk, a caspase 3/7 inhibitor, are used to generate H and L. Five compound concentrations: 0.01, 0.1, 1, 10, and 100 μM are tested to determine optimum concentration; 0.1 % DMSO is used as vehicle control. A kinetic study is performed using optimum volume of caspase 3/7 substrate. Optimum concentrations for staurosporine and z-VAD-fmk are 6 and 10 μM, respectively, which are then used as H and L signal generator, respectively. For determining the Z' factor, ten replicates are used for each group ($n = 10$). Substrate enzyme interaction stabilizes at ~60 min (manufacture's notes); data (Fig. 1) confirms the linear range between 45 and 90 min. Therefore, 45, 60, and 90 min incubation times are used to calculate Z' factors, which are 0.37, 0.37, and 0.27, respectively (Table 2); these are considered to be within the acceptable range (at 99.72 % confidence) for an assay [22].

Table 2
Mean, SD, and Z′ factor for determining microplate-based whole zebrafish caspase 3/7 assay robustness

Incubation time (min)	Z-VAD-fmk		Staurosporine		
	Mean	SD	Mean	SD	Z′ factor
45	5,343	792	23,190	2,986	0.37
60	6,962	1,026	20,807	1,898	0.37
90	8,349	1,047	18,158	1,383	0.27

To assess assay reproducibility, CV of baseline caspase 3/7 activity in untreated zebrafish (Fig. 2) is calculated; CV is 16.5 % and 14.5 % for 60 and 90 min incubation, respectively; this is within the acceptable range for a whole animal bioassay.

3.3.2 Optimum Incubation Time for Caspase 3/7 Inhibition and Activation

Since the assay is considered acceptable for 45–90 min incubation, this range is used to determine optimum incubation time to assess compound effects on caspase 3/7 inhibition and activation. To assess compound inhibition, both untreated and vehicle control zebrafish are expected to exhibit high signal prior to reaching plateau. To assess compound activation, controls are expected to exhibit low signal. Therefore, based on kinetic curves (Fig. 2), 90 and 60 min are considered optimum incubation times for assessing inhibition and activation, respectively.

Optimum assay conditions for this acceptable and reproducible assay are: (1) dechorionated 1 dpf zebrafish; (2) 24 h compound treatment; (3) three zebrafish per well; (4) 25 μl caspase 3/7 substrate volume; and (5) 90 and 60 min substrate incubation times for inhibition and activation, respectively.

3.4 Microplate-Based Whole Zebrafish Caspase 3/7 Assay Validation

To demonstrate that this zebrafish caspase 3/7 assay predicts compound effects in humans, validation can be done using: four caspase 3/7 inhibitors (ac-DNLD-cho, ac-DEVD-cho, z-VAD-fmk, and q-VD-oph), three characterized apoptosis activators (staurosporine, gambogic acid, borrelidin), and one control compound (buthionine sulfoximine), shown to have no effect on zebrafish apoptosis [14]. Untreated zebrafish are used as assay control and 0.1 % DMSO is used as vehicle control. Untreated zebrafish are incubated with buffer without substrate to measure background signal, which is subtracted from total signal in each well before calculating compound effects.

3.4.1 Compound Effects and Statistics

Data is normalized and % inhibition and % activation are calculated following formulas (2) and (3), respectively:

$$\%\,\text{inhibition} = \left(1 - \text{RLU}\left(\text{compound}\right) / \text{RLU}\left(\text{control}\right)\right) \times 100\% \quad (2)$$

$$\%\,\text{activation} = \left(\text{RLU}\left(\text{compound}\right) / \text{RLU}\left(\text{control}\right) - 1\right) \times 100\% \quad (3)$$

Statistical analysis can be done using GraphPad Prism 5 (GraphPad Software, San Diego, CA). To identify concentrations that cause significant effects, comparison of means among three or more groups is done using ANOVA, followed by Dunnett's test. Statistical significance is defined as $P < 0.05$.

Five compound concentrations (0.01, 0.1, 1, 10, and 100 μM) are tested; ten replicates are used per condition. Untreated zebrafish and zebrafish treated with vehicle solution (0.1 % DMSO) are used to identify baseline caspase 3/7 activity levels and to demonstrate that vehicle solution does not have an effect on baseline activity. As shown in Table 3, all inhibitors and activators cause significant effects, and the control compound buthionine sulfoximine causes essentially no effects. These results validate that the microplate-based whole zebrafish caspase 3/7 assay correctly predicts compound effects in mammals.

Concentration response curves for each compound are generated by plotting concentration against % inhibition and % activation (Fig. 3). ANOVA followed by Dunnett's *t*-test are used to analyze data and determine concentrations that caused significant effects ($P < 0.05$).

3.5 Confirmation of Compound Effects

This microplate-based whole zebrafish assay quantitates caspase 3/7 enzyme activity but not cell death. Since apoptosis is the end result of caspase 3/7 activation, to confirm quantitative results of the microplate-based whole zebrafish caspase 3/7 assay, and to assess the site of cell death, AO staining in whole zebrafish is visually assessed by fluorescence microscopy.

3.5.1 Whole Mount Acridine-Orange Staining

Using optimum compound concentrations (Table 3), 1 dpf zebrafish are treated for 24 h. After compound treatment, zebrafish are washed and incubated with AO (1 mg/ml) for 30 min. Zebrafish are then washed three times with fish water, and anesthetized in MESAB (0.5 mM 3-aminobenzoic acid ethyl ester, 2 mM Na_2HPO_4). Zebrafish are then placed on depression slides in 2.5 % methylcellulose, and examined using fluorescence microscopy. During development, at 1–2 dpf, apoptosis occurs in both nasal placodes and hatching gland [14] but not in the tail [14]. Therefore, these sites are defined as regions of interest (ROI), and images are captured using a constant exposure time and gain. Compound effects are compared with effects for vehicle control.

Untreated 2 dpf zebrafish exhibit developmentally regulated apoptosis in the nasal placodes (Fig. 4a, green arrow) and hatching gland (Fig. 4a, light blue arrows) but not in the tail (Fig. 4b, blue

Table 3
Summary of compound effects on caspase 3/7 activity in whole zebrafish[a]

Compound	Test conc. range (μM)	Range (%) for inhibition effect	Range (%) for activation effect	Optimum conc. (μM)	% inhibition	% activation	P[b]
Ac-DNLD-cho	0.01–100	16.4 ± 10.8 to 95.2 ± 1.1		10	86.6 ± 4.0		**<0.0001**
Ac-DEVD-cho	0.01–100	51.0 ± 3.2 to 95.6 ± 0.2		100	95.6 ± 0.2		**<0.0001**
Z-VAD-fmk	0.01–100	14.6 ± 15.7 to 53.3 ± 7.2		1	53.3 ± 7.2		**<0.0001**
Q-VD-oph	0.01–100	12.6 ± 12.8 to 34.5 ± 8.4		100	34.5 ± 8.4		**<0.0001**
Staurosporine	1–8		21.1 ± 26.1 to 267.0 ± 94.2	6		267.0 ± 94.2	**<0.0001**
Gambogic acid	0.01–1.5		−13.3 ± 13.8 to 125.0 ± 38.0	1.5		125.0 ± 38.0	**<0.0001**
Borrelidin	0.01–2		28.3 ± 26.4 to 140.7 ± 67.1	2		140.7 ± 67.1	**<0.0001**
Buthionine sulfoximine	0.01–100		−8.9 ± 11.6 to −7.6 ± 12.6	100		−7.6 ± 12.6	0.4740

[a]Compound effects express as mean ± SD ($n = 10$)
[b]Bolded P values indicate significant effects

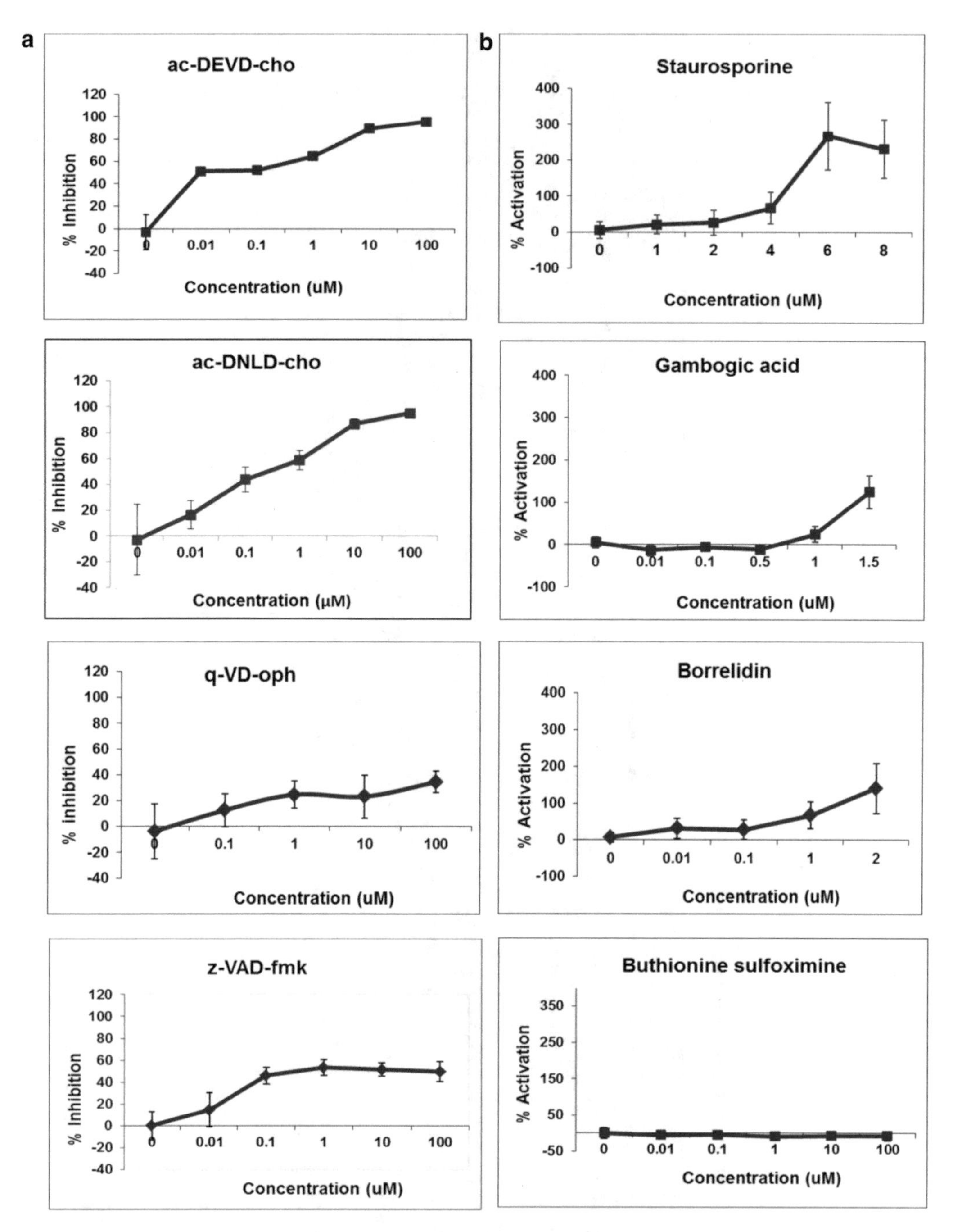

Fig. 3 Concentration response curves for caspase 3/7 inhibitors (a) and activators (b). *X* axis represents compound concentrations. Log scale is used for all inhibitors and control compound. In contrast, in order to clearly display concentration effects, equal spacing scale is used for all activators. *Y* axis represents % inhibition (**a**) or % activation (**b**). Buthionine sulfoximine, which exhibited a straight line near baseline throughout the concentration range, indicating no effect, is used as a control. Each point represents mean ± SD ($n = 10$)

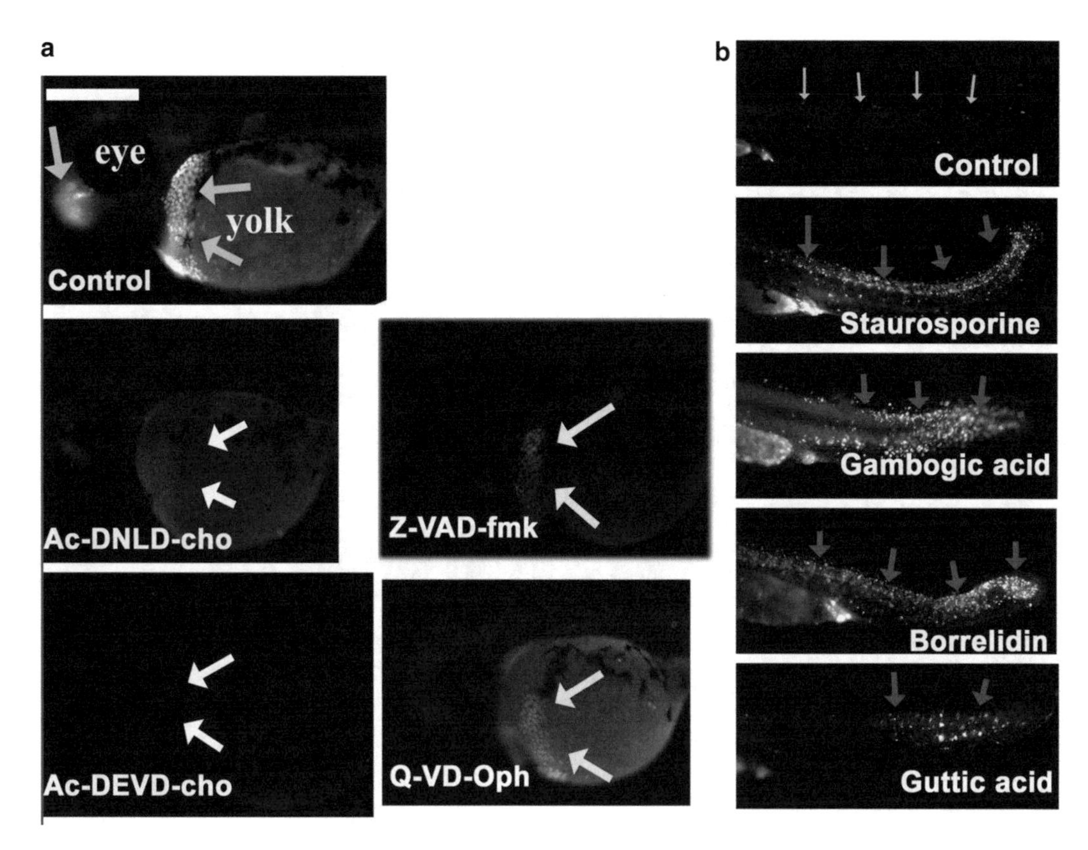

Fig. 4 Whole mount AO staining in control and compound treated zebrafish. 1 dpf zebrafish are treated with caspase inhibitors (**a**) and activators (**b**) for 24 h. Live zebrafish are stained with AO. (**a**) Strong AO staining in the nasal placodes (*green arrow*) and hatching gland (*light blue arrows*) is observed in control zebrafish. No AO staining in the nasal placodes and no (*white arrows*) or decreased (*yellow arrows*) AO staining in the hatching gland (*white arrows*) are observed in ac-DNLD-cho, ac-DEVD-cho, z-VAD-fmk, and q-VD-oph treated zebrafish, indicating apoptosis inhibition. (**b**) No AO staining (*blue arrows*) is observed in the tail of control zebrafish. In contrast, AO staining (*red arrows*) is observed in the tails of: staurosporine, gambogic acid, and borrelidin treated zebrafish, indicating apoptosis activation. Head: *left*; tail: *right*. Scale bar: ~250 μm

arrows). Effects for compound treated zebrafish are compared with vehicle control. As shown in Fig. 4a, zebrafish treated with all inhibitors exhibit no AO staining in the nasal placodes and no (white arrows) or decreased (yellow arrows) AO staining in the hatching gland, indicating decreased apoptosis. In contrast, zebrafish treated with activators exhibit increased AO staining in the tail (Fig. 4b, red arrows), indicating increased apoptosis.

3.6 Comparing Results in Zebrafish and Mammals

Results in zebrafish are then compared with published results in mammalian cell based assays and animal models (Table 4). Compound effects on zebrafish caspase activity are comparable to cell death results reported in mammalian cell based assays and animal models. The overall prediction success rate is 100 %, which, according to the European Centre for the Validation of Alternative Methods (ECVAM) [23], is considered "excellent."

Table 4
Comparison of compound effects on apoptosis/caspase activity in zebrafish and mammals

Compound	Effect in zebrafish	Effect in mammalian cell based or animal models	Correct prediction
ac-DEVD-cho[a]	Inhibition (89 %)	Inhibition (54 %) of apoptosis in infant rat hippocampal in pneumococcal meningitis model	Yes
ac-DNLD-cho[b]	Inhibition (89 %)	Inhibition (100 %) of human recombinant caspase-3 enzyme	Yes
q-VD-oph[c]	Inhibition (35 %)	Inhibition (48 %) of apoptosis in infant rat brain infarct volume in focal ischemia (stroke) model	Yes
z-VAD-fmk[d]	Inhibition (46 %)	Inhibition (90 %) of rat heart caspase-3 activity in endotoxin induced myocardial dysfunction model	Yes
Staurosporine[e]	Activation (267 %)	Activation (58 %) of cell death in human monocyte-like cells (THP-1)	Yes
Gambogic acid[f]	Activation (125 %)	Activation (31 %) of apoptotic cells in human tumor tissue transplanted in mice	Yes
Borrelidin[g]	Activation (141 %)	Activation (200 %) of caspase-3 activity in HUVEC	Yes

[a][28], [b][29], [c][30], [d][31], [e][32], [f][33], [g][34]

3.7 Microplate-Based Whole Zebrafish Caspase 3/7 Assay Adapted for HTS

Although in this study compound treatment is performed in 6-well microplates, and treated zebrafish are subsequently transferred to 96-well plates to quantitate chemiluminescent signal, using filter membranes, the entire assay can be performed in 96-well microplates. To further increase automation, a robotic liquid handler can also be adapted to perform washing and reagent dispensing steps (our unpublished data). Although cell based HTS assays now include 384- and 1536-well formats, based on our previous studies (our unpublished data), these formats are not suitable for all microplate based whole zebrafish assays due to potential crowding and oxygen depletion which can arrest development [9].

4 Notes

1. *Human specific caspase 3/7 specifically cross-reacts with zebrafish enzyme.* Caspase 3/7 is highly conserved among species, providing the opportunity to develop a zebrafish animal model to identify potential apoptosis modulators using a commercially available test kit developed for human cell based assays. These results (Figs. 1 and 2) confirm that human caspase 3/7 specific substrate cross reacts with zebrafish enzyme making it possible to develop a microplate-based whole animal assay format.

2. *Microplate-based whole zebrafish caspase 3/7 assay is acceptable and reproducible.* For most assays, in order to distinguish signal from noise, S/N ratio ≥3 is required [21]. Using incubation times ranging from 15 to 195 min and one to four zebrafish per well, S/N ranged from 15 to 118 (Table 1), indicating that signal is generated specifically by the caspase 3/7 enzyme. In addition, the Z' factor range is: $0.2 < Z' < 0.5$ during 45–90 min incubation, indicating that low and high signals are distinguishable and not overlapping; thus the assay is useful for assessing compound effects, confirming that the assay is acceptable. In contrast, an assay exhibiting $Z' < 0$ indicates the SDs are overlapping and the assay is not acceptable [22]. Further supporting assay reproducibility, variation observed for baseline caspase 3/7 activity in 2 dpf zebrafish is approximately 15 %, considered "good" reproducibility for a whole animal assay [21].
3. *Microplate-based whole zebrafish caspase 3/7 assay validation using characterized apoptosis modulating compounds.* The overall goal for this assay format is identification of potential caspase 3/7 inhibitors and activators. Therefore, the validation study is performed using characterized caspase 3/7 modulators. Since environmental factors, such as humidity and temperature, can affect RLU signal in a microplate-based format [24], compound effects are normalized to either % inhibition or % activation using vehicle control as the standard. These data show compound effects vary widely from compound to compound (ranging from 34.5 ± 8.4 to 95.6 ± 0.2 % for inhibitors, and 125.0 ± 38.0 to 267.0 ± 94.2 % for activators). However, variation is lower for inhibitors (<10 %) than for activators (>30 %), a phenomenon previously observed for compound effects in whole zebrafish (our unpublished data). For the control compound buthionine sulfoximine, inhibition effect is 7.6 ± 12.6 % (negative value for % activation), close to the baseline value. This result clearly shows that this compound has no effect on caspase 3/7.
4. *Results for quantitative microplate-based whole zebrafish caspase 3/7 assay are comparable to mammalian cell based assays and animal models.* Four inhibitors and three activators induce similar effects to effects observed in mammalian cell based assays and animal models (Fig. 3 and Table 4). Although published methods for these approaches differ widely, overall effects for mammals and zebrafish are in good agreement, further validating that the microplate-based whole zebrafish caspase 3/7 assay is a predictive model for identifying potential drug candidates for several diseases, including cancers [25], neurodegeneration [26], and Alzheimer's [27].
5. *Visual assessment of apoptosis confirmed quantitative microplate based results.* Our microplate based whole zebrafish caspase 3/7 assay measures enzyme activity, but not cell death. In

order to determine if compound effects on caspase 3/7 also inhibited or activated the final stages of apoptosis, and to assess the site of cell death, after compound treatment, AO staining in live zebrafish was visually assessed. At 2 dpf, developmentally regulated apoptosis is observed most prominently in the nasal placodes and hatching gland, but not in the tail; these parameters were used to visually confirm apoptosis inhibition and activation. These results demonstrate that zebrafish treated with four inhibitors show no or reduced AO staining in both the nasal placodes and hatching gland, indicating apoptosis inhibition. In addition, 2 dpf zebrafish treated with three activators exhibit increased AO staining in the tail region, indicating apoptosis activation. These visual results confirm quantitative microplate-based whole zebrafish caspase 3/7 assay results.

Acknowledgement

The research was supported in part by National Institutes of Health grants: National Institute of Medical Sciences: 1R43GM087754.

Conflict of Interest

Both Patricia McGrath and Wen Lin Seng are employees and shareholders of Phylonix.

References

1. Hetts SW (1998) To die or not to die: an overview of apoptosis and its role in disease. JAMA 279(4):300–307
2. Lamkanfi M, Declercq W, Kalai M et al (2002) Alice in caspase land. A phylogenetic analysis of caspases from worm to man. Cell Death Differ 9(4):358–361
3. Uren AG, O'Rourke K, Aravind LA et al (2000) Identification of paracaspases and metacaspases: two ancient families of caspase-like proteins, one of which plays a key role in MALT lymphoma. Mol Cell 6(4):961–967
4. Chakraborty C, Nandi SS, Sinha S et al (2006) Zebrafish caspase-3: molecular cloning, characterization, crystallization and phylogenetic analysis. Protein Pept Lett 13(6):633–640
5. Agniswamy J, Fang B, Weber IT (2007) Plasticity of S2–S4 specificity pockets of executioner caspase-7 revealed by structural and kinetic analysis. FEBS J 274(18):4752–4765
6. Fang B, Boross PI, Tozser J et al (2006) Structural and kinetic analysis of caspase-3 reveals role for s5 binding site in substrate recognition. J Mol Biol 360(3):654–666
7. Granato M, Nusslein-Volhard C (1996) Fishing for genes controlling development. Curr Opin Genet Dev 6(4):461–468
8. National Research Council (2007) Application of toxicogenomic technologies to predictive toxicology and risk assessment. National Academy of Sciences, Washington, DC, p 301
9. Westerfield M (1993) The zebrafish book: a guide for the laboratory use of zebrafish. The University of Oregon Press, Eugene
10. Seng WL, Eng K, Lee J et al (2004) Use of a monoclonal antibody specific for activated endothelial cells to quantitate angiogenesis in vivo in zebrafish after drug treatment. Angiogenesis 7(3):243–253
11. Serbedzija G, Semino C, Frost D et al (2003) Methods of screening agents for activity using teleosts. US Patent 6,656,449
12. Li C, Luo L, Awerman J et al (2012) Whole zebrafish cytochrome P450 assay for assessing drug metabolism and safety. In: McGrath P (ed) Zebrafish, methods for assessing drug safety and toxicity. Wiley, Hoboken, NJ, pp 103–116

13. Li C, Luo L, McGrath P (2012) Zebrafish xenotransplant cancer model for drug screening. In: McGrath P (ed) Zebrafish, methods for assessing drug safety and toxicity. Wiley, Hoboken, NJ, pp 219–232
14. Parng C, Ton C, Lin YX et al (2006) A zebrafish assay for identifying neuroprotectants in vivo. Neurotoxicol Teratol 28(4):509–516
15. Daroczi B, Kari G, McAleer MF et al (2006) In vivo radioprotection by the fullerene nanoparticle DF-1 as assessed in a zebrafish model. Clin Cancer Res 12(23):7086–7091
16. Geiger GA, Parker SE, Beothy AP et al (2006) Zebrafish as a "biosensor"? Effects of ionizing radiation and amifostine on embryonic viability and development. Cancer Res 66(16): 8172–8181
17. Negron JF, Lockshin RA (2004) Activation of apoptosis and caspase-3 in zebrafish early gastrulae. Dev Dyn 231(1):161–170
18. Yamashita M (2003) Apoptosis in zebrafish development. Comp Biochem Physiol B Biochem Mol Biol 136(4):731–742
19. Sorrells S, Toruno C, Stewart RA et al (2013) Analysis of apoptosis in zebrafish embryos by whole-mount immunofluorescence to detect activated caspase 3. J Vis Exp (82):e51060
20. Parng C, Anderson N, Ton C et al (2004) Zebrafish apoptosis assays for drug discovery. Methods Cell Biol 76:75–85
21. Iversen PW, Beck B, Chen Y-F et al (2012) HTS assay validation. In: Sittampalam GS, Gal-Edd N, Arkin MEA (eds) HTS assay validation. Eli Lilly & Company and the National Center for Advancing Translational Services, Bethesda, MD
22. Zhang JH, Chung TD, Oldenburg KR (1999) A simple statistical parameter for use in evaluation and validation of high throughput screening assays. J Biomol Screen 4(2):67–73
23. Genschow E, Spielmann H, Scholz G et al (2002) The ECVAM international validation study on in vitro embryotoxicity tests: results of the definitive phase and evaluation of prediction models. European Centre for the Validation of Alternative Methods. Altern Lab Anim 30(2):151–176
24. Michael S, Auld D, Klumpp C et al (2008) A robotic platform for quantitative high-throughput screening. Assay Drug Dev Technol 6(5):637–657
25. Chang H, Schimmer AD (2007) Livin/melanoma inhibitor of apoptosis protein as a potential therapeutic target for the treatment of malignancy. Mol Cancer Ther 6(1):24–30
26. Graham RK, Ehrnhoefer DE, Hayden MR (2011) Caspase-6 and neurodegeneration. Trends Neurosci 34(12):646–656
27. Jana K, Banerjee B, Parida PK (2013) Caspase: a potential therapeutic targets in the treatment of Alzheimer's disease. Trans Med S2:006
28. Gianinazzi C, Grandgirard D, Imboden H et al (2003) Caspase-3 mediates hippocampal apoptosis in pneumococcal meningitis. Acta Neuropathol 105(5):499–507
29. Yoshimori A, Sakai J, Sunaga S et al (2007) Structural and functional definition of the specificity of a novel caspase-3 inhibitor, Ac-DNLD-CHO. BMC Pharmacol 7:8
30. Renolleau S, Fau S, Goyenvalle C et al (2007) Specific caspase inhibitor Q-VD-OPh prevents neonatal stroke in P7 rat: a role for gender. J Neurochem 100(4):1062–1071
31. Fauvel H, Marchetti P, Chopin C et al (2001) Differential effects of caspase inhibitors on endotoxin-induced myocardial dysfunction and heart apoptosis. Am J Physiol Heart Circ Physiol 280(4):H1608–H1614
32. Voth DE, Howe D, Heinzen RA (2007) Coxiella burnetii inhibits apoptosis in human THP-1 cells and monkey primary alveolar macrophages. Infect Immun 75(9):4263–4271
33. Yang Y, Yang L, You QD et al (2007) Differential apoptotic induction of gambogic acid, a novel anticancer natural product, on hepatoma cells and normal hepatocytes. Cancer Lett 256(2):259–266
34. Kawamura T, Liu D, Towle MJ et al (2003) Anti-angiogenesis effects of borrelidin are mediated through distinct pathways: threonyl-tRNA synthetase and caspases are independently involved in suppression of proliferation and induction of apoptosis in endothelial cells. J Antibiot (Tokyo) 56(8):709–715

Chapter 12

Targeting Cancer Cell Death with Small Molecule Agents for Potential Therapeutics

Lan Zhang, Yaxin Zheng, Mao Tian, Shouyue Zhang, Bo Liu, and Jinhui Wang

Abstract

Time has come to switch from morphological to molecular definitions of cell death subroutines, due to substantial progress in biochemical and genetic exploration. Currently, cell death subroutines are defined by a series of precise, measurable biochemical features; these include apoptosis, autophagic cell death and necroptosis. Accumulating evidence has gradually revealed the core molecular machinery of cell death in carcinogenesis; the intricate relationships between cell death subroutines and cancer, however, still need to be clarified. Cancer drug discovery, in particular, has benefitted significantly from a rapid progress in utilization of several small molecule compounds to target different cell death modularity. Thus, this review provides a comprehensive summary of the interrelationships between the cell death subroutines (e.g., apoptosis and autophagic cell death) and relevant anticancer small molecule compounds (e.g., Oridonin and Rapamycin). Moreover, these interconnections between different cell death subroutines may be integrated into the entire cell death network. This would be regarded as a potential cancer target for more small molecule drug discovery. Taken together, these findings may provide new and emerging clues that fill the gap between cell death subroutines and small molecule drugs for future cancer therapy.

Key words Apoptosis, Autophagy, Necroptosis, Small molecule compounds, Cancer therapy

1 Introduction

Cell death plays crucial roles in regulating embryonic development, tissue homeostasis, immune function, tumor suppression, and infection resistance. Regulated cell death is essential in order to keep balance between cell survival and cell death for multicellular organisms. In fact, studies have shown that cells can undergo various cell death subroutines, including apoptosis, autophagic cell death, necroptosis, mitotic catastrophe, anoikis, cornification, entosis, netosis, parthanatos, pyroptosis, paraptosis and pyronecrosis, according to different microenvironments (Fig. 1).

The best known cell death form is apoptosis, which proceeds through two signaling pathways named as extrinsic and intrinsic

Perpetua M. Muganda (ed.), *Apoptosis Methods in Toxicology*, Methods in Pharmacology and Toxicology,
DOI 10.1007/978-1-4939-3588-8_12, © Springer Science+Business Media New York 2016

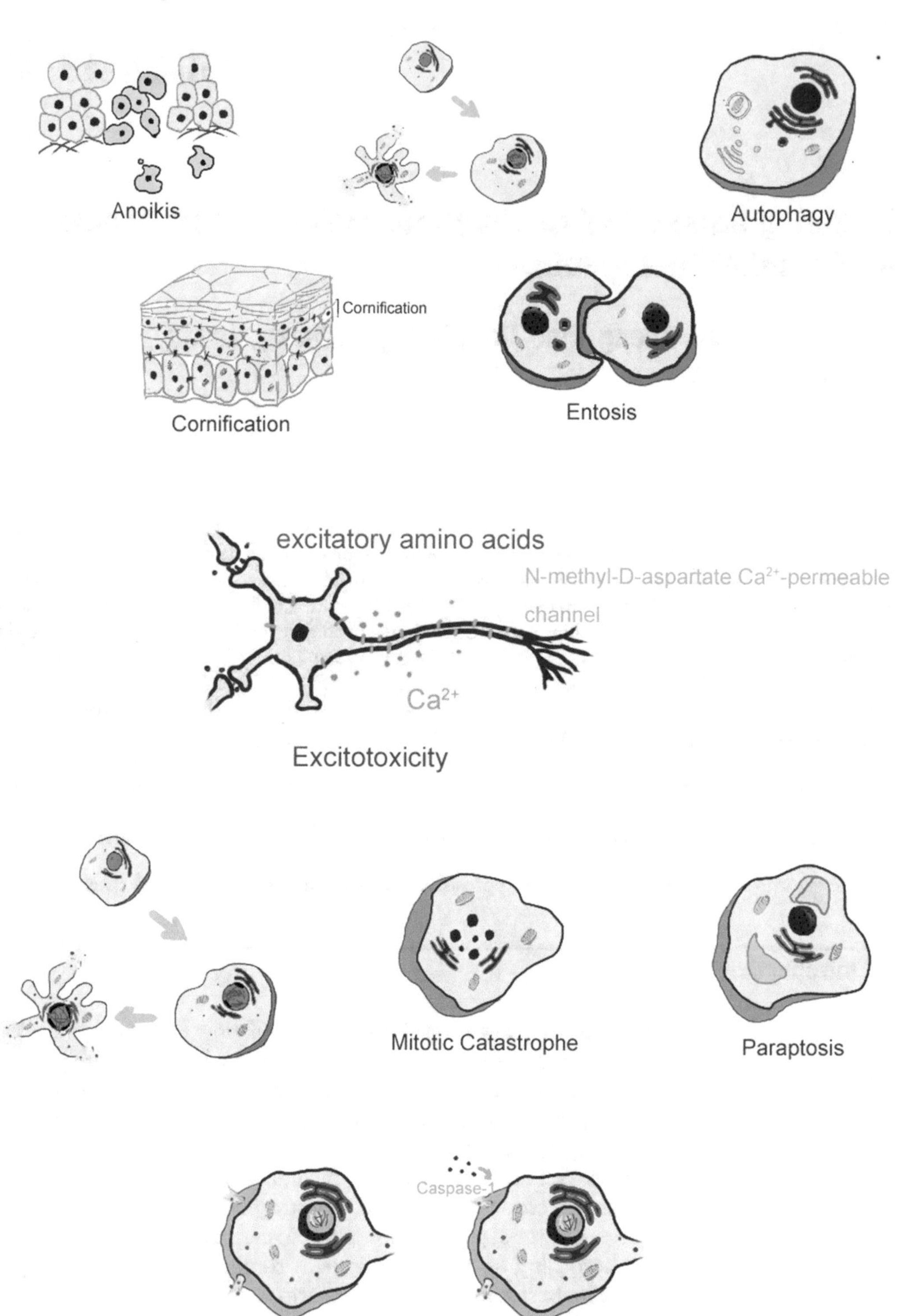

Fig. 1 Various cell death subroutines in different microenvironments

pathways. Autophagy is another cell death subroutine characterized by the appearance of large autophagic vacuoles in the cytoplasm. Impaired apoptosis is frequently associated with hyperproliferative conditions, such as autoimmune diseases and cancer, while defective autophagy is associated with developmental disorders and muscular dystrophy [1]. Both apoptosis and autophagy influence the antitumor effects of chemotherapeutic agents. Necrosis was previously considered to be an "accidental" cell death that occurred in response to physiochemical insults. However, it has been shown recently that necrosis can be a genetically regulated form of cell death that results in cellular leakage.

The existence of multiple cell death subroutines protects the cells against abnormalities in a single or multiple pathways, thus minimizing disease rates. Cross talk also exists between these cell death subroutines, making cell death a complicated complex. Many key proteins involving different pathways can be evaluated as targets for drug discovery. This evaluation process requires understanding of how proteins interact in a cellular signaling scheme. It also requires investigating the clinical relevance of the target, as well as the feasibility of creating small molecule antagonists or agonists [2]. Many small molecule compounds targeting cell death pathways have recently been tested in vitro and in vivo. These compounds have different functions, with some act as inhibitors while others act as inducers. Research into the molecular mechanisms of these compounds promises to provide new strategies for anticancer drug discovery by exploiting these cell death pathways [3].

2 Cell Death Classifications and Cancer

Programmed cell death (PCD) can occur both normally or under pathological conditions, and may involve multiple pathways, such as apoptosis, autophagy, and paraptosis. Indeed, considerable amount of cross talk exists between different cell death subroutines. More than one cell death subroutine can be activated at the same time, and this regulation is toxicant and cellular context-dependent. Cell death subroutines are divided into different subtypes according to various criteria, such as caspases activities and nuclear morphology.

2.1 Cell Death Classification—Caspases Activities

In the past decades, PCD was held synonymous with apoptosis characterized by typical morphological features such as chromatin condensation, phosphatidylserine exposure, cytoplasmic shrinkage, zeiosis, and the formation of apoptotic bodies. Caspases, a specific family of proteases, are indispensable in regulating the initiation of apoptosis. However, it has been shown in recent experiments that caspases are not the sole determinant of survival and death switch.

Various caspase-independent cell death forms cannot simply be classified as apoptosis or necrosis. Therefore, definitions of alternative death subroutines such as autophagy and paraptosis have been created [4]. The advantage for organisms to adopt various death subroutines is to protect themselves from abnormalities, since it is dangerous to depend on a single protease family for the clearance of unwanted and potential harmful cells.

2.2 Cell Death Classification—Nuclear Morphology

In contrast to the definitions above, another description model classifies cell death into four subtypes according to their nuclear morphology. Apoptosis can be distinguished from apoptosis-like PCD using chromatin condensation as a criterion. It is evident that during apoptosis chromatin condenses to compact and displays simple geometric (globular, crescent-shaped) figures in the nucleus. However, for apoptosis-like PCD, chromatin condenses into less geometric shapes, and phagocytosis markers on the plasma membrane can be seen before cell rupture. Furthermore, necrosis-like PCD is a cell death subroutine that occurs without chromatin condensation, or at best with chromatin clustering to speckles and often classified as "aborted apoptosis." Accidental necrosis or cell rupture is another death subtype stimulated by high concentrations of detergents, oxidants, ionophore or high intensities of pathologic insult and can be prevented only by the removal of stimulus [5]. Remarkably, recent genetic evidence, as well as discovery of chemical inhibitors reveals that necrosis is not an "accidental" cell death, and that it is actually regulated by multiple pathways [6]. Defined as a genetically controlled cell death process, regulated necrosis cause morphological changes, like cellular swelling, leakage, and cytoplasmic granulation.

2.3 Different Cell Death Subroutines in Cancer

For multicellular organisms, the balance between cell division and cell death is tightly controlled during different stages of development and normal physiology. Apoptosis, or programmed cell death, is the major cell death mechanism. Evidences have shown that accelerated apoptosis leads to acute and chronic degenerative disease, immunodeficiency, and infertility; conversely, insufficient apoptosis are usually associated with cancer and autoimmunity [7].

Cell death dysfunction is a hallmark of cancer. Harmful tumor cells proliferate at an abnormal high level, while their removal rate decreases. Suppression of apoptosis is thought to play a critical role in carcinogenesis via both the intrinsic and extrinsic signaling defects of the apoptosis pathways. Disruption of the intrinsic apoptosis signaling pathway is extremely common in cancer cells. Indeed, the p53 tumor suppressor gene is frequently mutated, and loss of p53 function disrupts apoptosis and promotes cancer development. Upstream regulators (ATM, Chk2, Mdm2, and p19ARF) or downstream effectors (PTEN, Bax, Bak, and Apaf-1) of p53 may also display functional mutations and changed expression level. Overexpression of Bcl-2 is seen in various cancers and can accelerate tumorigenesis. Pro-apoptotic Bcl-2, however, may be inactivated in certain cancers [8]. In contrast, the extrinsic apoptosis signaling pathway is less disrupted in tumorigenesis, and tumor cells are often resistant to extrinsic apoptosis. The extrinsic signaling pathways are mediated by cell death receptors and their ligands, such as tumor necrosis factor (TNF), CD95, and TNF-related apoptosis inducing ligand (TRAIL). In fact defects of apoptosis in cancer cells are detected at different stages, including initiation, transduction, amplification, and execution. Mutated apoptosis components and their frequency, however, are distinct in different cancer types; this indicates that the critical control point of apoptosis may be context dependent. Furthermore, tumor cells can undergo other cell death subroutines, such as autophagy and necrosis. Interestingly, autophagy plays a Janus role in different tumor progression stages and cancer types; this can either function as part of tumor-suppression machinery or may contribute to cancer development. Although necrosis has always been considered to be almost random and uncontrolled process, it may also promote rapid tumor growth by disrupting the integrity of cancer cells and inducing the inflammatory responses [9].

3 Apoptosis and Small Molecule Compounds in Cancer Therapy

3.1 Anticancer Compounds Targeting Extrinsic Apoptosis by Death Receptors or Dependence Receptors

The extrinsic apoptosis pathway is initiated through proapoptotic membrane death receptors, such as CD95 (APO-1/Fas), TNF receptor 1 (TNFR1), TNF-related apoptosis-inducing ligand-receptor 1 (TRAIL-R1/DR4), TRAIL-R2 (DR5), DR3 (TRAMP/Apo-3/WSL-1/LARD), and DR6. The corresponding ligands of TNF superfamily is composed of death receptor ligands, such as CD95 ligand (CD95L), TNFα, lymphotoxin-α, TRAIL, TWEAK, APO2L, and TNFSF10 [10].

Attempts have been made to exploit death receptors for cancer drug discovery. However, toxic side effects are observed with these recombinant agents; this limits their therapeutic use. For example, TNFα is unsuitable for systemic administration due to its pro-inflammatory activities. Hepatotoxic side effects are also seen in Fas-targeted therapy. By contrast, targeting soluble recombinant human Apo2L/TRAIL represents a more promising approach. TRAIL is able to induce apoptosis by binding to DR4 and DR5 receptors, with a minimal cytotoxicity to normal cells. Several anticancer compounds targeting extrinsic apoptosis have been tested in experiments; these compounds enhance the extrinsic apoptosis pathway through different mechanisms [11]. The main subtypes are shown in Fig. 2: these include drugs targeting the TRAIL and TRAIL/DR receptor system, monoclonal antibodies to DR4 and

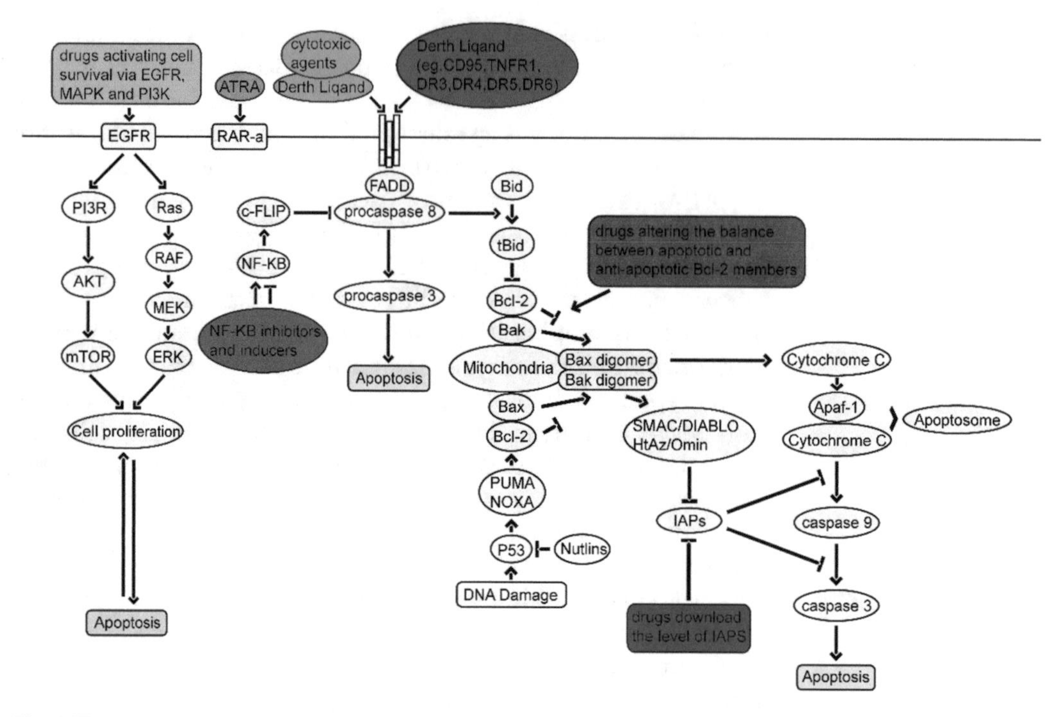

Fig. 2 The main subtypes of small molecule compounds that act at various points within the extrinsic and intrinsic apoptosis pathways

DR5, NF-κB inhibitors and inducers, as well as all trans retinoic acid (ATRA).

It has been reported that numerous conventional and investigational anticancer drugs, like the proteasome inhibitor bortezomib, can bind to and serve as agonists of TRAIL/DR. A combination of TRAIL with a variety of cytotoxic agents (including irinotecan, camptothecin, 5-fluorouracil, carboplatin, paclitaxel, doxorubicin, and gemcitabine) triggers the extrinsic apoptosis pathway in similar modes. Smac peptides strongly enhanced the antitumor activity of TRAIL in an intracranial malignant glioma xenograft model in vivo [12]. Thus, Smac agonists are also promising candidates for cancer therapy by potentiating cytotoxic therapies. In addition, a series of investigational anticancer drugs (such as histone deacetylase (HADC) inhibitors, rituximab, synthetic triterpenoids, and sorafenib) shows synergy with TRAIL. In preclinical trials, recombinant TRAIL induces apoptosis in various cancer cell lines, including cells with p53 mutations. In addition, TRAIL induces apoptosis with varying sensitivity. Chemotherapeutic drugs significantly augment TRAIL-induced apoptosis in cancer cells through upregulation of DR4, DR5, Bax, and Bak, and induction of caspase activation [13].

Agonistic monoclonal antibodies targeting death receptors 4 (DR4) and 5 (DR5) have been developed and evaluated at phase

1 and phase 2 trails. DR4 and DR5 recruit adaptor proteins via death domain interactions, and initiate the formation of the death inducing signaling complex (DISC), which leads the initiation of apoptosis. Dulanermin, an optimized zinc-coordinated homotrimeric recombinant Apo2L/TRAIL protein, is the only agent that serves as an agonist to both Apo2L/TRAIL death receptors, DR4 and DR5. Monoclonal antibodies selectively targeting DR5 have been tested in clinical trials; these include conatumumab, drozitumab, tigatuzumab, LBY135, and lexatumumab, whereas mapatumumab is a fully human agonistic antibody only against DR4. In several phase 1b safety studies, death receptor agonists in combination with chemotherapy and/or targeted agents enhanced antitumor activities of apoptosis without sensitizing normal cells to apoptosis [14].

Acquired resistance of tumor cells emerges as a significant impediment to effective cancer therapy when chemotherapeutic agents are used clinically. Activation of NF-κB pathway is common in many malignant tumor cells; this suppresses the apoptotic potential of chemotherapeutic agents and contributes to resistance. Intriguingly, several anticancer agents stimulate NF-κB activation, potentially leading to chemoresistance. Therefore, inhibitors of NF-κB are in great need to overcome resistance. The application of several NF-κB inhibitors (such as BAY 11-7085, BAY 11-7082, soy isoflavone genistein, parthenolide, CHS828, flavopiridol, and gliotoxin), blocks NF-κB activation and promotes apoptosis. It has been demonstrated that steroids and nonsteroidal anti-inflammatory drugs, like cyclooxygenase-2 inhibitors, also enhance the efficacy anticancer agents by disrupting the regulation of NF-κB. Notably, the caspase 8 homologue FLICE-inhibitory protein (cFLIP) is an important NF-κB-dependent regulator in death receptor-induced apoptosis; cFLIP levels are upregulated in some cell lines. Drugs, like the proteasome inhibitor N-benzoyloxycarbonyl (Z)-Leu-Leu-leucinal (MG-132) or geldanamycin, are able to interfere with TNF-induced NF-κB activation to inhibit the upregulation of cFLIP, promoting the extrinsic apoptosis pathway [15].

Acute promyelocytic leukemia (APL) is distinct from the myeloid leukemias (AML). Patients with APL have a characteristic translocation on chromosome 17, in the region of retinoic acid receptor-alpha (RAR-α). This receptor is involved in growth and differentiation of myeloid cells. Retinoic acid (RA), an analog of vitamin A, has the ability to trigger differentiation and terminal cell death of leukemic cells in vitro. All-trans retinoic acid (ATRA) has been reported as an effective inducer in attaining complete remission in APL by modulating RAR-α. In addition, ATRA is thought to induce the paracrine release of membrane-bound TRAIL, leading to extrinsic apoptosis both in ARAR-treated APL cells and in adjacent non-ATRA responsive and non-APL cells [16].

3.2 Anticancer Compounds Targeting Caspase-Dependent/Independent Intrinsic Apoptosis

The intrinsic apoptosis signaling pathway is triggered by non-receptor mediated stimuli under a series of intracellular stress conditions. All of these stimuli cause changes in the inner mitochondria that results in dissipation of mitochondrial transmembrane potential, release of intermembrane mitochondria proteins into the cytosol, and respiratory chain inhibition. Two main groups of proteins are released from the mitochondria. One group is composed of cytochrome c, Smac/DIABLO, and serine protease HtA2/Omi, which activate the caspase-dependent mitochondrial pathway. The other protein group consists of pro-apoptotic proteins AIF, endonuclease G, and ICAD are released later when cell has committed to death. The Bcl-2 family of proteins, which are controlled by p53, play a vital role in these apoptotic mitochondria events by regulating the release of cytochrome c. Notably, Bcl-2 family proteins can either be anti-apoptotic like Bcl-2, Bcl-x, Bcl-xL, Puma, Noxa, Bcl-XS, Bcl-w, and Bag or pro-apoptotic like Bcl-10, Bax, Bid, Bad, Bim, Bik, and Blk [9].

The intrinsic apoptosis pathway plays an important role in cancer development. Therefore, small molecule compounds that can selectively induce apoptosis are potentially useful in cancer therapy. In fact, those compounds used in cancer therapy function in three main categories: (1) Altering the balance between pro-apoptotic and anti-apoptotic members of the Bcl-2; (2) Downregulating the level of inhibitors of apoptosis proteins (IAPs), such as XIAP, survivin, and c-IAP; (3) Other cancer therapy mechanisms. These approaches are described in detail in the following passage [17].

Pro-apoptotic and anti-apoptotic Bcl-2 proteins have generated great interest as drug targets in cancer therapy. The increased expression of Bcl-2, Mcl-1, and Bcl-w occurs significantly in common cancer types with poor response to apoptosis. Consequently, several strategies have been developed; these include antisense techniques that utilize BH3-domain peptides or synthetic small molecule drugs interfering with Bcl-2-like protein function. Specific small molecule inhibitors targeting anti-apoptotic Bcl-2 family includes Gossypol, Obatoclax (GX15-070), ABT737, HA14-1, 2-Methoxy Antimycin A, Chelerythrine, and Sanguinarine. Meanwhile, pro-apoptotic Bcl-2 family of proteins are also targeted by small molecule compounds; these include 3, 6-dibromocarbazole piperazine derivatives of 2-propanol, 4-phenylsulfanyl-phenylamine derivatives, humanin peptides, and Ku70 peptides. Application of small molecule compounds towards pro-apoptotic and anti-apoptotic Bcl-2 proteins disrupt the balance between them, leading to apoptosis. Notably, Mcl-1 is a highly expressed pro-survival protein in human malignancies. Inhibition of its expression and/or neutralization of its anti-apoptotic function does rapidly render Mcl-1 dependent cells more susceptible to apoptosis, providing an opportunity in cancer therapy [18, 19].

Another group of small molecule compounds involved in targeting the intrinsic apoptosis pathway interact with inhibitors of apoptosis proteins (IAPs). IAPs are a family of proteins characterized by the baculoviral IAP repeat (BIR) domain; they include XIAP, cIAP1, cIAP2, NAIP, ML-AIP, livin, survivin, and apollon. IAPs have been reported to directly inhibit active caspase-3 and caspase-7 as well as block caspase-9 activation, thus suppressing apoptosis. Various IAP inhibitors have been identified and tested in clinical trials. For example, both of the 19-mer phosphorothioate antisense oligonucleotide, AEG-35156(Aegera), and Smac/Diablo peptides-embelin can inhibit XIAP in cancer cells. Other studies also show that down regulation of XIAP sensitizes colon cancer cells to PPARγ ligand-induced apoptosis in vitro and in vivo [17]. Therefore, targeting IAPs is a promising therapeutic approach in cancer treatment [10].

Many other small compounds for various cancer therapy strategies have been reported. Some compounds, like Nutlins, restores the function of the tumor suppressor transcription factor p53, while others inhibit the activities of proteasome like Bortezomib [20]. These small compounds target the intrinsic apoptosis pathway creates promising approaches in cancer therapy.

4 Autophagic Cell Death and Small Molecule Compounds in Cancer Therapy

Autophagy is an evolutionarily conserved and multistep lysosomal process, in which cellular materials are delivered to lysosomes for degradation and recycling. It has been demonstrated that autophagy is modulated by a limited number of autophagy-related genes (ATGs). The autophagy process can be dissected into five different steps; these include induction, vesicle nucleation, elongation & completion, docking & fusion, and degradation & recycling [21]. Therefore, autophagy can help cells respond to a wide range of extracellular and intracellular stresses.

Autophagy has been demonstrated to act as either a guardian or executioner in cancer, a multistep process caused by genetic mutations in oncogenes and tumor suppressors. The self-eating property of autophagy can function as a tumor-promoting mechanism, while the cell death characteristic may help autophagy execute self-destructive function. Both oncogenesis and tumor suppression are influenced by perturbations of the molecular machinery that controls autophagy. Numerous oncogene proteins (including Bcl-2, Ras, mTOR, PI3KCI, AKT, and MAPKs) may suppress autophagy, while several tumor suppressor proteins (such as Beclin-1, DAPK1, LKB1/STK11, and PTEN) promote autophagy [22]. It is surprising, however, to find that p53, one of the most important tumor suppressor proteins, regulates autophagy based on its different subcellular localization. Small molecule

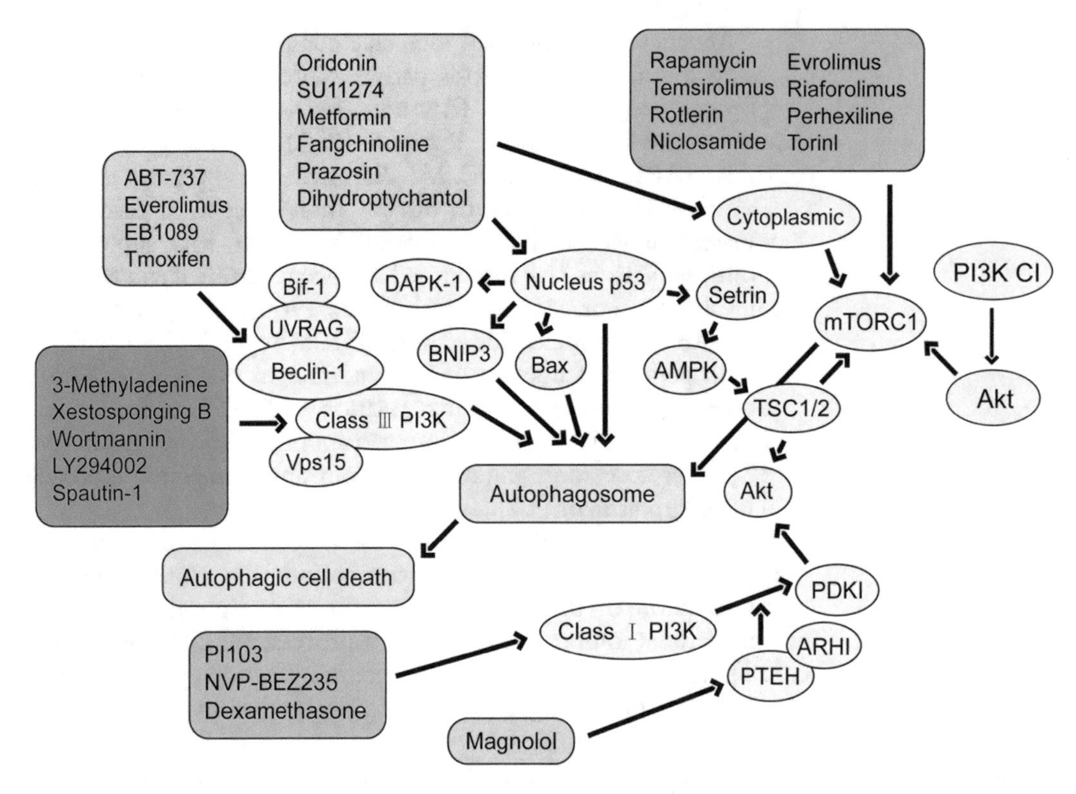

Fig. 3 The autophagy pathway with relevant small molecule compounds

compounds that regulate protein function and affect biological processes have been extensively employed to dissect biological pathways and to study different diseases. With the deepening research for cancer treatment via different autophagic signaling pathways, several kinds of small molecule compounds have been revealed to promote or suppress autophagy by targeting corresponding autophagy-related signaling pathways. Here we outline specific mechanisms of some widely accepted autophagic modulators, such as PI3KCI/AKT/mTORC1 signaling pathway, Beclin-1 interactomes, and p53. We also summarize some known small molecule compounds, such as Rapamycin and its derivatives, Rottlerin, PP242, AZD8055, Spautin-1, Tamoxifen, Chloroquine, Oridonin, and Metformin (Fig. 3). This demonstrates the potential of small molecule compounds targeting autophagy for future cancer treatment [23, 24].

4.1 PI3KCI/AKT/mTORC1 Pathway with Relevant Small Molecule Compounds

The mTOR-dependent signaling pathway, induced by insulin or amino acid depletion, is the major mechanism which may control autophagic activity. As a principal regulator of cell growth, mTORC1 is deregulated in most human cancers. The activation of the class I phosphatidylinositol 3-kinase (PI3KCI) and downstream components, such as the kinase AKT, can help promote mTORC1. The enhanced activity of PI3KCI is often achieved via activating kinase mutations or gene amplification in cancers, thus inhibiting autophagy. By mediating protein translational modification, such as ULK1 phosphorylation, AMPK phosphorylation, mTORC1 can negatively regulate autophagy, and mTORC1 subnetwork may occupy a central position in autophagic pathways. Thus, targeting the PI3KCI/AKT/mTORC1 pathway, an important oncogenic signaling pathway that is involved in many human cancers, can control the level of autophagy in cancer cells and exert potential therapy capacities [25].

Our understanding of the interplay between autophagy and cancer first benefited from the availability of Rapamycin, a natural product that can inhibit mTORC1 by dissociating raptor from mTOR, thus limiting the access of mTOR to some substrates. With its exquisite selectivity, Rapamycin can be used as an indispensable pharmacological probe for elucidating biological functions of mTOR serine/threonine kinase in governing cell growth and proliferation [26]. Since Rapamycin has significant therapeutic effects, its synthetic analogs including temsirolimus (CCI-779), everolimus (RAD001), and ridaforolimus (AP23573), have also been developed to improve pharmacokinetic properties and produce advantageous intellectual property positions. Temsirolimus is a prodrug of Sirolimus, which is hydrolyzed quickly after intravenous administration, thus treating different cancers such as acute myeloid leukemia (AML), nonsmallcell lung cancer (NSCLC), and renal cell carcinoma (RCC). Besides, Everolimus may enhance antitumor effect of the oncolytic adenovirus Delta24RGD by inducing autophagy in glioma. It can also increase autophagy and reduce tumor size in acute lymphoblastic leukemia, prostate cancer, and NSCLC. With a dimethyl phosphate group at C-40-O position, Ridaforolimus can aim at various cancers, such as advanced malignancies and relapsed hematological malignancies. Four additional mTORC1 inhibitors, including rottlerin, niclosamide, perhexiline, and Torin1, can promote autophagy by directly inhibiting mTORC1 function or inhibiting proteins in the upstream of mTOR pathway in cancer cells under nutrient-rich conditions. Interestingly, Rottlerin can also target TSC2, a negative regulator of mTORC1, thereby suppressing mTORC1 signaling [27].

Moreover, ATP-competitive inhibitors can directly prevent mTOR function, and suppress AKT phosphorylation in primary leukemic cells and stromal cells cultured alone or in combination with leukemic cells. As ATP-competitive inhibitors, PP242 and AZD8055 can block phosphorylation of mTORC1 substrates and AKT, while AZD8055, also shows excellent selectivity against all PI3K isoforms and other members of PI3K-like kinase family. Both of these two inhibitors can greatly inhibit cell proliferation and suppresses tumor growth by inducing autophagy in cancers, such as head and neck squamous cell carcinoma, as well as acute myeloid leukemia. Furthermore, Phenethyl isothiocyanate (PEITC), a promising cancer chemopreventive agent in human prostate cancer cells, induces autophagic cell death by suppressing phosphorylation of both AKT and mTORC1 in prostate cancer. The compounds 2-deoxyglucose and glucose 6-phosphate also reduce mTORC1 and AKT phosphorylation; these have been used in the treatment of acute lymphoblastic leukemia and other lymphoid malignancies. As a main active component of marijuana, THC may induce autophagic cell death and enhance endoplasmic reticulum (ER) stress via inhibition of AKT and mTORC1 in cancer cells. Resveratrol, a polyphenol present in grapes, peanuts and other plants, is known to display antitumor activities, since it may promote cell death by triggering autophagy and thereby activating AKT and mTORC1 in ovarian cancer cells. The standard first-line systemic drug Sorafenib is used for advanced hepatocellular carcinoma (HCC); it can activate AKT through the feedback loop of mTOR, switching protective autophagy to autophagic cell death.

A class of PI3KCI inhibitors with multiple-target abilities has also emerged. The 1H-imidazo [4, 5-c] quinoline derivative NVP-BEZ235 inhibits activities of PI3KCI/AKT/mTOR cascade by binding to the ATP-binding cleft of both PI3CKI and mTOR in glioma cells. PI103 is a potent inhibitor of both PI3KCI and mTOR, thereby preventing AKT phosphorylation and resulting in autophagy enhancement. This compound may also inhibit proliferation and invasion of cancer cells, since it displays suppression of tumor growth in different human cancers with genetic abnormalities and PI3KCI activation. With a morpholino pyridinopyrimidine core structure, KU0063794 (AstraZeneca) is presumably developed using PI103 as a lead compound, and has high potential and selectivity as an mTOR inhibitor. Besides, the synthetic glucocorticoid dexamethasone potently inhibits PI3KCI/AKT/mTOR signaling pathway in acute lymphoblastic leukemia cells, and is used clinically as a chemotherapeutic agent in many hematologic malignancies [28].

In addition, PTEN, a tumor suppressor frequently mutated in human tumors, can induce autophagy by inhibiting the PI3KCI/AKT/mTOR signaling cascade. It has been found that Magnolol blocks PI3K/PTEN/AKT and induces H460 autophagic cell death, underlining the potential utility of its induction as a new cancer treatment (Fig. 3).

4.2 Beclin-1 Interactome with Relevant Small Molecule Compounds

Beclin-1, the mammalian homolog of Atg6, can enhance autophagy by combining with PI3KIII/Vps34 in the induction of autophagy, since the evolutionarily conserved domain (ECD) of Beclin-1 interacts with PI3KIII/Vps34. Beclin-1 is a direct substrate of caspase-3/7/8 in apoptosis, whose caspase cleavage is sufficient to suppress autophagy in cancer. Beclin-1 can positively regulate autophagy by coupling with PI3KCIII/Vps34, and other positive and negative cofactors, like Vps15, ATG14L/Barkor, UVRAG, Bif-1, Rubicon, Ambra1, high mobility group box 1 (HMGB1), Survivin, AKT, and Bcl-2/Bcl-xL to form the Beclin-1 interactome. UVRAG interacts with Beclin-1 and PI3KCIII/Vps34, while Bif-1 may fuse with Beclin-1 through UVRAG; thereby enhancing autophagy. AMBRA1 promotes Beclin-1 interaction with its target PI3KCIII/Vps34 and mediates autophagosome nucleation. As an extracellular damage-associated molecular pattern molecule, HMGB1 disrupts interaction between Beclin-1 and its negative regulator Bcl-2 by competitively binding to Beclin-1. Another positive modulator is the anti-apoptotic protein survivin, which may present a possible mechanism in the cross talk between autophagy and apoptosis by Beclin-1-mediated degradation. PINK1, a serine/threonine protein kinase in mitochondria, can also interact with Beclin-1 and thus inducing autophagy. Moreover, Beclin-1 contains a BH-3 motif which is necessary for binding to Bcl-2, Bcl-xL, and Mcl-1. Bcl-2 can block Beclin-1 interaction with PI3KCIII/Vps34 and decrease PI3KCIII activity, while Bcl-xL and Mcl-1 inhibit Beclin-1 activity by stabilizing Beclin-1 homodimerization. Different from UVRAG and Bif-1, RUBICON prevents kinase activity of PI3KCIII/Vps34 and block autophagosome maturation, and is also involved in the endocytic pathways by inducing aberrant endosomes and blocking EGFR degradation. Moreover, interaction of Bcl-2/Bcl-xL with the Beclin-1/Vps34 complex reduces PI3KCIII/Vps34 activity, and this interaction can be competitively disrupted by BH3-only proteins, such as Bad. Therefore, Beclin-1 interactcomes may enhance autophagy and inhibit tumorigenesis through mediation of positive and negative regulators of its interactcomes in cancers [29, 30].

Although currently no Beclin-1-related drugs have been applied for clinical use, several small molecule compounds can target Beclin-1 in different cancers. As a possible lead compound for development of anticancer drugs, Spautin-1 is also a potent inhibitor of autophagy. It can promote degradation of PI3KCIII/Vps34 complexes by inhibiting two ubiquitin-specific peptidases, USP10 and USP13, which target the Beclin-1 subunit of Vps34 complexes. Xestospongin B disrupts Beclin-1 through an indirect link established by Bcl-2, and then interferes with the molecular complex formed by PI3KIII and Beclin-1. The BH3 mimetic ABT-737 specifically decreases the interaction between Bcl-2 and Bcl-xL with the BH3 motif of Beclin-1, which stimulates Beclin-1-dependent activation of PI3KCIII. A chemotherapeutic vitamin D analog, EB1089, promotes massive autophagy by disrupting the inhibitory interaction between two BH3 domains of Beclin-1 and Bcl-2. Another small molecule inhibitor is Gossypol, which interrupts interaction between Beclin-1 and Bcl-2/Bcl-xL at ER and induces autophagy in a Beclin-1-/Atg5-dependent manner. Tamoxifen, a well-recognized antitumor drug for breast cancer treatment, is able to increase the level of Beclin-1 to stimulate autophagy. Generally regarded as an mTORC1 inhibitor, RAD001 (Everolimus) has also been found to increase Beclin-1 expression and induce autophagy in leukemia.

Moreover, some small molecule compounds can directly inhibit PI3KIII, thereby influencing the expression of Beclin-1. Chloroquine (CQ), a previously recognized antimalarial agent, can effectively sensitize cell-killing effects by ionizing radiation and chemotherapeutic agents in a cancer-specific manner. It may act as a competitive PI3KIII inhibitor and promote degradation of the PI3KCIII/Vps34 complex by inhibiting two ubiquitin specific peptidases, USP10 and USP13. 3-Methyladenine is also used as a specific inhibitor of autophagic sequestration, inhibiting PI3KCIII and thus providing a target for its action. Furthermore, the anti-proteolytic effect of Wortmannin (IC50 = 30 nM) and LY294002 (IC50 = 10 μM) has been found to be accompanied not by an increase in lysosomal pH or a decrease in intracellular ATP, but by inhibition of autophagic sequestration. Thus, PI3KCIII inhibitors wortmannin and LY294002 can inhibit autophagy in isolated hepatocytes [31] (Fig. 3).

4.3 p53 with Relevant Small Molecule Compounds

Embedded within a highly interconnected signaling pathway, p53 regulates key cellular processes, such as DNA repair, metabolism, development, inflammation, endocytosis, and cell death. It has been demonstrated that p53 can positively or negatively control autophagy. The role of p53 in autophagy, however, is paradoxical depending on its subcellular location, primarily present in two forms: cytoplasmic p53 and nuclear p53. During response of p53 to DNA damage, a number of cell death genes are transcriptionally activated, some of which play important roles in autophagy. When exposed to stress, nuclear p53 may induce autophagy by acting at multiple levels of the AMPK-mTOR axis. It may downregulate mTOR through transcriptional regulation of main activators of AMPK-Sestrin1/2, and target it to phosphorylate tuberous sclerosis 2 (TSC2) [32]. Damage-regulated autophagy modulator (DRAM) can also be regarded as a transcriptional target of nuclear p53 in autophagy modulation, explaining the complexity of mutual regulation in apoptosis and autophagy. Regarding this cross talk, the pro-autophagic role of p53 may influence the expression of Bcl-2 protein family members, including Bcl-2, Bcl-xL, Mcl-1, and Bad, while the reduction of p53 induces the release of Beclin-1 from sequestration. The p53-inducible BH3-only protein PUMA induces mitochondrial autophagy, but Bax alone can induce mitochondria-selective autophagy in the absence of PUMA activation. Following activation, cytoplasmic p53 translocates to the nucleus and regulates expressions of several target genes. Under this condition, PUMA may enhance cell death by targeting mitochondria to autophagy. In response to mitochondrial dysfunction, p53 can also induce DRAM1-dependent autophagy. As a target of cytoplasmic p53, TP53-induced glycolysis and apoptosis regulator (TIGAR) can inhibit autophagy by negatively modulating glycolysis and suppressing reactive oxygen species (ROS). However, divergent effects of nuclear and cytoplasmic p53 are still controversial. The p53 variants in the cytoplasm significantly inhibit autophagy, whereas those in the nucleus may fail to suppress autophagy. Therefore, p53 plays a role in controlling the basal level of autophagy, as nuclear p53 boost autophagy and cytoplasmic p53 decreases autophagy independent on its transcriptional activity [33].

There are several small molecule compounds that have been identified to target p53 in different cancer cells. The tetracycline diterpenoid natural product Oridonin can induce autophagy in L929 cells via the NO-ERK-p53 positive-feedback loop. As a selective Met tyrosine kinase inhibitor, SU11274 leads to autophagic cell death in non-small-cell lung cancer (NSCLC) A549 cells. It has been demonstrated that p53 can be activated after SU11274 treatment, while interruption of p53 activity decreases SU11274-induced autophagy. Treatment with the polyphenolic compound, Galangin, may inhibit cell proliferation and induce autophagy, particularly leading to an accumulation of autophagosomes and an increase of p53 expression; this mediates autophagy through a p53-dependent pathway in HCC HepG2 cells. Additionally, it has been demonstrated that anti-melanoma effects of Metformin are mediated through autophagy associated with p53/Bcl-2 modulation, mitochondrial damage, and oxidative stress. Fangchinoline is a highly specific antitumor agent which induces autophagic cell death via p53/Sestrin2/AMPK signaling in HCC HepG2 and PLC/PRF/5 cells. Previously used to treat high blood pressure and anxiety, Prazosin has been found to induce patterns of autophagy via a p53-mediated mechanism in HPC2 cells, since cells exposed to prazosin may increase levels of phospho-p53 and phospho-AMPK. Dihydroptychantol A, a macrocyclic bisbibenzyl derivative, also increases p53 expression, induces p53 phosphorylation, and upregulates p21 (Waf1/Cip1); this is responsible for mediating autophagy associated with p53 in human osteosarcoma U2OS cells [34].

5 Necroptosis, Necrosis, and Small Molecule Compounds in Cancer Therapy

Unlike the features of apoptosis and autophagy, "necrotic cell death" or "necroptosis" were initially only viewed merely as an accidental subroutine of cell death. Currently, "necroptosis" is considered as a form of programmed necrosis whose molecular effects are partially shared with apoptosis. A series of researches showed that necroptosis can be avoided by inhibiting RIP1 through genetic or pharmacological methods [35]; additionally, it can be initiated by the members of the tumor necrosis factor (TNF) families (TNFR1, TNFR2, TRAILR1, and TRAILR2), Fas ligand, toll-like receptors (TLRs), lipopolysaccharides (LPS), and genotoxic stress. Different kinds of physical-chemical stress stimuli can also induce necroptosis, including anticancer drugs, ionizing radiation, photodynamic therapy, glutamate, and calcium overload. RIPK1 functions to limit caspase-8-dependent, TNFR-induced apoptosis, whereas TNFR-induced RIPK3-dependent necroptosis requires RIPK1, and cells lacking RIPK1 were sensitized to necroptosis triggered by stimuli or interferons. Necroptosis is generally dependent on RIP3, which is activated under conditions that are insufficient to trigger apoptosis [36].

Necrostatin-1 (Nec-1), a small molecule inhibitor of RIP1 kinase, is identified to selectively target the kinase activity of RIP1, a key mediator of necroptosis, and thus induce necroptosis [37]. Moreover, Bcr-Abl inhibitor Ponatinib in leukemia therapy is also identified as a dual inhibitor of RIPK1 and RIPK3. Based on the structure of Nec-1 and Ponatinib, optimized compound PN10 has shown great efficacy to dual target RIPK1 and RIPK3, and is a powerful blocker of TNF-induced injury, whether under inflammation or cancer [38]. Neoalbaconol, a constituent extracted from *Albatrellus confluens*, abolished the ubiquitination of RIPK1 by downregulating E3 ubiquitin ligases; it also induced necroptosis, including RIPK1/NF-κB-dependent expression of TNFα and RIPK3-dependent generation of ROS [39]. In addition, it has been demonstrated that targeting human phosphatidylethanolamine-binding protein 4 (hPEBP4), an anti-apoptotic protein, can increase the sensitivity of cancer cells to TNFα or TRAIL-induced apoptosis and necroptosis. IOI-42 can increase hPEBP4 expression, promote TNFα-mediated growth inhibition, and suppress cell growth in MCF-7 cells [40]. Inhibitors of glucose transporter 1 have been shown to reduce glycolysis, and induce cell-cycle arrest as anticancer agents. They also induce necroptosis by targeting the signaling pathway of necroptosis. For instance, WZB117 can increase AMPK and reduce cyclin E2 by inhibiting glucose transport to induce necroptosis [41].

6 The Cross-Talks Between Cell Death Subroutines and Cancer Drug Discovery

The death of cancer cells is a complicated but well controlled process with various cell death subroutines. The best characterized cell death subroutines are apoptosis, autophagy and necrosis. Apoptosis, autophagy and necrosis bear distinct morphological characteristics and physiologically processes, however, the intricate interrelationships between them remain uncover.

Apoptosis and autophagy engage in complex interplay with each other. Under various cellular conditions, autophagy can function to promote cell survival or cell death. In fact, apoptosis and autophagy are highly interconnected and share many key signal transduction pathways. The general cellular stress mediator ROS and increased cytosolic free Ca^{2+} concentration not only activate autophagy but also induce apoptosis [42]. Although, the prominent apoptosis inducer, sphingolipid ceramide also mediates autophagy processes. Notably, the sphingolipid sphingosine-1-phosphate appears to stimulate autophagy and antagonize ceramide-induced apoptosis.

The transcription factor p53 acts as a tumor suppressor that detects stressful conditions within cells, and induces cells senescence or apoptosis. A number of studies have demonstrated that p53 can stimulate autophagy by inhibiting mTOR or acting on damage-regulated autophagy modulator (DRAM). The pharmacological BH3 mimetics, ABT737 and HA14-1, are able to induce autophagy in cells without causing the cells to undergo apoptosis. Autophagy inhibition by knockdown of Beclin-1 promotes apoptosis, while inhibition of caspases promotes autophagy. Similarly, proteins, such as members of the death-associated protein kinase (DAPK) family, are able to induce both autophagy and apoptosis, depending on the cell types, by acting as molecular switches to modulate cell to programmed death between apoptosis and autophagy [43]. In summary, apoptosis and autophagy may be induced by common stressors in a context-dependent manner, and with different thresholds. Necrosis was always regarded as an accidental and uncontrolled process. However, accumulating evidence suggests that necrosis is also a regulated and programmed form of cell death distinct from apoptosis. As is shown in recent studies, a number of cell death receptors (including TNFR1, FAS, TNFR2, TRAILR1 and TRAILR2) which typically affect apoptosis, can also induce necrosis in different cell types [44]. Thus, the interplay between apoptosis, autophagy and necrosis is complex, and these events can be activated in parallel or sequentially [45].

The multiple layers of interaction between various cell death pathways indicate that there is a balance between life and death in response to a given cellular stress condition. The disruption of this balance may cause pathological symptoms, such as cancer. Since many key proteins and pathways play a critical role in cell death interactions, discovery of drugs targeting these proteins and pathways may create more chances for the treatment of cancer now and in the future. Many small molecule compounds aimed at multiple targets and functions may be able to work on various types of cancer. The characterization of cell death pathways, as well as the understanding of pathological conditions will form the basis for discovering novel context-dependent drugs in cancer therapy.

7 Conclusions and Perspectives

The development and progression of cancer is complicated process related to various cell death subroutines, including apoptosis, autophagy and necroptosis. Each cell death subroutine has its own distinct characteristics. Therefore, cell death can be classified into different categories according to morphological features or molecular mechanisms. Cross talks between cell death subroutines have also been observed in various cancer cells and some molecular machinery have been deciphered, whereas knowledge of how such combined activity occurs is fragmented and incomplete. In fact, the selection of death subroutine by cancer cells is a context-dependent process that is largely dependent on the microenvironment. Many small molecule compounds targeting cell death pathways have been discovered recently, and tested for use in cancer therapy both in vitro and in vivo. These novel drugs can be used either alone or in combination with safer doses of conventional anticancer therapies to enhance their efficiency. Therefore, future cancer therapy will pay more attention to personalized treatment and selection of targeted drugs. Further research aimed at understanding the molecular mechanisms of all kinds of cell death subroutines and the dynamic drug induced signaling pathways is needed. The availability of improved experiment approaches to test the efficacy of small molecule compounds will significantly facilitate the identification of more drug targets, drug combination methods, as well as the design of clinical trials.

Acknowledgments

This work was supported by grants from the Key Projects of the National Science and Technology Pillar Program (No. 2012BAI 30B02), National Natural Science Foundation of China (Nos. U1170302, 81402496, 81260628, 81303270 and 81172374).

References

1. Kepp O, Galluzzi L, Lipinski M, Yuan J, Kroemer G (2011) Cell death assays for drug discovery. Nat Rev Drug Discov 10(3): 221–237
2. Huang P, Oliff A (2001) Signaling pathways in apoptosis as potential targets for cancer therapy. Trends Cell Biol 11(8):343–348
3. Kim R (2004) Recent advances in understanding the cell death pathways activated by anticancer therapy. Cancer 103(8):1551–1560
4. Jäättelä M (2004) Multiple cell death pathways as regulators of tumour initiation and progression. Oncogene 23(16):2746–2756
5. Leist M, Jäättelä M (2001) Four deaths and a funeral: from caspases to alternative mechanisms. Nat Rev Mol Cell Biol 2(8):589–598
6. Vanden Berghe T, Linkermann A, Jouan-Lanhouet S, Walczak H, Vandenabeele P (2014) Regulated necrosis: the expanding network of non-apoptotic cell death pathways. Nat Rev Mol Cell Biol 15(2):135–147
7. Danial NN, Korsmeyer SJ (2004) Cell Death: Critical Control Points. Cell 116(2):205–219
8. Johnstone RW, Ruefli AA, Lowe SW (2002) Apoptosis: a link between cancer genetics and chemotherapy. Cell 108(2):153–164

9. Ouyang L, Shi Z, Zhao S, Wang FT, Zhou TT, Liu B, Bao JK (2012) Programmed cell death pathways in cancer: a review of apoptosis, autophagy and programmed necrosis. Cell Prolif 45(6):487–498
10. Fulda S, Debatin KM (2006) Extrinsic versus intrinsic apoptosis pathways in anticancer chemotherapy. Oncogene 25(34):4798–4811
11. Sayers TJ (2011) Targeting the extrinsic apoptosis signaling pathway for cancer therapy. Cancer Immunol Immunother 60(8):1173–1180
12. Fulda S, Wick W, Weller M, Debatin KM (2002) Smac agonists sensitize for Apo2L/TRAIL- or anticancer drug-induced apoptosis and induce regression of malignant glioma in vivo. Nat Med 8(8):808–815
13. Fulda S (2015) Targeting extrinsic apoptosis in cancer: Challenges and opportunities. Semin Cell Dev Biol 39:20–25
14. Holland PM (2014) Death receptor agonist therapies for cancer, which is the right TRAIL? Cytokine Growth Factor Rev 25(2):185–193
15. Nakanishi C, Toi M (2005) Nuclear factor-κB inhibitors as sensitizers to anticancer drugs. Nat Rev Cancer 5(4):297–309
16. Huang ME, Ye YC, Chen SR, Chai JR, Lu JX, Zhao L, Gu LJ, Wang ZY (1989) Use of all-trans retinoic acid in the treatment of acute promyelocytic leukemia. Haematol Blood Transfus 32:88–96
17. Ralph SJ, Neuzil J (2009) Mitochondria as targets for cancer therapy. Mol Nutr Food Res 53(1):9–28
18. Hockenbery DM (2010) Targeting mitochondria for cancer therapy. Environ Mol Mutagen 51(5):476–489
19. Akgul C (2009) Mcl-1 is a potential therapeutic target in multiple types of cancer. Cell Mol Life Sci 66(8):1326–1336
20. Vangestel C, Van de Wiele C, Mees G, Peeters M (2009) Forcing cancer cells to commit suicide. Cancer Biother Radiopharm 24(4):395–407
21. Klionsky DJ (2007) Autophagy: from phenomenology to molecular understanding in less than a decade. Nat Rev Mol Cell Biol 8(11):931–937
22. Liu B, Wen X, Cheng Y (2013) Survival or death: disequilibrating the oncogenic and tumor suppressive autophagy in cancer. Cell Death Dis 4, e892
23. Liu B, Cheng Y, Liu Q, Bao JK, Yang JM (2010) Autophagic pathways as new targets for cancer drug development. Acta Pharmacol Sin 31(9):1154–1164
24. Maiuri MC, Tasdemir E, Criollo A, Morselli E, Vicencio JM, Carnuccio R, Kroemer G (2009) Control of autophagy by oncogenes and tumor suppressor genes. Cell Death Differ 16(1):87–93
25. Zoncu R, Efeyan A, Sabatini DM (2011) mTOR: from growth signal integration to cancer, diabetes and ageing. Nat Rev Mol Cell Biol 12(1):21–35
26. Hartford CM, Ratain MJ (2007) Rapamycin: something old, something new, sometimes borrowed and now renewed. Clin Pharmacol Ther 82(4):381–388
27. Liu Q, Thoreen C, Wang J, Sabatini D, Gray NS (2009) mTOR Mediated Anti-Cancer Drug Discovery. Drug Discov Today Ther Strateg 6(2):47–55
28. Apsel B, Blair JA, Gonzalez B, Nazif TM, Feldman ME, Aizenstein B, Hoffman R, Williams RL, Shokat KM, Knight ZA (2008) Targeted polypharmacology: discovery of dual inhibitors of tyrosine and phosphoinositide kinases. Nat Chem Biol 4(11):691–699
29. McKnight NC, Zhenyu Y (2013) Beclin 1, an Essential Component and Master Regulator of PI3K-III in Health and Disease. Curr Pathobiol Rep 1(4):231–238
30. Kang R, Zeh HJ, Lotze MT, Tang D (2011) The Beclin 1 network regulates autophagy and apoptosis. Cell Death Differ 18(4):571–580
31. Fu LL, Cheng Y, Liu B (2013) Beclin-1: Autophagic regulator and therapeutic target in cancer. Int J Biochem Cell Biol 45(5):921–924
32. van Veelen W, Korsse SE, van de Laar L, Peppelenbosch MP (2011) The long and winding road to rational treatment of cancer associated with LKB1/AMPK/TSC/mTORC1 signaling. Oncogene 30(20):2289–2303
33. Ryan KM (2011) p53 and autophagy in cancer: guardian of the genome meets guardian of the proteome. Eur J Cancer 47(1):44–50
34. Sui X, Jin L, Huang X, Geng S, He C, Hu X (2011) p53 signaling and autophagy in cancer: a revolutionary strategy could be developed for cancer treatment. Autophagy 7(6):565–571
35. Golstein P, Kroemer G (2007) Cell death by necrosis: towards a molecular definition. Trends Biochem Sci 32(1):37–43
36. Dillon CP, Weinlich R, Rodriguez DA, Cripps JG, Quarato G, Gurung P, Verbist KC, Brewer TL, Llambi F, Gong YN, Janke LJ, Kelliher MA, Kanneganti TD, Green DR (2014) RIPK1 blocks early postnatal lethality mediated by caspase-8 and RIPK3. Cell 157(5):1189–1202

37. Degterev A, Hitomi J, Germscheid M, Ch'en IL, Korkina O, Teng X, Abbott D, Cuny GD, Yuan C, Wagner G, Hedrick SM, Gerber SA, Lugovskoy A, Yuan J (2008) Identification of RIP1 kinase as a specific cellular target of necrostatins. Nat Chem Biol 4(5):313–321
38. Najjar M, Suebsuwong C, Ray SS, Thapa RJ, Maki JL, Nogusa S, Shah S, Saleh D, Gough PJ, Bertin J, Yuan J, Balachandran S, Cuny GD, Degterev A (2015) Structure Guided Design of Potent and Selective Ponatinib-Based Hybrid Inhibitors for RIPK1. Cell Rep 10(11):1850–1860, pii:S2211-1247(15)00210-7
39. Yu X, Deng Q, Li W, Xiao L, Luo X, Liu X, Yang L, Peng S, Ding Z, Feng T, Zhou J, Fan J, Bode AM, Dong Z, Liu J, Cao Y (2015) Neoalbaconol induces cell death through necroptosis by regulating RIPK-dependent autocrine TNFα and ROS production. Oncotarget 6(4):1995–2008
40. Qiu J, Xiao J, Han C, Li N, Shen X, Jiang H, Cao X (2010) Potentiation of tumor necrosis factor-alpha-induced tumor cell apoptosis by a small molecule inhibitor for anti-apoptotic protein hPEBP4. J Biol Chem 285(16):12241–12247
41. Liu Y, Cao Y, Zhang W, Bergmeier S, Qian Y, Akbar H, Colvin R, Ding J, Tong L, Wu S, Hines J, Chen X (2012) A small-molecule inhibitor of glucose transporter 1 downregulates glycolysis, induces cell-cycle arrest, and inhibits cancer cell growth in vitro and in vivo. Mol Cancer Ther 11(8):1672–1682
42. Nishida K, Yamaguchi O, Otsu K (2008) Crosstalk between autophagy and apoptosis in heart disease. Circ Res 103(4):343–351
43. Solarewicz-Madejek K, Basinski TM, Crameri R, Akdis M, Akkaya A, Blaser K, Rabe KF, Akdis CA, Jutel M (2009) T cells and eosinophils in bronchial smooth muscle cell death in asthma. Clin Exp Allergy 39(6):845–855
44. Bialik S, Kimchi A (2006) The death-associated protein kinases: structure, function, and beyond. Annu Rev Biochem 75:189–210
45. Nikoletopoulou V, Markaki M, Palikaras K, Tavernarakis N (2013) Crosstalk between apoptosis, necrosis and autophagy. Biochim Biophys Acta 1833(12):3448–3459

Chapter 13

Liposomes in Apoptosis Induction and Cancer Therapy

Magisetty Obulesu and Magisetty Jhansilakshmi

Abstract

Cancer is the leading cause of death with multiple obstacles in therapeutic arsenals employed to date. Apoptosis induction in cancer cells has hitherto been a prominent unresolved obstacle for a few decades. Liposomes with multiple merits were extensively employed to entrap several types of anticancer agents, biomolecules and imaging agents to achieve substantial therapeutic effect for various types of cancers. Multifunctional liposomes with enhanced biocompatible properties were designed to enhance the therapeutic effect. Despite the promising drug delivery strategies and significantly reduced toxicity of the liposomal formulations a few demerits still limit their success considerably. This chapter reviews recent advances in liposomal formulations, methods of therapeutic loaded liposomal preparation, their merits and demerits. A few challenges associated with liposomal drug delivery and apoptosis induction are also summarized.

Key words Apoptosis, Liposomes, Cancer, Multidrug resistance

1 Apoptosis

Apoptosis, a biological process of cell death, primarily facilitates growth by removing some unwanted cells. Mitochondria and endoplasmic reticulum are the major platforms for the process of apoptosis. Mitochondria integrate various stimuli and execute cell death through a few common effector caspase pathways [1]. In addition, Fenton chemistry with imperfectly liganded iron provokes apoptotic events in several neurodegenerative diseases, such as Alzheimer's disease [1]. Caspases such as caspase-9, caspase-3, and caspase-8 also play a pivotal role in execution of apoptosis [1]. Although apoptosis plays an essential role in the physiology, when uncontrolled it leads to several deleterious effects [1]. Apoptosis induction in cancer cells is of significant importance currently, and has remained a Herculean task for a few decades.

Perpetua M. Muganda (ed.), *Apoptosis Methods in Toxicology*, Methods in Pharmacology and Toxicology,
DOI 10.1007/978-1-4939-3588-8_13, © Springer Science+Business Media New York 2016

2 Liposomes

Liposomes were first discovered in 1961 by Alec D Bangham, a British hematologist [2]. Liposomes are vesicular structures ranging in the size from 50 to 1000 nm [3–5]; they have been extensively used in drug delivery for a few decades [6]. Primarily liposomes are amenable carriers of water soluble drugs in their core and lipid-soluble drugs in their corona without affecting the physicochemical properties of drugs [7]. Liposomes carry nucleic acids such as DNA, RNA, and genes [8], and have been extensively used in the induction of apoptosis in cancer cells. In fact, epidermal growth factor receptor targeted immunoliposomes showed pronounced delivery of celecoxib to cancer cells [9].

3 Liposome Mediated Apoptosis Induction

3.1 Liposomes for Natural Compounds

Curcumin, a popular ingredient of Indian curry spice, comprises ample therapeutic benefits. Its poor aqueous solubility, however, considerably hinders its progress as an amenable therapeutic agent [10]. To address these issues, solid lipid curcumin nanoparticles were designed and delivered orally to the patients with osteosarcoma. These particles showed substantial therapeutic effect without any toxic effects [10]. Other curcumin liposomal formulations were also extensively studied and thoroughly reviewed [11–14].

Specifically designed mitochondria targeted resveratrol liposomes succeeded in overcoming multidrug resistance in human lung adenocarcinoma A549 cells and resistant A549/cDDP cells [15]. Dequalinium polyethylene glycoldistearoylphosphatidylethanol amine (DQA-PEG(2000)-DSPE) conjugate facilitated mitochondrial targeting and internalization of nanoparticles into the tumor core [15]. Liposomal associated lysosomal Saposin C–dioleoylphosphatidylserine nanovesicles showed substantial tumor vessel accumulation and demonstrated the robust therapeutic efficacy against wide variety of cancers without affecting normal cells [16]. Bufalin liposomes conjugated with anti-CD40 antibody demonstrated significant toxicity in a mouse B16 melanoma model through the co-delivery and synergistic therapeutic effect of bufalin and CD-40 [17]. Therefore, liposomes serve as appropriate carriers of natural compounds with apoptotic activity.

3.2 Drug Loaded Liposomes

Several drug loaded liposomes also show better therapeutic effect. More recently, it has been found that photo-provoked tumor vascular treatment ameliorated the therapeutic efficacy of paclitaxel loaded liposomes in a mouse model [18]. Robust liposomes targeting mitochondria improved drug uptake in the mitochondria which further provoked all apoptotic events in MCF-7 cancer stem cells. These apoptotic events include initiation of the pro-apoptotic Bax protein, degeneration of the mitochondrial membrane potential, opening of the mitochondrial permeability transition pores, translocation of cytochrome C, and stimulation of a cascade of caspase 9 and caspase 3 reactions [19]. Significant therapeutic effect of Irinophore, a liposomal irinotecan, was observed in patient's xenografts of primary human colorectal tumors grown in NOD-SCID mice [20].

Doxorubicin, the robust anticancer drug has several toxicity issues despite the promising therapeutic effect. Doxorubicinol, a pegylated liposomal formulation developed at a later date showed substantial inhibition of doxorubicin induced toxicity [21, 22] as well as a substantial therapeutic effect in multifarious cancer models [21]. The basic underlying mechanism of the formulation was to enhance apoptosis and attenuate anti-apoptotic pathways [21]. Although doxil, a pegylated liposomal doxorubicin, has currently been extensively used, its release mechanism is yet to be unraveled. Recent reports proposed an ammonia provoked release [23].

Loading of two therapeutic molecules and their co-delivery is also an appropriate therapeutic strategy to induce apoptosis in cancer cells [17]. In line with this, encapsulation and co-delivery of antagomir 10b and paclitaxel in antimicrobial peptide liposomes remarkably hampered the migration of 4T1 cells and provoked apoptosis [24]. In another study, epigallocatechin gallate (EGCG) and paclitaxel (PTX) co-loaded liposomes were designed to induce significant matrix metalloproteinase inhibition and apoptosis in vitro [25].

3.3 Biomolecule Loaded Liposomes

Cationic liposomes with the appropriate DNA–lipid ratio (1:6) and diameter of 143.3 ± 5.7 nm showed substantial apoptosis in lung cancer model [26]. In another study significant apoptosis and inhibition of proliferation was observed in STAT3 induced ovarian cancer mouse model by liposomal delivery of short hairpin RNA [27]. Sterically stable liposomes were found to show significant therapeutic effect in hepatic fibrosis mouse model via the interferon delivery [28]. Therefore, the liposomes play an essential role in the successful delivery of the promising loaded therapeutics to the target sites. The corresponding therapeutic such as drug or RNA or DNA can induce apoptosis in cancer cells.

More recently, epirubicin and/or antisense oligonucleotide loaded PEGylated liposomes exhibited appreciable apoptosis in cancer cells. They could substantially circumvent the multidrug resistance through the attenuation of multidrug resistance transporters [29]. Costa et al [30, 31] designed anti mi-RNA oligonucleotide (AMO) loaded liposome using (DODAP): cholesterol (CHOL): 1,2-distearoyl-sn-glycero-3-phosphocholine (DSPC): Ceramide C16-polyethylene glycol 2000(CerC16-PEG2000) (25:49:22:4, % molar ratio to total lipid). Interestingly, these liposomes showed appreciable therapeutic effect against glioblastoma cell lines. Taken together these findings open new avenues to overcome cancer through successful internalization of loaded therapeutics. Kogure et al. [32] developed innovative MITO-porter with plasmidic DNA in the center and a lipid outer layer which successfully delivers encapsulated drugs to the mitochondria by a membrane fusion mechanism in the disguise of viruses. Dihydropyridopyrazoles encapsulated prodrug liposomes exhibited antiproliferative activity and apoptosis-mediated cancer cell death in HeLa and Jurkat cancer cell lines thus overcoming the aqueous solubility of substantial anticancer agents [33].

Nonviral gene delivery has been made possible by the Trojan Horse Liposome (THL) Technology, which effectively targets brain unlike other delivery systems [34, 35]. Weekly treatment with such liposomal formulation showed a significant amelioration in mice with brain tumors [35, 36]. APRPG-PEG liposomes exhibited intense accumulation in tumor site and ameliorated pancreatic cancer by inhibiting angiogenesis [37]. Liposomes have been found to inhibit proliferation and avert apoptosis by targeted in vivo delivery of STAT-3 siRNA plasmid or scramble plasmid, thus preventing restenosis of vein graft in a mouse model [38].

3.4 Multifunctional Liposomes

Multifunctional liposomes comprise a nanocarrier, therapeutic drugs/biomolecules, protective polymer, such as PEG, to avoid its removal from systemic circulation, and a targeting ligand, to reach the definite receptor on the target tissue [39]. They encompass multifarious therapeutic properties—distribution of drug or imaging agent in vivo, ability to target specific tissues, and extended circulation. Their promising therapeutic benefit includes the delivery of cargo not only to specific cells but also to cell organelles [40]. They induce enhanced therapeutic effect in target cells through enhanced uptake of therapeutic into specific organelles, such as mitochondria.

With a view to achieving the targeted delivery of the liposomes, they are commonly conjugated to driving moieties such as antibodies without affecting the properties of liposomes or attached moieties [41]. Multifunctional target specific daunorubicin in association with quinacrine liposomes showed significant therapeutic effect against brain glioma and glioma stem cells [42]. Surprisingly, these liposomes played a pivotal role in apoptosis induction through caspase-mediated signaling pathways both in vitro and in vivo by spanning blood–brain barrier. The method of preparation of these liposomes followed by drug encapsulation involves a few simple steps such as dissolution of lipids in chloroform, hydration, sonication, dialysis and incubation with drugs.

3.5 Pegylated Liposomes

Significance of targeted liposomes is further accentuated by PEGylated liposomal antisense oligonucleotides (ASOs), which enhanced cytotoxicity in Caco-2 cells through epirubicin-provoked apoptosis [43]. The primary role of PEG in this liposomal formulation is to enhance the half-life in systemic circulation and to surpass the reticuloendothelial system. Given the cardinal role of these pegylated liposomes in circumventing the multidrug resistance, they can be successfully employed to cure multifarious cancer by inducing apoptosis.

4 Demerits of Liposomes

Despite the numerous advantages of liposomes, a few disadvantages impede their success. They include feeble stability in vivo, vulnerability to peripheral stimuli such as difference in temperature, pH and osmotic pressure and inability to facilitate transmembrane permeability [44, 45]. Liposomes demonstrated cargo leakage in a few formulations and probable detection and deletion from the circulatory system through the reticuloendothelial system (RES) [21, 46, 47]. Although doxorubicinol significantly deteriorated doxorubicin uptake in the heart compared to doxorubicin alone, its uptake in liver and skin augmented 4–48 times, thus impeding the success of this formulation [21].

5 Conclusion

Liposomes have been extensively used to encapsulate and deliver several types therapeutic drugs, natural compounds and biomolecules to achieve enhanced therapeutic efficacy. Despite the demerits associated with a few liposomal formulations, a few FDA approved liposomal formulations such as liposomal bupivacaine, liposomal vincristine (Marqibo®) [48, 49] are extensively used to enhance therapeutic effect of loaded cargo. Additionally, a few multifunctional liposomes targeted specifically towards the mitochondrial apoptotic events in several cancer cell lines and animal models showed appreciable success. Therefore, they become corner stone in the treatment of several types of cancers. Furthermore, it can be concluded that an appropriate multifunctional liposome with several amenable properties can be a substantial therapeutic arsenal when a few challenges associated with them are absolutely resolved.

Acknowledgements

The authors sincerely thank Professor Dr. Yukio Nagasaki, Department of Materials Science, Graduate School of Pure and Applied Sciences, University of Tsukuba, Tennodai 1-1-1, Tsukuba, Ibaraki, Japan for this generous support.

References

1. Obulesu M, Lakshmi MJ (2014) Apoptosis in Alzheimer's disease: an understanding of the physiology, pathology and therapeutic avenues. Neurochem Res 39:2301–2312
2. Sateesh Madhav NV, Ojha A, Saini A (2015) A platform for liposomal drug delivery. Int J Pharm Drug Anal 3:6–11
3. Musacchio T, Torchilin VP (2010) Recent developments in lipid-based pharmaceutical nanocarriers. Front Biosci (Landmark Ed) 16:1388–1412
4. Bitounis D, Fanciullino R, Iliadis A et al (2012) Optimizing druggability through liposomal formulations: new approaches to an old concept. ISRN Pharm 2012:738432
5. Sawant RR, Torchilin VP (2012) Challenges in development of targeted liposomal therapeutics. AAPS J 14:303–315
6. Mallick S, Choi JS (2014) Liposomes: versatile and biocompatible nanovesicles for efficient biomolecules delivery. J Nanosci Nanotechnol 14:755–765
7. Apostolova N, Victor VM (2014) Molecular strategies for targeting antioxidants to mitochondria: therapeutic implications. Antioxid Redox Signal 22(8):686–729
8. Farooqi AA, Rehman ZU, Muntane J (2014) Antisense therapeutics in oncology: current status. Onco Targets Ther 7:2035–2042
9. Limasale YD, Tezcaner A, Ozen C et al (2015) Epidermal growth factor receptor-targeted immunoliposomes for delivery of celecoxib to cancer cells. Int J Pharm 479:364–373
10. Yallapu MM, Jaggi M, Chauhan SC (2012) Curcumin nanoformulations: a future nanomedicine for cancer. Drug Discov Today 17:71–80
11. Li L, Braiteh FS, Kurzrock R (2005) Liposome-encapsulated curcumin: in vitro and in vivo effects on proliferation, apoptosis, signaling, and angiogenesis. Cancer 104:1322–1331
12. Pandelidou M, Dimas K, Georgopoulos A et al (2011) Preparation and characterization of lyophilized EGG PC liposomes incorporating curcumin and evaluation of its activity against colorectal cancer cell lines. J Nanosci Nanotechnol 11:1259–1266
13. Narayanan NK, Nargi D, Randolph C et al (2009) Liposome encapsulation of curcumin and resveratrol in combination reduces prostate cancer incidence in PTEN knockout mice. Int J Cancer 125:1–8
14. Mulik RS, Monkkonen J, Juvonen RO et al (2010) Transferrin mediated solid lipid nanoparticles containing curcumin: enhanced in vitro anticancer activity by induction of apoptosis. Int J Pharm 398:190–203
15. Wang XX, Li YB, Yao HJ et al (2011) The use of mitochondrial targeting resveratrol liposomes modified with a dequalinium polyethylene glycol-distearoylphosphatidyl ethanolamine conjugate to induce apoptosis in resistant lung cancer cells. Biomaterials 32:5673–5687
16. Qi X, Chu Z, Mahller YY et al (2009) Cancer-selective targeting and cytotoxicity by liposomal-coupled lysosomal saposin C protein. Clin Cancer Res 15:5840–5851
17. Li Y, Yuan J, Yang Q et al (2014) Immunoliposome co-delivery of bufalin and anti-CD40 antibody adjuvant induces synergetic therapeutic efficacy against melanoma. Int J Nanomedicine 9:5683–5700
18. Araki T, Ogawara KI, Suzuki H et al (2015) Augmented EPR effect by photo-triggered tumor vascular treatment improved therapeutic efficacy of liposomal paclitaxel in mice bearing tumors with low permeable vasculature. J Control Release 200C:106–114
19. Zhang L, Yao HJ, Yu Y et al (2012) Mitochondrial targeting liposomes incorporating daunorubicin and quinacrine for treatment of relapsed breast cancer arising from cancer stem cells. Biomaterials 33:565–582
20. Neijzen R, Wong MQ, Gill N et al (2014) Irinophore C™, a lipid nanoparticulate formulation of irinotecan, improves vascular function, increases the delivery of sequentially administered 5-FU in HT-29 tumors, and controls tumor growth in patient derived xenografts of colon cancer. J Control Release 199C:72–83
21. Tacar O, Sriamornsak P, Dass CR (2013) Doxorubicin: an update on anticancer molecular action, toxicity and novel drug delivery systems. J Pharm Pharmacol 65:157–170
22. Gabizon A, Shmeeda H, Barenholz Y et al (2003) Pharmacokinetics of pegylated liposomal doxorubicin: review of animal and human studies. Clin Pharmacokinet 42:419–436
23. Silverman L, Barenholz Y (2015) In vitro experiments showing enhanced release of doxorubicin from Doxil® in the presence of ammonia may explain drug release at tumor site. Nanomedicine 11(7):1841–1850
24. Zhang Q, Ran R, Zhang L et al (2015) Simultaneous delivery of therapeutic antagomirs with paclitaxel for the management of metastatic tumors by a pH-responsive antimicrobial peptide-mediated liposomal delivery system. J Control Release 197:208–218

25. Ramadass SK, Anantharaman NV, Subramanian S et al (2015) Paclitaxel/Epigallocatechin gallate coloaded liposome: a synergistic delivery to control the invasiveness of MDA-MB-231 breast cancer cells. Colloids Surf B: Biointerfaces 125:65–72
26. Kwon OJ, Kang E, Kim S et al (2011) Viral genome DNA/lipoplexes elicit in situ oncolytic viral replication and potent antitumor efficacy via systemic delivery. J Control Release 155:317–325
27. Jiang Q, Dai L, Cheng L et al (2013) Efficient inhibition of intraperitoneal ovarian cancer growth in nude mice by liposomal delivery of short hairpin RNA against STAT3. J Obstet Gynaecol Res 39:701–709
28. Li Q, Yan Z, Li F et al (2012) The improving effects on hepatic fibrosis of interferon-γ liposomes targeted to hepatic stellate cells. Nanotechnology 23:265101
29. Lo YL, Liu Y, Tsai JC et al (2013) Overcoming multidrug resistance using liposomal epirubicin and antisense oligonucleotides targeting pump and nonpump resistances in vitro and in vivo. Biomed Pharmacother 67:261–267
30. Costa PM, Cardoso AL, Mendonca LS et al (2013) Tumor-targeted Chlorotoxin coupled nanoparticles for nucleic acid delivery to glioblastoma cells: a Promisingsystem for glioblastoma treatment. Mol Ther Nucleic Acids 2, e100
31. Costa PM, Cardoso AL, Custodia C et al (2015) MiRNA-21 silencing mediated by tumor-targeted nanoparticles combined with sunitinib: A new multimodal gene therapy approach for glioblastoma. J Control Release 207:31–39
32. Kogure K, Akita H, Yamada Y et al (2008) Multifunctional envelope-type nano device (MEND) as a non-viral gene delivery system. Adv Drug Deliv Rev 60:559–571
33. Mikhalin AA, Evdokimov NM, Frolova LV et al (2014) Lipophilic prodrug conjugates allow facile and rapid synthesis of high-loading capacity liposomes without the need for post-assembly purification. J Liposome Res 1–29
34. Boado RJ (2007) Blood-brain barrier transport of non-viral gene and RNAi therapeutics. Pharm Res 24:1772–1787
35. Collet G, Grillon C, Nadim M et al (2013) Trojan horse at cellular level for tumor gene therapies. Gene 525:208–216
36. Zhang Y, Zhang YF, Bryant J et al (2004) Intravenous RNA interference gene therapy targeting the human epidermal growth factor receptor prolongs survival in intracranial brain cancer. Clin Cancer Res 10:3667–3677
37. Asai T (2012) Nanoparticle-mediated delivery of anticancer agents to tumor angiogenic vessels. Biol Pharm Bull 35:1855–1861
38. Sun J, Zheng J, Ling KH et al (2012) Preventing intimal thickening of vein grafts in vein artery bypass using STAT-3 siRNA. J Transl Med 10:2
39. Torchilin V (2009) Multifunctional and stimuli-sensitive pharmaceutical nanocarriers. Eur J Pharm Biopharm 71:431–444
40. Deshpande PP, Biswas S, Torchilin VP (2013) Current trends in the use of liposomes for tumor targeting. Nanomedicine (Lond) 8: 1509–1528
41. Torchilin VP (2007) Targeted pharmaceutical nanocarriers for cancer therapy and imaging. AAPS J 9:E128–E147
42. Li XT, Ju RJ, Li XY et al (2014) Multifunctional targeting daunorubicin plus quinacrine liposomes, modified by wheat germ agglutinin and tamoxifen, for treating brain glioma and glioma stem cells. Oncotarget 5:6497
43. Lo YL, Liu Y (2014) Reversing multidrug resistance in Caco-2 by Silencing MDR1, MRP1, MRP2, and BCL-2/BCL-xL using liposomal antisense oligonucleotides. PLoS One 9, e90180
44. Beloglazova NV, Goryacheva IY, Shmelin PS (2015) Preparation and characterization of stable phospholipid–silica nanostructures loaded with quantum dots. J Mater Chem B 3:180–183
45. Juliano R, Stamp D (1975) The effect of particle size and charge on the clearance rates of liposomes and liposome encapsulated drugs. Biochem Biophys ResCommun 63:651–658
46. Poste G, Bucana C, Raz A et al (1982) Analysis of the fate of systemically administered liposomes and implications for their use in drug delivery. Cancer Res 42:1412–1422
47. Yamada Y, Furukawa R, Yasuzaki Y et al (2011) Dual function MITO-Porter, a nano carrier integrating both efficient cytoplasmic delivery and mitochondrial macromolecule delivery. Mol Ther 19:1449–1456
48. Leonetti C, Scarsella M, Semple SC et al (2004) In vivo administration of liposomal vincristine sensitizes drug-resistant human solid tumors. Int J Cancer 110:767–774
49. Wang X, Song Y, Su Y et al. (2015) Are PEGylated liposomes better than conventional liposomes? A special case for vincristine. Drug Deliv 1–9.

INDEX

Perpetua M. Muganda (ed.), *Apoptosis Methods in Toxicology*, Methods in Pharmacology and Toxicology,
DOI 10.1007/978-1-4939-3588-8,

G

H

I

L

Printed in the United States
By Bookmasters